Water Societies and Technologies from the Past and Present

Edited by
Yijie Zhuang
Mark Altaweel

First published in 2018 by
UCL Press
University College London
Gower Street
London WC1E 6BT

A CIP catalogue record for this book is available from The British Library.

Preface

Inspiration for this volume came from extensive fieldwork in Asia conducted by us, the editors, as well as discussions about establishing a new course on 'Rivers and Civilisations'. However, rather than focusing on a course, we thought it would be a good idea to establish a small conference on water-related issues that covered a variety of settings and periods, to see what commonalities might be determined across time. A benefit of archaeology is that it provides a long-term perspective on the use of critical resources such as water. We thought a conference that looked at different cases, applying archaeological methods focused on water, while also inviting noted scholars of current issues relating to water, could provide a longer view of the major issues debated, such as water sustainability, technology and management.

In choosing who to invite to participate, our intention was to incorporate cutting-edge research conducted by various scholars and see how their research addresses the main areas of focus on the topic of water. The main issues we cover are how water relates to societies across time, the technologies of water, and how we can understand water-related issues across time through models (conceptual and quantitative) of water use and management. By the nature of these topics, the scholars' work had to be interdisciplinary, and often collaborative.

The intention is still to gather momentum for an interdisciplinary teaching initiative on water that aims to provide a coherent overview of how water technologies and societies have co-evolved. The concept of sustainability, we thought, also required investigation, since past and even modern cases have begun to challenge it. A comparative conference could test these ideas and indicate the feasibility of theme-based approaches that are not wedded to disciplines such as archaeology but look at a critical resource across time.

Initial invitations were sent to potential participants, and we received very positive feedback from a variety of scholars in China, the United States, the UK and elsewhere in Europe. This helped us put forward a strong application for funding, which was ultimately accepted. We are therefore grateful to the Institute of Archaeology and the Institute of Advanced Studies at UCL for their generosity in funding the conference, which was entitled 'Comparative Water Technologies and Management: Pathways to Social Complexity and Environmental Change'.

This conference proved successful; it focused on the main water-related issues outlined above. Particularly stimulating was the post-presentation group discussion in a seminar room, which not only addressed the achievements of the conference but helped develop the foundations of this volume. We had the opportunity to share, in a relaxed manner, common concerns, our vision – both humble and grand – of future research directions, and anecdotes about fieldwork and interdisciplinary collaborations. Perhaps what was telling was that, outside the building on Gower Street where the conference was held, anti-austerity protestors were marching. We gave sandwiches, which could be considered delicious by UK conference standards, to the upbeat protestors.

Ultimately, we wanted our ideas to be heard and shared more widely, hence this edited open access volume you are reading. We have not made this book simply a proceedings volume. Rather, authors were asked to refine their work to focus on the key takeaway themes that were developed in the conference, so that a larger, more coherent picture could be developed of the long-term use of water and of how different case studies could inform us about how societies have integrated water resources across time.

We thank our contributors for their quick response to emails and for producing their work to a tight production schedule. We are also grateful to UCL Press for making this book happen. In a radically changing publication climate, UCL Press has become a pioneer by making all its publications open access. We are confident that this volume will be a small and stimulating start to a long academic journey that puts water, one of our most critical resources, at the centre of an interdisciplinary and long-term understanding of how this critical resource has been used across time, and the lessons this use has for us today.

Acknowledgements

We are grateful for the generous support of the UCL Institute of Archaeology and the UCL Institute of Advanced Studies.

Contents

List of figures

List of tables

List of contributors

Fawzi Abudanah is an associate professor in the Department of Archaeology at Al-Hussein Bin Talal University. Dr Abudanah received his PhD from the School of Historical Studies, University of Newcastle, where his studies focused on the Roman period. Additionally, Dr Abudanah's research interests include water systems and landscape archaeology in Jordan.

Mark Altaweel obtained his PhD from the University of Chicago and is now a reader in Near East Archaeology at UCL. He has conducted fieldwork in various parts of the Middle East and North America, focusing on environmental and social-environmental interactions in modern and ancient societies. In addition to fieldwork, he uses quantitative and analytical methods to gain insight into water-management and water-use issues.

Sarah Bell is a chartered engineer and a professor of environmental engineering at UCL. She is the director of the Engineering Exchange, which supports community engagement with engineering and built environment research. She is an EPSRC Living with Environmental Change research fellow, researching community engagement with infrastructure projects.

Judith Bunbury obtained her PhD from the University of Cambridge and has worked as a geoarchaeologist based at the Department of Earth Sciences there. She has studied the interaction between people and the waterscapes of Egypt throughout the Holocene. Her current work is an investigation of the movements of the Nile and the effects of both human intention and natural processes.

Peter D. Clift is a marine and sedimentary geologist who has worked in Asia for 27 years. Educated at Oxford and Edinburgh Universities, he is now the Charles T. McCord Endowed Professor at Louisiana State

University. He is a visiting professor at the Chinese Academy of Sciences and Nanjing Normal University.

Mark Driessen is an assistant professor in the Archaeology Group of the Roman Provinces, Middle Ages and Modern Period (Eurasia) in the Faculty of Archaeology, Leiden University. His fields of research relate to ancient harbours, ancient water management and agriculture, Roman logistics and trade, the Roman army, landscape archaeology, and community-based programming.

Maurits W. Ertsen is an associate professor in the Water Resources group at the Delft University of Technology in the Netherlands. He studies how irrigation realities emerge from short-term actions of (non-)human agents. His work spans current, historical and archaeological times. He is one of two main editors of the journal *Water History*.

Liviu Giosan obtained his PhD from Stony Brook University, New York. He is now a geoscientist with the Woods Hole Oceanographic Institution (USA). His recent work has focused on fluvial and marine morphodynamics, the future of deltas, the effects of historic land use on marine environments, and the role of monsoons as a driver of civilisation florescence or collapse in Asia.

Jaafar Jotheri obtained his PhD from Durham University in 2016; his thesis title is 'Holocene avulsion history of the Euphrates and Tigris rivers in the Mesopotamian floodplain'. He has a general interest in the geoarchaeology of the southern Mesopotamian floodplain, with a special focus on how the changing of rivers' courses impacted on human activities and settlement patterns in the past. He also researches the development of irrigation systems in Mesopotamia and around the world. He currently works as an assistant professor in the Department of Archaeology at the University of Al-Qadisiyah, Iraq, where he teaches remote sensing, GIS, field archaeology and geoarchaeology.

Heejin Lee obtained her PhD from the Department of Archaeology, University of Cambridge. She is now a lecturer in the Division of Cultural Heritage Convergence, Korea University. Her research focuses on ancient soils associated with agricultural management and ancient use of space, and on understanding the complex relationship between humans and the environment. She has conducted geoarchaeological research on ancient settlements and agricultural sites in the Korean peninsula and Jeju Island.

Duowen Mo is a professor at the College of Urban and Environmental Sciences, Peking University. His main research interests include surface geomorphological processes, sedimentary processes, and how these environmental processes are related to social changes. He is the director of the Association for the Environmental Archaeology of China. He has directed numerous environmental archaeological research projects across China.

Julia Shaw obtained her PhD from the University of Cambridge. She is now a lecturer in South Asian Archaeology at the UCL Institute of Archaeology. She has been carrying out fieldwork in India for the last 20 years. Her current research interests include: South Asian environmental and socio-religious history; religion, medical knowledge and attitudes towards food and the body; and global ecological discourse, environmental health and archaeology as environmental humanities.

Janice Stargardt is a professor and senior fellow of Sidney Sussex College, University of Cambridge. She is also a senior fellow of the McDonald Institute for Archaeological Research and the Department of Geography. She works on the historical geography and archaeology of South and South East Asia. The overarching theme of her research has been the transition of societies in south-east India, Burma and Thailand from Iron Age villages to complex, literate and urbanised communities.

Tim Williams, FSA, is a reader in Silk Roads Archaeology at the UCL Institute of Archaeology. He worked in the Department of Urban Archaeology (Museum of London) and at English Heritage before joining UCL in 2002. His current research is focused on urbanism, especially Islamic and central Asian, along the Silk Roads. He has undertaken major projects in Beirut (Lebanon) and Merv (Turkmenistan), and the International Council on Monuments and Sites (ICOMOS) thematic study of the Silk Roads, which was the basis for the UNESCO World Heritage nomination strategy.

Yijie Zhuang obtained his PhD from the Department of Archaeology, University of Cambridge. He is now a senior lecturer in Chinese Archaeology at the UCL Institute of Archaeology. He has conducted intensive fieldwork in different parts of East and South East Asia. He is primarily interested in the reconstruction of the water-management systems and agricultural ecologies of both the monsoonal and arid regions of Asia and in how these related to long-term social evolution.

1 Introduction: Interdisciplinary research into water management and societies

Yijie Zhuang and Mark Altaweel

Background to the book

This book presents a broad, interdisciplinary discussion of water systems and their use. It investigates a variety of case studies from different periods, ancient and modern, and from different regions, to explore how water relates to issues of social complexity, scale and organisation. Water sustainability is an active topic of interest today for major cities and countries with large populations (see, e.g., Liu and Yang, 2012). While this issue is not new, as societies have had to address water issues over millennia, such research becomes more relevant today as we try to understand how past societies may present lessons and case studies that are useful for understanding current issues. We see that both past and present societies have faced great challenges concerning water solutions; however, states and cities have often found ways to thrive even in challenging environments. Societies have adapted in very arid regions and have dealt with considerable environmental constraints, sometimes showing an ability to create adequate or even innovative responses. However, evidence also exists of major environmental damage such as salinisation, crop failure and water pollution that resulted in diseases or had more profound, far-reaching socio-environmental consequences (Mithen, 2012; Norrie, 2016: 15).

As societies have encountered a diversity of environmental constraints, over time and space, we have found that a conference on how societies tackle their challenges in relation to water is timely and contributes to a wider theoretical knowledge about this critical resource that is relevant today. Based on this premise, a conference entitled 'Comparative Water Technologies and Management: Pathways to Social Complexity and Environmental Change' was held at University College London in April 2016. Participants with diverse backgrounds were asked to contribute to an interdisciplinary dialogue on water issues that would help to connect the past and give insights into the present. The first theme is understanding the past through modelling or reconstructing water systems in order to better understand their process, system and management. The second theme is the technologies of water. It looks at how societies co-evolved with water systems so that those systems both served societies and enabled new forms of societies to emerge across time. The last theme addresses the relationship of water and societies in different times and places in order to understand the mechanism of sustainable water use; several speakers investigated this theme by looking at critical examples that allow us to investigate commonalities of water-management practices under different socio-environmental conditions.

Few volumes take a long-term and cross-regional perspective that presents different case studies together to demonstrate the variety of ways in which societies have evolved alongside water. This volume is not just the direct output of that conference; it also seeks to understand how societies have dealt with water resource challenges, or used water in ways that allowed them to evolve or persist, or drastically alter their environment.

Commonalities of water management

Technologies and water

Chapters presented in this volume identify the commonalities of water management in societies across a wide range of ecological and climate zones. This constructs a solid step towards the development of effective solutions to some acute water-related problems. A constant theme across time is the need for water, to fulfil which technologies have changed to make obtaining and controlling it easier. Technologies evolve accordingly and become closely intertwined with social and developmental issues. Water technologies include irrigation canals, underground *qanats*, aqueducts, dams, and modern water technologies that desalinate or clean polluted water. Driessen and Abudanah investigate the chronology and extension of surviving ancient water-harvesting and -distribution systems in southern Jordan, some of which were still functioning not long ago. Their careful dating exercise contributes to a more reliable chronological reconstruction of some water-management features from Nabataean to Islamic times. The Udhruh *qanat*, a subterranean water system consisting of 'vertical *qanat* shafts and horizontal underground water conduits', might be dated to around 700–600 BC. While it undoubtedly required great innovation and investment to design and build such water-management systems, the authors point out that maintaining knowledge and experience was key to their successful functioning. This point is further emphasised by Bunbury in her geoarchaeological survey of and textual information about water-management systems and ecological habitats in the Lower Nile. She proposes that the development of the *qanat*, and of other technologies, benefited from this knowledge accumulation. Knowledge about ecological cycles and past experiences of water management could be inherited and passed on to subsequent generations. Innovations that lead to intensification of water management (and agricultural production) often result from legacy derived from earlier societies (Bunbury, Chapter 3 in this volume) or are shaped by a collective memory of water management.

Distinguishing water-related features of the past is an important aspect of understanding past water use. Jotheri's chapter looks at key features that differentiate natural and anthropogenic channels in ancient Mesopotamia, where irrigation features such as dams, canals and conduits of early hydraulic systems have received much scholarly attention in the past. On the other hand, the functions of prehistoric and early historical irrigation features in East and South East Asia, often comprised of moats and dams, are less well understood by Western scholars. This contrast arises from an important ecological distinction between rice and wheat farming. For the former, drainage is as important as irrigation. The two chapters on China investigate the construction and sustainment of moat-and-ditch-based water-management systems and how they are constrained by the ecological and environmental conditions of the Middle and Lower River Yangtze. While knowledge and engineering of micro-topography in and around the fields may have been minimal, these moat-and-ditch-based systems maximise the benefits for rice farming and flood prevention. The scales dealt with in these two chapters are different: Mo and Zhuang's chapter focuses on the long-term developmental trajectories of large moated sites and their socio-economic consequences, and Zhuang's emphasises the effect of local soil and hydrological conditions on the expansion of paddy fields. Thus, these western and eastern Asian examples are significant in demonstrating the technological evolution of ancient water management, with a remarkably similar emphasis on the importance of knowledge to successful water management.

While technologies do provide benefits, often immediate, by making it easier to obtain potable or usable water, or even to store it, they also create potential long-term problems unless there is an adequate social adaptation and response to them. Some technologies, such as the enormous hydraulic enterprise in the Liangzhu culture (3300–2300 BC) of the Lower Yangtze of China, show that complex systems can emerge quickly, as threats such as flooding develop and create the need for the technology (Liu et al., 2017).

Moreover, technologies require long-term maintenance. The costs of modern and ancient technologies, such as sophisticated irrigation works, dam construction and desalination projects, sometimes make them vulnerable to such problems as insufficient funds for maintenance or high labour requirements that are not always met, and less adaptable to major environmental change. Technologies that are low in cost, and adaptable to different forms of societies (large or small) and to different environments, may prove more useful in the long term, as societies undergo inevitable change in their economic and social structures. Resilience is not a characteristic simply of the technology but also of society's ability to use that technology effectively regardless of social change. In effect, what we see is that dependence on technology without adequate social adaptation to using and sustaining it makes its use vulnerable. New technological achievements even today in many cases parallel similar developments in the past, such as when *qanats* or waterwheels were made to facilitate the extraction of clean water. It is the parallel development of management strategies and social adaptation to new water use possibilities in a given social structure that has been among the more successful historical adaptations to technologies (Scarborough, 2017; Scarborough et al., 2012; Angelakis et al., 2005).

For modern societies, low-cost, easy-to-make solutions have been found to be more suitable, particularly in regions vulnerable to drought (Lasage et al., 2008). The benefit of simple, adaptive methods is that they allow communities to form strong social bonds and networks through the maintenance of these systems. These solutions also avoid measures that require a lot of resources. Effective systems enable new social norms to develop that revolve around water maintenance and use, in environmental conditions such as drought. We see similar cases in the past, where technologies that are closely linked with communities and are relatively easy to maintain have often had very long-term durability and use. Stargardt's chapter demonstrates, similarly, how ancient technology adapted to different environmental and social settings. In effect, technologies that divorce the environment from the social setting may become more vulnerable by ignoring local challenges, social needs and environmental history. The water-management practices of various past societies, developed to consider the social, physical and economical needs for water concurrently, show modern water managers the benefits of bringing communities, cultures and water needs closer together (Keen, 2003).

Intensification of water management and economic production

The maintenance and enhancement of large cities and agricultural systems have been primary goals of modern (Pinderhughes, 2004) as well as past societies. Cabrera and Arregui (2012) examine a variety of management and other patterns related to large-scale agricultural systems and urban water provision. In addition to this focus on engineering and technical development, some of the chapters included in our volume show the socio-economic dimension of intensified water management and economic production.

Merv, or Sultan Kala, in central Asia witnessed different episodes of urbanisation through history. In the Hellenistic period the city expanded to around 340 ha, and it continued to grow through different periods on the two banks of the local river. The Islamic period saw the construction of large-scale water-management facilities, coinciding with the building of religious buildings in the city. Drawing upon his team's survey and excavation data, Williams provides a detailed calculation of the increasing water demand from daily activities, industrial production and, especially, religious purposes at a time when the city was booming. This flourishing was closely related to the careful planning and construction of the water supply systems, mainly consisting of canals, conduits and water pipes inside and surrounding the city, and to water usage and disposal aspects. Williams discusses how these systems contribute to the functional-civil dichotomy of Islamic cities proposed by Paul Wheatley (see Williams, Chapter 8 in this volume), and reflects on the classic theory of the centralised state management of hydro-infrastructures and its impact.

A top-down process of intensifying water management can also be seen in ancient Korea. Lee scrutinises archaeological evidence and historical documents from the proto- and Three Kingdoms periods in ancient Korea and demonstrates the close combination of agricultural revolution and intensified irrigation that drove the development of large-scale agricultural production. Clear archaeological evidence shows that technological innovations, including the increasing use of iron farming tools and the adoption of traction animals for ploughing, enabled the emerging state society to establish crop rotation and intensify agricultural production. A key aspect of this intensification process was the construction of large-scale dams organised by the state.

As Stargardt demonstrates, while the construction of hydro-infrastructures may be organised by states, their functioning and maintenance, and water distribution, do not always follow a top-down, institutional process (also see Zhuang et al., 2017). The central Asian and ancient Korean cases nonetheless illustrate that states could initiate the large-scale construction of hydro-infrastructures, even though it is clear that the incentives for such intensification, and its socio-economic consequences, vary greatly from case to case. Equally clear is that the long-term development of water management is not always sustainable, and ecological outcomes sometimes affect society negatively (Zhuang et al., 2017).

Ecological consequences

The ecological consequences of agricultural intensification are a shared concern of societies which practise diverse farming under different environmental conditions. Salinisation, for instance, derives from long-term, intensified, irrigated agriculture that could occur not only in semi-arid environments but also in monsoonal areas. Poor land use practices accelerate the salinisation process and aggravate land degradation. Many societies have developed mechanisms to mitigate this problem, some very simple (for example fallowing), others complicated and advanced such as developing salt-tolerant crops (Connor et al., 2011; Hillel, 2000). Archaeologists have long been aware of the occurrence of salinisation in different ancient environmental and socio-economic contexts (Stargardt, 1976; Wilkinson, 2003). Altaweel's chapter demonstrates the 'optimal strategies for irrigation' adopted in ancient Mesopotamia, with updated evidence. The delicate decision of 'too much or too little' water facing local farmers on a daily or short-term basis fundamentally defined the drainage and fallowing patterns. Such carefully planned fallowing patterns remained crucial to long-term sustainable farming, as can be supported by anthropological surveys in the region (Fernea, 1970).

As noted by Altaweel, greater pressure arises 'where irrigation salinity is more substantial'. The responses, both from agriculturalists and from states or higher-level organisations, become vital to the maintenance of ecological equilibrium and long-term sustainability. Decision making involves mediating potential conflicts of interest between different stakeholders, an issue that is also discussed by Fernea.

Driessen and Abudanah, as well as discussing similar issues of overwatering causing salinisation of soils in the functioning of *qanats*, observe the so-called 'hit-and-run agriculture' practised by contemporary farmers who 'seem to be interested only in a quick profit'. It leads to 'an enormous waste of precious water' and deteriorating soil conditions. Such a problem is illustrative as well as alarming, as it shows how quickly soil can degrade if ill-grounded policies focused on short-term yields are implemented. This brings us to the topic of water and societies, to which many of the chapters contribute significant theoretical insights.

Theoretical underpinnings: Water and societies

Classical theory on the close relationship between water management and social evolution, proposed by Wittfogel (1981), has been disputed and modified by archaeologists and anthropologists. Many archaeological and anthropological examples have demonstrated dynamic relationships between social complexity, local environmental conditions and the development of hydraulic engineering. These challenge Wittfogel's idea of the 'hydraulic society' whose agricultural production depended on large-scale irrigation. Wittfogel further proposed that the oriental type of despotism, different from that in Western societies, stemmed from an 'absolutist managerial' structure. This theory was strongly criticised by Adams (1981) for ancient Mesopotamia, and by Butzer (1976), who worked on ancient Egyptian water management. The contribution by Stargardt (Chapter 12 in this volume) further criticises the Wittfogelian model by providing five different cases of ancient hydraulic systems from societies with diverse ecological settings and socio-economic institutions in South and South East Asia. Continuous research in some early urban centres, such as Angkor Wat, and in its hydraulic landscape has further illustrated that an emerging pattern characterises the relationship between organisation of water management, local environments and social evolution. However, as suggested by Stargardt, there is a need to examine small-scale hydraulic systems, such as village-based water-management infrastructures. She argues that the 'vital technical knowledge, the ability to assemble and employ the required materials and coordinate irrigation labour' lie in the achievement of a fine balance of these aspects by village groups. To a large extent, Zhuang's chapter echoes this important point.

Another salient characteristic some of the chapters deal with is the close association between ideological change and the formation of urban and rural hydraulic landscapes. Building on earlier work by Geertz (1972), Lansing (2007) has vividly demonstrated an alternative trajectory to the intensification of agricultural production and water management in Bali that was centred on the management of *Subak* (terraced paddy fields) around water temples.

Several chapters explore how water is incorporated into different social systems. Themes explored include the economic, ideological and philosophical roles of water in society. Indeed, water systems, and irrigation in particular, co-evolve with social development: religion, language, customs and social norms often reflect a close coupling with the natural systems they develop in. Julia Shaw's research on early Buddhism and water management in India has brought together the strengths of field survey data, archaeological data and textual evidence to provide a coherent understanding of this important topic. Her chapter expands this theme to introduce environmental ethics to the debate. As acute readers will notice when reading some of the relevant chapters, the dynamics between water management and society are much more complicated and discursive than the one-sided perspective adopted by Wittfogel suggests. Religion, ideology and other 'spiritual' aspects of humans could also act as agents in shaping past hydraulic landscapes, in either a positive or a negative manner (see also the issue of *World Archaeology* on 'Archaeology and Environmental Ethics', 2016, volume 48).

While these new theoretical contributions have aimed to adopt a more balanced perspective on the causal relationship between water and societies, religion has always been placed in a predominant position in this transforming scholarship. A more dynamic and discursive perspective on the interactions between water and societies is needed. Williams's chapter looks at this question from a different angle. He argues that water systems and technologies shaped the identity of communities in the early Islamic city of Sultan Kala in present Turkmenistan. Societies' decision-making process regarding water management is constrained by many factors other than religion. Spiritual needs and use of water intertwine and develop with water systems, which also fulfil the basic human need for water. The physical and spiritual needs for water are related, and the spiritual reinforces the importance and relevance of water to societies.

Zhuang and colleagues (Zhuang and Kidder, 2014; Zhuang et al., 2017) provide cases in ancient China and Korea that address the delicate choice facing states and societies with growing populations and expanding territories. Communities along the Yellow River in Henan province had been reclaiming lands from river floodplains for farming during the Han period (206 BC–AD 220). The government was aware of the potential consequences but had to succumb to this movement because of the increasing conflict between the unprecedented growth of population and the amount of arable land available. Their solution was to build levees along the river, but this led to a breach that caused a massive flood in about AD 14–17, which submerged the village compounds and the farming landscape (Kidder et al., 2012; Zhuang and Kidder, 2014). Lee's contribution explores this dilemma in the form in which it faced the Three Kingdoms of ancient Korea. By showing how agricultural and technological innovations combined to foster societal reform in the Three Kingdoms–Unified Silla period of the Korean peninsula, she rethinks the relationship between social resilience and environmental vagaries in the shaping of agrarian societies in East Asia.

There is a legend about the Great Yu's taming of floods, which was written down in the Han (2nd century BC) period. The Great Yu was believed to have founded the Xia Dynasty (c. 2000 BC). Though the veracity of the legend has long been disputed, inscriptions on a Western Zhou-era (c. 1040–771 BC) bronze called *He zun* confirm that, at least by the beginning of the first millennium BC, the legend about the Great Yu taming water was already popular, at least in elite circles. Later, the Great Yu's image repeatedly appeared in Han-period pictorial evidence, and his 'association' with water was further popularised. Similarly, there are many stories about water, religion and society (see also Shaw, Chapter 11 in this volume). Water might appear in different forms in society, but water issues should not be scrutinised without investigation of the social interactions and developments that may relate to it.

Methodological contributions

The recurrence of water issues, such as water shortage, pollution and floods, across space and time and the diversity of adaptations and responses to water conditions have shaped the approaches used in understanding water-management practices, in which methods often become interdisciplinary by nature. However, while interdisciplinary research has become fashionable, dealing with the complex nature of the material and interpretative methods remains challenging (Lach, 2014). The conference that inspired this volume was dedicated to creating a platform for interdisciplinary dialogue, and we have seen very lively communication among the conference participants, who are field archaeologists, civil engineers, geologists and environmental scientists.

Water remains a crucial interface at which physical environmental conditions are transformed into hydro-infrastructures for water management. Most of the chapters included in this volume are based on surveys and excavations of water-management facilities and the environments in which they are built. These environments generate different fossils or sediments, both susceptible to taphonomic issues; caution is needed when dealing with a wide array of evidence for water-management practices. As well as presenting first-hand data and illustrating cases, the chapters aim to develop holistic approaches with strict criteria for multidisciplinary enquiry into past and present water management, including the integration of modern and ancient environmental data.

The range of evidence of water management that is often under scrutiny by scholars includes biotic, abiotic, ethnographic and archaeological data, including their spatial patterns. For biotic data, one of the most persisting and challenging issues is taphonomy. How were plant remains, for instance, brought back to archaeological sites and how did they get preserved? Was it via charring or waterlogging processes? Such issues should be examined before plant remains can be used as a reliable proxy to reconstruct past water regimes, growing environments, and ecological modification by water-management strategies. Abiotic proxies, albeit less vulnerable to post-depositional alteration, should be used with caution in archaeological investigations of past water management. The main issue lies in the business of making correlations between environmental data and archaeological evidence. Data correlation is often rendered difficult by a lack of high-resolution chronological data, and by the fact that, even when it is available, it may have been obtained by different parties, who have used different processing procedures or different sample selection criteria. In effect, this data, which can contribute to long-term perspectives on water management, may complicate, indeed cloud, our understanding of water management.

The interpretation of ethnographic, archaeological, archival and interview data is even more complex, primarily because different researchers adopt different data-retrieval techniques and because the stakeholders involved in interviews and surveys are intrinsically complicated. A simple example of the latter point is the managerial coordination of water use upstream and downstream of a river watershed (Cai et al., 2003; Jønch-Clausen and Fugl, 2001; Richter et al., 2003). People living in different watersheds might react differently to the same policy. The study of water-related issues is thus diachronic in nature. On the one hand, water issues remain critical to communities across time and space. On the other, there is often a disconnect between the present and the distant past, or at least the latter might be perceived as having little to do with the present situation. It is this alerting disassociation, which sometimes serves to escalate water shortage, pollution and many other problems, that calls for more interdisciplinary research towards a holistic understanding of ancient and contemporary water issues.

Similarly important to such an approach is the *longue durée* perspective. This is evident, for instance, in irrigation projects in traditionally long-occupied and heavily farmed areas, such as the Yellow, Yangtze, Ganges, Euphrates and Tigris rivers, which are an outcome of long-term technological and economic developments that sometimes replace, but are also integrated with, older irrigation practices. While modern technology has distanced societies in some cases from water (e.g., Bone et al., 2011), in other cases it has facilitated new solutions that promote efficient and novel methods of conservation and improve water quality and monitoring, for example technologies that convert polluted water to reusable water, or unmanned vehicles for better monitoring and decision making (Chaabane et al., 2017; Stagakis et al., 2012). A *longue durée* perspective is important for an understanding of how technologies and economic conditions accumulated that led to the construction of large-scale irrigation infrastructures, a process often defined as 'intensification', and why these were used sustainably over a long period.

Another useful approach to examining the long-term trajectories of water management is to disentangle the seasonality, or, as Bunbury puts it in Chapter 3, cyclicity, of water (climate) and society. Cycles of local ecosystems follow annual, decadal or longer scales, and researchers may sometimes be fortunate enough to capture the moment at which such cycles begin or finish, for instance when irrigation ditches are covered by construction material during repair, or when irrigation systems are sealed by flooding deposits that cause their abandonment (Liu et al., 2017). Societies have often developed ways of adapting to short-term cyclical changes such as drought. However, sometimes evidence indicates risk trade-offs between long-term ecological equilibrium and short-term outcomes due to population pressure or other societal reasons. It is an important fact that ecological feedback often presents a time lag, which means that societies often have to bear the ecological consequences of the socio-economic actions of their predecessors.

Another focus for this *longue durée* perspective is computational modelling of long-term change to water-based systems. A benefit of archaeological data is that they provide a deep temporal understanding of past social change. Archaeological data can be used to explore how societies developed but also changed their social and environmental systems through long-term use. Issues such as salinisation, which is investigated in Altaweel's chapter (9), are affected by long-term water use in arid regions susceptible to high salinity.

Integrating past, present and prospective water practices

As briefly mentioned above, many water-related issues, such as shortage, uneven distribution, and conflicts arising from unfair laws or improper management, recur throughout history. Not only are water-management facilities used continuously for very long periods in regions such as those of the River Yangtze and the River Ganges, knowledge of sustainable water management and local ecosystems has been accumulated and transmitted through the generations. Thus, not only is the connection between past and present water practices crucial to the survival and development of local communities, but an understanding of these practices is also indispensable to the future sustainability of these regions. This latter point is, unfortunately, often neglected, because local communities are often unaware of such issues. Bunbury's contribution touches upon such topics (Chapter 3). One of the challenges in contemporary water management is how to reconnect past and present water practices in order to maximise the ecological and economic benefits for local areas of sustainable water management. Some argue that knowledge of traditional and sustainable water management 'hibernates' in local communities, and that it is the researcher's responsibility to reawaken this knowledge in different stakeholders.

This ecological continuity is advocated by Stargardt in the concluding statements of Chapter 12, where she shows that knowledge could be 'transmitted orally over long periods'. Maurits Ertsen and Sarah Bell make this point even more strongly. Ertsen's chapter blends an in-depth study of the Hohokam culture in America, which built canals on an impressive scale, with theoretical considerations of the human and non-human agencies in water management. It confirms the complex interlink between social and environmental factors, and between natural and cultural agencies, in our research approach to water management. His chapter examines the 'materiality' of water. Though this somewhat radical proposal needs more theoretical construction, and the backup of more practical examples, it is absolutely necessary to contextualise water (and decision making about water) in its natural and cultural backgrounds.

A fallacy of sustainability is that it can be achieved without adaptation. On the one hand, social systems and water technologies develop in tandem, but changes to one can cause disruption in, or require multiple adaptive responses from, the other. In the modern world, population and water demand pressures do this, while in the past, similarly, water and social systems continually changed, or had to adapt to, the other. In effect, a fully sustainable model is not achievable where systems are constantly evolving. What has been required are social systems that can rapidly modify and adapt as new disturbances to the balance between social and water systems emerge. Sarah Bell, as a civil engineer, presents the urgent need to establish a holistic, integrated approach to understanding past and present water management for the benefit of the future. Sustainability, she says, lies in an understanding of the multiple discourses about it.

Another past–present perspective associated with water is that modern political controversy often haunts scientific research and biases archaeological investigations. It is easy to imagine this affecting scholarly enquiries into the alluvial history of major rivers that cross national borders and are intertwined with ancient and modern cultural and political situations in the countries through which they run. This is a sensitive field that most scholars are either too immersed in to realise its problems, or too afraid or biased to have an objective view. Chapter 2, by Clift and Giosan, touches upon this by presenting unambiguous evidence of the Holocene evolution of alluvial systems in the Indus basin and showing how this is crucial to our understanding of societal evolution in the area. Undoubtedly, there is still a long way to go in this research direction, and it requires interdisciplinary endeavour.

References

Adams, Robert McCormick. 1981. *Heartland of Cities: Surveys of Ancient Settlement and Land Use on the Central Floodplain of the Euphrates*. Chicago: University of Chicago Press.

Angelakis, A. N., D. Koutsoyiannis and G. Tchobanoglous. 2005. 'Urban wastewater and stormwater technologies in ancient Greece.' *Water Research* 39 (1), 210–20.

Bone, Christopher, Lilian Alessa, Mark Altaweel, Andrew Kliskey and Richard Lammers. 2011. 'Assessing the impacts of local knowledge and technology on climate change vulnerability in remote communities.' *International Journal of Environmental Research and Public Health* 8 (3), 733–61.

Butzer, Karl W. 1976. *Early Hydraulic Civilization in Egypt: A Study in Cultural Ecology*. Prehistoric Archaeology and Ecology Series. Chicago: University of Chicago Press.

Cabrera, Enrique and Francisco Arregui. 2012. 'Engineering and water management over time: Learning from history.' In Enrique Cabrera and Francisco Arregui (eds), *Water Engineering and Management through Time: Learning from History*, 3–28. Boca Raton, FL: CRC Press.

Cai, X., D. C. McKinney and M. W. Rosegrant. 2003. 'Sustainability analysis for irrigation water management in the Aral Sea region.' *Agricultural Systems* 76 (3), 1043–66.

Chaabane, Safa, Khalifa Riahi, Hédi Hamrouni and Béchir Ben Thayer. 2017. 'Suitability assessment of grey water quality treated with an upflow-downflow siliceous sand/marble waste filtration system for agricultural and industrial purposes.' *Environmental Science and Pollution Research* 24 (11), 9870–85.

Connor, D., R. S. Loomis and K. G. Cassman. 2011. *Crop Ecology: Productivity and Management in Agricultural Systems*. 2nd edn. Cambridge: Cambridge University Press.

Geertz, Clifford. 1972. 'The wet and the dry: Traditional irrigation in Bali and Morocco.' *Human Ecology* 1 (1), 23–39.

Fernea, Robert A. 1970. *Shaykh and Effendi: Changing Patterns of Authority among the El Shabana of Southern Iraq*. Middle Eastern Studies. Cambridge, MA: Harvard University Press.

Hillel, Daniel. 2000. *Salinity Management for Sustainable Irrigation: Integrating Science, Environment, and Economics*. Washington, DC: World Bank.

Jønch-Clausen, Torkil and Jens Fugl. 2001. 'Firming up the conceptual basis of integrated water resources management.' *International Journal of Water Resources Development* 17 (4), 501–10.

Keen, Meg. 2003. 'Integrated water management in the South Pacific: Policy, institutional and socio-cultural dimensions.' *Water Policy* 5 (2), 147–64.

Kidder, Tristram R., Haiwang Liu, Qinghai Xu and Mingin Li. 2012. 'The alluvial geoarchaeology of the Sanyangzhuang site on the Yellow River floodplain, Henan Province, China.' *Geoarchaeology* 27, 324–43.

Lach, Denise. 2014. 'Challenges of interdisciplinary research: Reconciling qualitative and quantitative methods for understanding human–landscape systems.' *Environmental Management* 53 (1), 88–93.

Lansing, J. Stephen. 2007. *Priests and Programmers: Technologies of Power in the Engineered Landscape of Bali*. Princeton, NJ: Princeton University Press.

Lasage, R., J. Aerts, G.-C. M. Mutiso and A. de Vries. 2008. 'Potential for community based adaptation to droughts: Sand dams in Kitui, Kenya.' *Physics and Chemistry of the Earth* 33 (1–2), 67–73.

Liu, Bin, Ningyuan Wang, Minghui Chen, Xiaohong Wu, Duowen Mo, Jianguo Liu, Shijin Xu and Yijie Zhuang. 2017. 'Earliest hydraulic enterprise in China, 5,100 years ago.' *Proceedings of the National Academy of Sciences* 114 (52), 13637–42.

Liu, Jianguo and Wu Yang. 2012. 'Water sustainability for China and beyond.' *Science* 337 (6095), 649–50.

Mithen, Steven. 2012. *Thirst: Water and Power in the Ancient World*. London: Phoenix.

Norrie, Philip. 2016. *A History of Disease in Ancient Times: More Lethal than War*. New York: Palgrave Macmillan.

Pinderhughes, Raquel. 2004. *Alternative Urban Futures: Planning for Sustainable Development in Cities throughout the World*. Lanham, MD: Rowman & Littlefield.

Richter, Brian D., Ruth Mathews, David L. Harrison and Robert Wigington. 2003. 'Ecologically sustainable water management: Managing river flows for ecological integrity.' *Ecological Applications* 13 (1), 206–24.

Scarborough, Vernon L. 2017. 'The hydraulic lift of early states societies.' *Proceedings of the National Academy of Sciences* 114 (52), 13600–1.

Scarborough, Vernon L., Nicholas P. Dunning, Kenneth B. Tankersley, Christopher Carr, Eric Weaver, Liwy Grazioso, Brian Lane, John G. Jones, Palma Buttles, Fred Valdez and David L. Lentz. 2012. 'Water and sustainable land use at the ancient tropical city of Tikal, Guatemala.' *Proceedings of the National Academy of Sciences* 109 (31), 12408–13.

Stagakis, S., V. González-Dugo, P. Cid, M. L. Guillén-Climent and P. J. Zarco-Tejada. 2012. 'Monitoring water stress and fruit quality in an orange orchard under regulated deficit irrigation using narrow-band structural and physiological remote sensing indices.' *ISPRS Journal of Photogrammetry and Remote Sensing* 71, 47–61.

Stargardt, Janice. 1976. 'Man's impact on the ancient environment of the Satingpra Peninsula, South Thailand. I. The natural environment and natural change.' *Journal of Biogeography* 3 (3), 211–28.

Wilkinson, T. J. 2003. *Archaeological Landscapes of the Near East*. Tucson: University of Arizona Press.

Wittfogel, Karl A. 1981. *Oriental Despotism: A Comparative Study of Total Power*. 1st Vintage Books edn. New York: Vintage Books.

Zhuang, Yijie and Tristram R. Kidder. 2014. 'Archaeology of the Anthropocene in the Yellow River region, China, 8000–2000 cal. BP.' *The Holocene* 24 (11), 1602–23.

Zhuang, Yijie, Heejin Lee and Tristram R. Kidder. 2017. 'The cradle of heaven-human induction idealism: Agricultural intensification, environmental consequences and social responses in Han China and Three-Kingdoms Korea.' *World Archaeology* 48 (4), 563–85.

I

Modelling long-term change

2 Holocene evolution of rivers, climate and human societies in the Indus basin

Peter D. Clift and Liviu Giosan

Abstract

The Indus valley (or Harappan) civilisation is known to have experienced significant deterioration after around 2000 BC when many of the major urban centres were abandoned. This societal decay has been linked to a drying of the climate at the same time, when lakes in the Rajasthan area contracted and dried up, and discharge from the River Indus reduced. Although rivers may also have played a part in supplying water to the Harappan, attempts to reconstruct the Holocene drainage network show that most major reorganisation preceded human settlement by a significant amount of time. Any connection between the River Yamuna and the Indus basin was severed no later than 10,000 years ago. Instead, seasonal flooding by the River Ghaggar-Hakra, sourced in the Himalayan foothills, was key to successful agriculture during the Mature Harappan cultural phase, but declined as the monsoon rains weakened after 3000–2000 BC. Subsequently the Ghaggar-Hakra was overwhelmed by dunes in an expanding Thar Desert. Direct comparisons between the Ghaggar-Hakra and the mythical, glacier-fed River Saraswati are hard to support during the Bronze Age.

Introduction

The Harappan, or Indus valley, civilisation is one of the oldest known urban cultures and was located within the Indus drainage basin to the west and north-west of the Thar Desert in what is now north-western India and Pakistan (Figure 2.1). The Harappan society is known to have suffered deterioration after around 1900 BC, which resulted in the abandonment of the large urban centres and the dispersal of the population. The causes

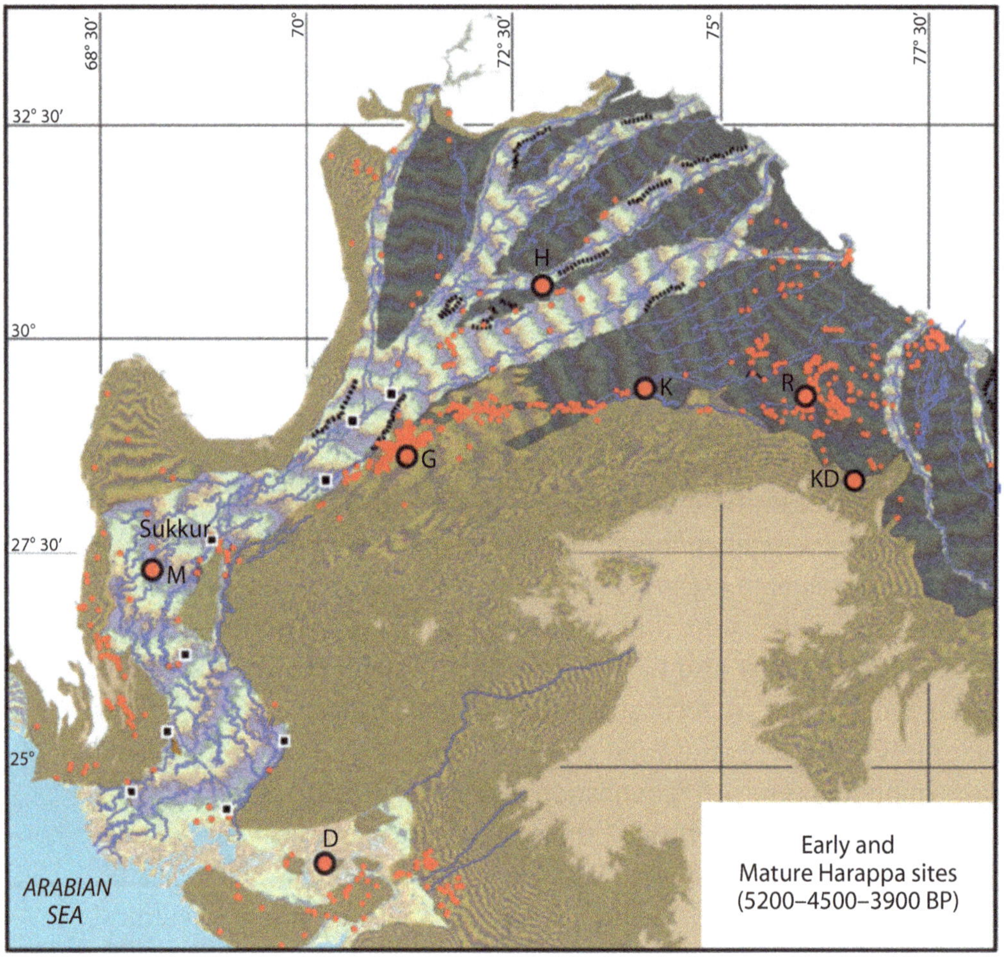

Figure 2.1 The main Harappan landscapes of the western Indo-Gangetic plain: large-scale morphology of the region (interfluves or bar uplands in grey mask; terrace edges in dashed black lines) with river channels, and chronological information. Early and Mature Harappan sites with the names of some major urban centres: D = Dholavira; M = Mohenjo-daro; G = Ganweriwala; H = Harappa; K = Kalibangan; KD = Kotla Dahar; R = Rakhigarhi. Modified from Giosan et al., 2012

of this change remain controversial but are often linked to climate change and especially to a weakening and a sustained low intensity of Asian summer monsoon rains at that time (Madella and Fuller, 2006; G. Singh et al., 1974). This hypothesis was widely accepted, especially outside the archaeology community (Possehl, 1999), in view of the similar declines seen in other early societies, such as in Mesopotamia (Akkadian Empire), in Egypt (Old Kingdom) and in northern China (Qijia Culture) (Lawler, 2008), together with the recognition that monsoon strength, as tracked by many proxies, appears to weaken during the Holocene. This rough synchroneity in societal change is suggestive of a link to wider, possibly global, climate change, impacting high-density centres.

However, river systems may also have been influential in the fate of the Harappan. Societies have shown an ability to survive and even thrive in otherwise arid environments when they have ready access to water from nearby rivers, which can be used to irrigate fields and supply domestic water to towns. The Nile has been critical to human society in Egypt despite being surrounded by the Sahara Desert. The possibility that the development of rivers might have played a part in cultural evolution is moreover hinted at by the appearance in satellite and aerial photographs of abandoned fluvial channels on the edge of the desert, as well as in the mythology of local religions.

The Rig Veda, a Sanskrit Hindu religious text, tells of a major river known as the Saraswati or Sarasvati that flowed in this part of the subcontinent, which was subsequently interpreted as a major stream originating from glacial sources in the Himalayas (see discussion in Aklujkar, 2014). Most recent scholarly attempts to locate this stream place it close to the modern River Ghaggar in Indian Punjab (Figure 2.2), although currently at least this river is ephemeral and only drains the foothills of the Himalayas. Downstream, where it flows through Pakistan, the Ghaggar is known as the Hakra. Presently the Hakra peters out in the dune fields of the northern Thar Desert shortly after crossing the border. Because of its religious, and potentially archaeological, significance there have been several attempts to define the path and age of the Ghaggar-Hakra, downstream of where its recent course is clear in northern India. Not all the proposed pathways are realistic: some involve crossing hilly country between the Ghaggar headwaters and the Rann of Kutch where several reconstructions have this stream entering the ocean. In this study we review the evolution of streams on the north and west sides of the Thar Desert in order to determine what influence they may have had on the documented settlements of the Harappan culture.

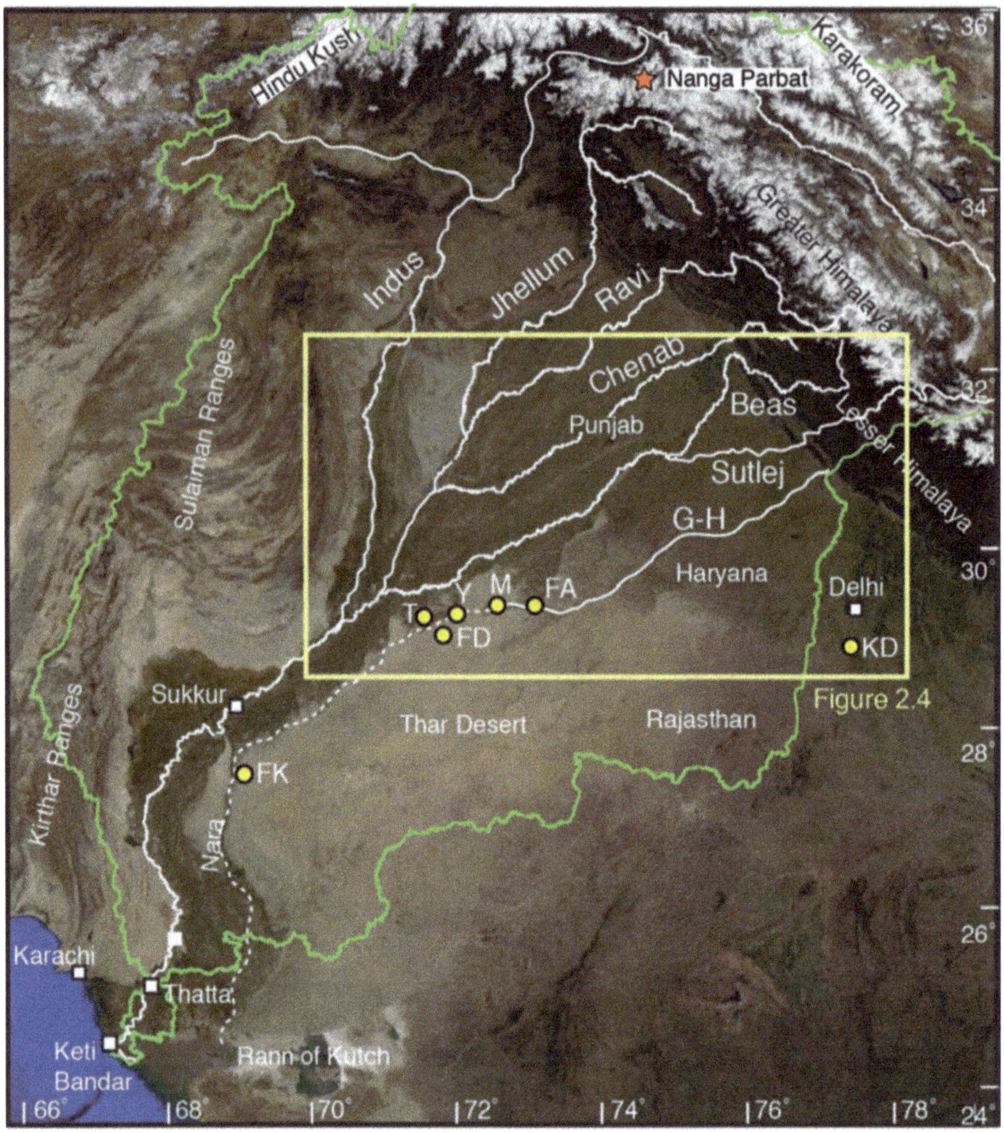

Figure 2.2 Satellite image of the Indus basin (outlined in green) showing the major sites and rivers discussed in the text. G-H = Ghaggar-Hakra; FA = Fort Abbas; M = Marot; FD = Fort Derawar; Y = Yazman; T = Tilwalla; FK = Fakirabad. (The area within the yellow rectangle is shown at a larger scale as Figure 2.4.). Source: authors

Figure 2.3 shows a reconstruction of the River Sarasvati which is typical in showing its course parallel to the Indus and its connection into High Himalayan terrain, where it is supposed to have drained a catchment with a glacial source. The lower reaches of the Sarasvati essentially parallel the Indus before reaching the ocean at the Rann of Kutch. It is debatable whether, in its upper reaches, the Sarasvati previously incorporated the headwaters of other major modern Indus tributaries. Some

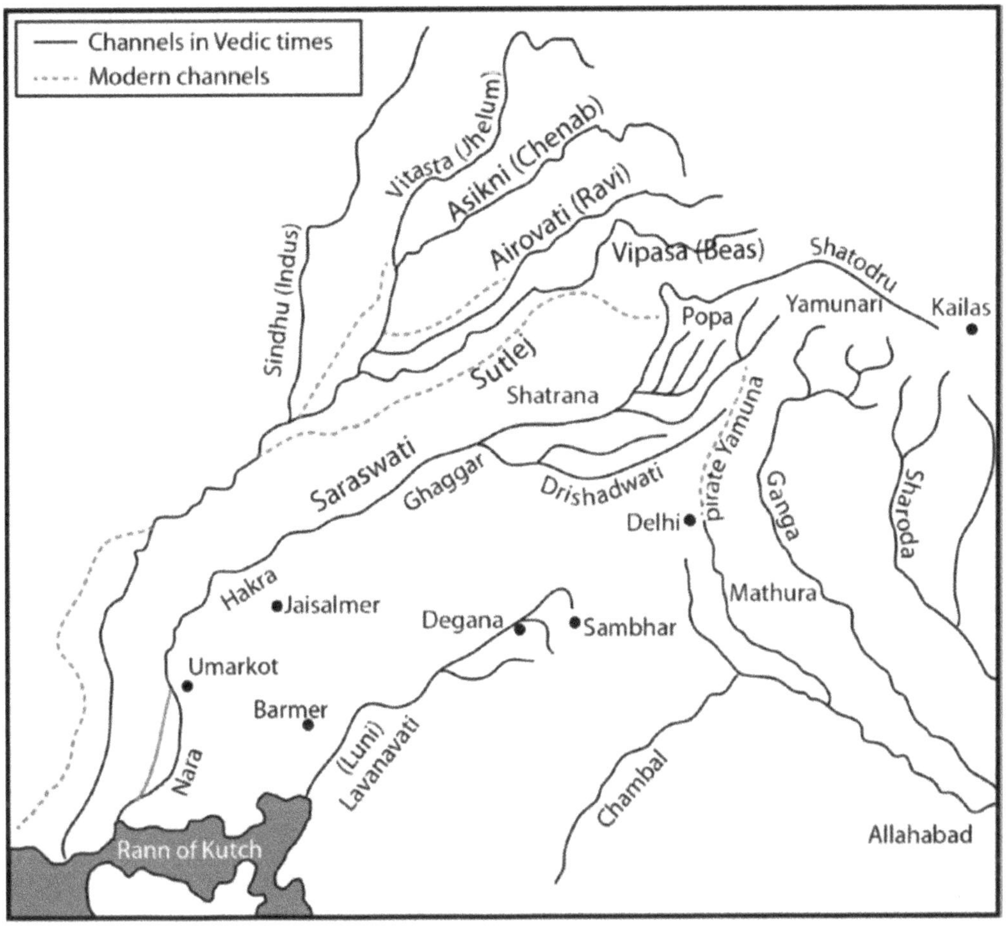

Figure 2.3 Suggested palaeo-drainage map of the western Ganga and Indus floodplains in the late Quaternary, redrawn from Valdiya, 2002. Note the proposed River Saraswati running sub-parallel to the Indus as far as the Rann of Kutch

interpretations of religious texts led to the view that the Sarasvati was a large river, potentially comparable to the Indus itself, yet the modern-day Ghaggar-Hakra is a relatively small stream that does not support such legends. In the view of Valdiya (2002) and others cited in his compilation, the river not only comprises the modern Ghaggar-Hakra but is also receiving water from the Sutlej and taps the Yamuna, which now flows towards the east, after being captured by the Ganges at an unspecified time. Certainly there were suggestions that such a connection might have occurred in the past. Coring work undertaken by Saini et al. (2009) and Sinha et al. (2013) in the Indian parts of Punjab and Haryana west of Delhi, and close to the modern course of the Ghaggar, have identified a series of essentially east–west-trending channels in the subsurface which may represent the now buried channels of the connection between the

Yamuna and the Ghaggar-Hakra (Figure 2.4). However, sediments were not fingerprinted to ascertain a Yamuna origin and the optically stimulated luminescence (OSL) dating of these channel materials indicated that flow in these courses far pre-dates the Harappan culture area further west. Indeed, this study favoured an integrated drainage system between the Yamuna and the Ghaggar-Hakra sometime between 28,000 and 18,000 BC, although there is some suggestion of a limited younger phase of river activity between 4000 and 900 BC.

Further constraints on the age of river activity around the northern, western and eastern sides of the desert came from an analysis of the river valleys and terraces around the primary tributaries of the modern-day Indus (Giosan et al., 2012). Figure 2.4 is an enhanced topographic map showing the network of the modern rivers and highlighting the fact that many of these channels are sitting in broad, albeit shallow, valleys separated by terraces or interfluves. Radiocarbon and OSL dating have demonstrated that the northern part of the floodplain, where these terraces are best defined today, was in a state of

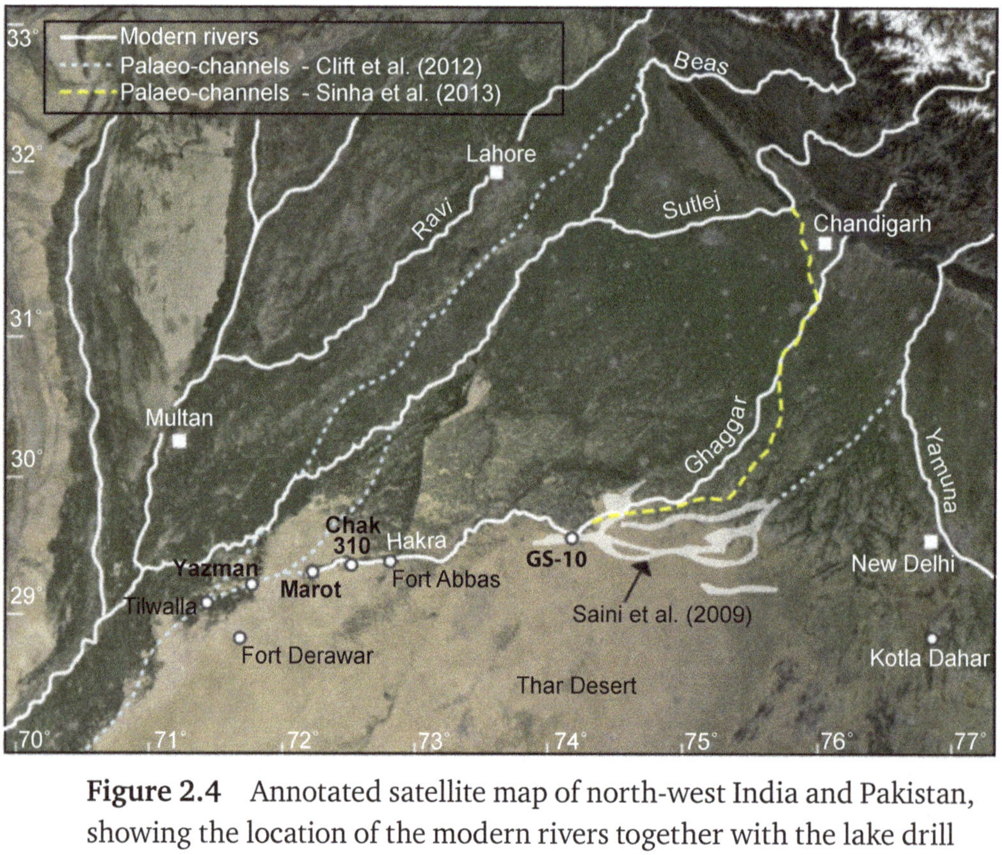

Figure 2.4 Annotated satellite map of north-west India and Pakistan, showing the location of the modern rivers together with the lake drill sites mentioned in the text, as well as the palaeo-channels proposed by Clift et al., 2012; Saini et al., 2009; Sinha et al., 2013. Source: authors

accretion until around 10,000 years ago (Alizai, Clift et al., 2011; Durcan et al., 2010; Giosan et al., 2012), after which the floodplain experienced incision resulting in the modern rivers being entrenched up to 15 m below the earlier Quaternary sediments in wide valleys that stretch to 30 to 50 km across (Giosan et al., 2012). This implied that the rivers have been in their present course since that time and to feed the Ghaggar-Hakra from their incised valley would have involved unrealistically large floods (Giosan et al., 2013). It was also noteworthy that the Ghaggar-Hakra is anomalous among Himalayan tributaries in not being located within an wide incised valley, which suggests that significant flow in this particular stream must have pre-dated the Holocene (Giosan et al., 2012). That is not to say that ephemeral or low-volume flows could not have passed along this channel, but it does mostly preclude the idea of a large-volume Ghaggar-Hakra flowing during the times of archaeological interest, as shown by the distribution of archaeological sites associated with the Early and Mature Harappan cultures. In fact, Maemoku et al. (2013) showed that sand dunes along the Ghaggar and the Chautang on either side of their valleys began to develop from *c.* 13,000 BC, with some forming as late as *c.* 3000 BC. Further south, East et al. (2015) used a combination of OSL and 14C dating to show dune advance towards the River Nara even more recently, between 1000 BC and AD 1, which suggests that the dunes may still be advancing westward, in that area at least. Thus, dunes with habitation layers already existed along the Ghaggar during the Harappan period and have not been washed away in any significant degree since then. Nonetheless, these sites cluster along the course of the Ghaggar-Hakra, which does indicate a link between settlement and water supply along this route.

Climatic evolution

The summer monsoon climate of South Asia is reasonably well constrained since the start of the Holocene (*c.* 8000 BC), as a result of a number of different proxy records that mostly support a consistent drying of the climate after 3000 BC (Ponton et al., 2012). Summer monsoon winds that blow from Arabia and towards the Indian subcontinent drive water away from the coast of Oman and generate upwelling and high productivity along that coast (Prell and Curry, 1981). Analyses of planktonic foraminifera in sediments sampled from the western Arabian Sea show an abundance of the upwelling-related form *G. bulloides* after around 9000 BC (Gupta et al., 2003), and while there is a weaker phase around 7000 BC it is clear that the upwelling entered a decreasing trend, with several weak stages at *c.* 5500 and 4000 BC, and especially after 3000 BC, and possible renewed activity over the last couple of thousand years (Figure 2.5B). Because this long-term trend correlates with a decrease in insolation at these latitudes it was concluded that solar forcing is the most important control on summer monsoon intensity, on millennial timescales.

The idea that the summer monsoon weakened progressively after *c.* 3000 BC is moreover consistent with data from other regions. Cave records from China show a steady decrease in summer monsoon intensity after 6000 BC, although the sharpest decrease appears to be somewhat younger at 1600 BC (Dykoski et al., 2005). In speleothem records, the oxygen isotope character of the calcite, which forms the deposit, is taken to preserve the composition of the water that was percolating through the rock at the time of precipitation and in turn to mirror the composition of the rainfall. As a general rule isotopically lighter compositions are interpreted to indicate heavier rainfall, and although some doubt has been cast on the reliability of these proxies (Clemens et al., 2010) they continue to be widely relied on to provide high-resolution records of rainfall across Asia during the Quaternary. Closer to the area of the Harappan culture, similar cave records at Qunf in southern Oman show a steady decrease in rainfall intensity through the Holocene, with a complete break in the record starting around 1000 BC (Fleitmann et al., 2003), which can be interpreted to indicate such dry conditions that speleothems did not grow at all. Further evidence of long-term drying of the climate in South West Asia has come both from lake records on the edge of the Tibetan Plateau and from those in the floodplains of north-west India. For example, Figure 2.5D shows a proxy based on pollen data extracted from lake sediments in the Ladakh, within the Indian Himalayas but on the edge of the plateau (Wünnemann et al., 2010). This showed a sharp drop in humidity from around 3000 BC and the establishment of the very dry climate seen in that region in the present day. In Rajasthan, on the eastern side of the Thar Desert, a number of lakes have been studied, and their sediments cored and dated, to derive a history of their development (Dixit et al., 2014; Enzel et al., 1999). In general, peak lake levels are seen in lowland India from 5000 BC to *c.* 2000 BC (Dixit et al., 2014; Swain et al., 1983). It is noteworthy, however, that lake-level rise and fall is not synchronous. Although many peak at 5000–4000 BC, Lake Nal fell sharply at 4000 BC, with Lake Lunkaransar drying up completely shortly after 3500 BC, while Didwana continued to be high until 2500 BC (Prasad and Enzel, 2006). Many of these have now completely dried up, so that there is no modern lake in existence at all (Deotare et al., 2004). A study of oxygen isotopes from the palaeo-lake Kotla Dahar in Rajasthan, very close to the Harappan sites (Figure 2.4), was able to pinpoint a date of 2100

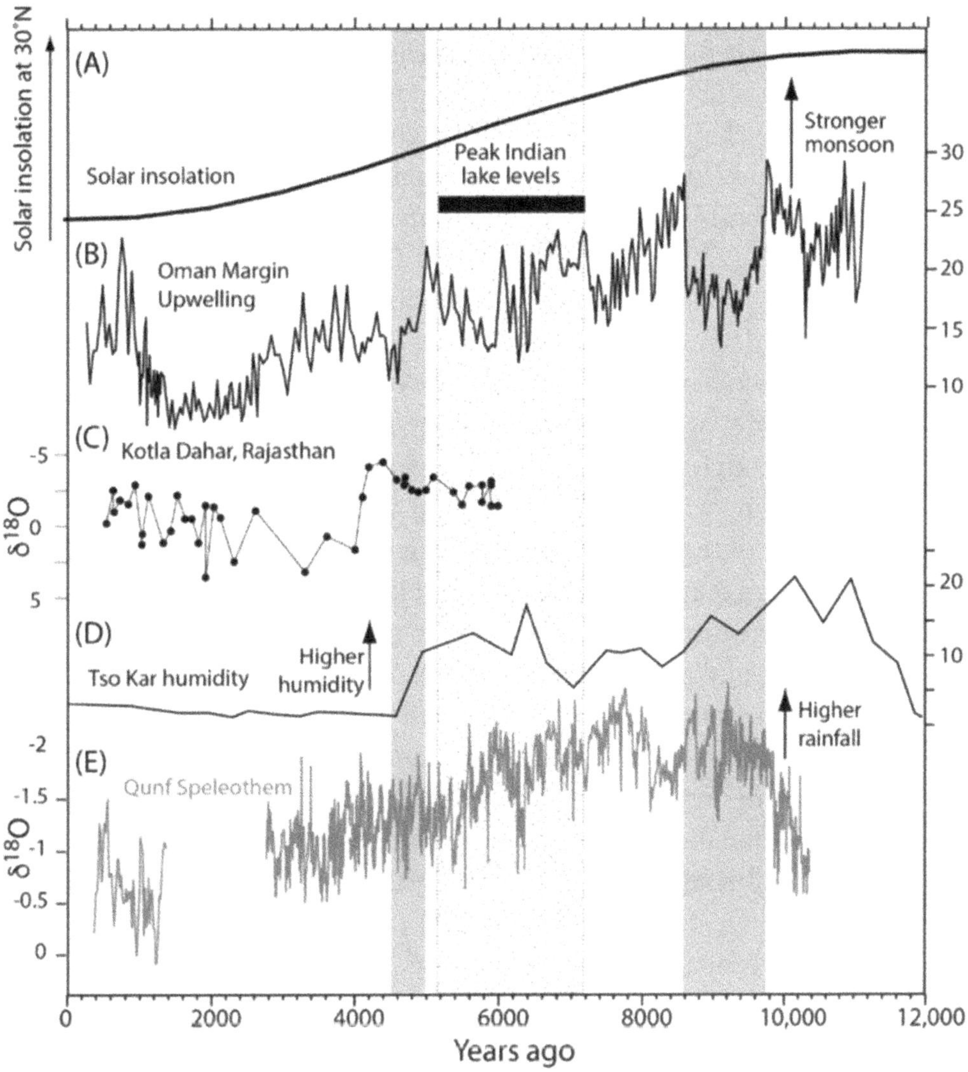

Figure 2.5 Compilation of monsoon intensity proxies showing the peak intensity in summer monsoon in 6000 BC, decreasing until around AD 1. (A) Solar insolation has been falling since its peak in 8000 BC. (B) Abundance of *G. bulloides* planktonic foraminifera in hemipelagic sediment on the Oman margin (Gupta et al., 2003). (C) Oxygen isotope data from Kotla Dahar, palaeo-lake in Rajasthan (Dixit et al., 2014). (D) Artemisia/Chenopodiaceae ratio (A/C) from a composite core from Tso Kar, Ladakh, India (Wünnemann et al., 2010). (E) Oxygen isotope data from stalagmites from Qunf Cave, southern Oman (Fleitmann et al., 2003). Grey bands show periods of rapid weakening in monsoon strength across several of the study regions. Source: authors

BC as a time of intense drought and a shift to drier climates lasting until the present day. This change is within error of the age of the Harappan deurbanisation, suggestive of a causative link (Dixit et al., 2014) (Figure 2.5(C)).

Although the timings of various lake disappearances span a longer time interval before the demise of the Harappan culture, this is consistent with the concept of a climate which was getting drier and in which, perhaps, the Harappan were increasingly struggling to maintain a viable agriculture to feed their population in their large urban centres. Indeed, the best-dated lake section from Kotla Dahar, which dated the desiccation at 2100 BC, is within error of the Harappan dispersal (Dixit et al., 2014). These climatic records are consistent with data interpreted as reflecting discharge from the Indus itself. In a study of oxygen isotopes in planktonic foraminifera offshore the Indus Delta mouth Staubwasser et al. (2003) proposed that trends in salinity link to discharge. Although the oxygen isotopic variability is rather small, a drop in discharge may have occurred around 2200 BC around the time of the termination of the Mature Harappan civilisation onshore. These workers concluded that it was drying of the climate and the establishment of late-Holocene drought cycles after that initial decline in rainfall which were principally responsible for the abandonment of the large urban centres seen along the Indus valley and especially in the region of the Ghaggar-Hakra. However, monsoon rainfall was not the principal agricultural mode for Harappans, who, in fact, depended on river floods for their inundation agriculture. Instead, based on the semi-fossilisation of the alluvial plain along the Indus, Giosan et al. (2012) proposed that the climate drying came with less extensive and less reliable river floods, which led to migration to monsoon-rain-fed areas and the abandonment of cities.

Palaeo-drainage systems

Satellite and aerial photography were used for identifying possible former drainage channels (Ghose et al., 1979; Gupta et al., 2004; Yashpal et al., 1980). Mehdi et al. (2016) identified three channel systems to the north of the Thar Desert in Rajasthan and Haryana. The two northern channels appear to flow from the Himalayas, while the southern one originates from the drainage divide between the Indus and Ganga basins. These workers argued that the channels were last active at 4000–1000 BC and could have supplied water to the Harappan settlement sites. Orengo and Petrie (2017) employed new analytical methods to confirm and extend earlier studies (Bhadra et al., 2009; Yashpal et al., 1980) that showed a complex array of shallow buried palaeo-channels in this region, between the Rivers Yamuna and Sutlej, around the

Ghaggar-Hakra channel. The age of these channels is of course unconstrained by remote sensing approaches. However, such data was integrated with satellite radar-based topography by Giosan et al. (2012) to make it possible to test connections between river systems in the past. Furthermore, the timing of flow was estimated through extensive coring and OSL dating of specific landscape features in this region. As noted above, coring in northern India has demonstrated that channels possibly connecting the Ghaggar-Hakra to the Yamuna mostly date from 28,000 to 18,000 BC (Saini et al., 2009). Tracing evidence of an independent river closer to the delta is more complicated, as a result of its proximity to the Thar Desert. An initial attempt to do this was made using radiogenic lead isotope measurements in grains of potassium feldspar, because this is a common mineral type in many continental rocks; it was felt to be ubiquitous, and therefore not necessarily biased against one or other possible source terranes (Tyrrell et al., 2006). More importantly, earlier work had established that different parts of the Himalayas had unique lead (Pb) isotopic signals, so that it would be possible to distinguish sediment derived from different areas and therefore resolve independent drainage systems (Clift et al., 2002). Lying directly along the course of the old Ghaggar-Hakra, the trench sites at Fort Abbas and Fort Derawar (Figures 2.2 and 2.4) established that there were active sandy river channels in that area dating from around 3700 BC, pre-dating the main Harappan culture (Alizai, Clift et al., 2011; Giosan et al., 2012). Figure 2.6A compares Pb isotopic ratios of sand grains from the fluvial sand bodies in both of these areas and compares them with data from dune sand in the adjacent desert, as well as with sand from the Sutlej River, which is the nearest large tributary of the Indus close to the sites (Alizai, Clift et al., 2011). If the Sutlej had flowed further south in the past then isotope values in the river and the cored channel sand might correspond, and therefore it is possible that the Sutlej was a contributor to a wider Sarasvati river system. These analyses showed that sand grains at Fort Abbas and Fort Derawar were very similar, and overlapped with both the Sutlej and the desert dune compositions. However, a small number of grains in the channels were characterised by relatively high isotope values, which were inconsistent with an origin in the Sutlej, but suggested derivation of material from the main River Indus that drains ancient continental crust in the region of Nanga Parbat, which is known to yield high $^{207}Pb/^{204}Pb$ and $^{206}Pb/^{204}Pb$ values (Clift et al., 2002). Since the main River Indus could not have flowed in the coring area in the recent past it was deduced that the source of these grains was in the desert (Alizai, Clift et al., 2011).

In practice, the sands that fill these late-stage river channels are isotopically indistinguishable from the desert dunes of the Thar Desert that have been analysed so far. Such dunes surround the channels today and provide clear opportunities for recycling. This relationship was interpreted to mean that the rivers, when they were active around 3700 BC, were recycling large amounts of material out of the dunes, so that they cannot readily be distinguished. We can rule out derivation of dune sediment from local river floodplains by using the existing isotopic and geochronology information, which points to a dominant influence from the Indus Delta (East et al., 2015). This provenance argues against the rivers being dominated by large volume flow from glacially fed sources in the High Himalayas, as suggested by the Saraswati legends (Valdiya, 2013). Nonetheless, there must have been an active flow of relatively high-energy water to transfer the relatively coarse material found in the channel, although the scale of the channel and the river system itself is not clear from the existing evidence. This work does, however, demonstrate that there were active river channels flowing along the edge of the desert before the establishment of the Harappan culture, and that subsequently they dried up and clogged with desert sands shortly before the Harappan sites were abandoned. Further to the west, trenching of sediments in the area of Yazman (Figure 2.4) showed river sands overlain by dune deposits, which argues for cessation of river flow shortly after 2500 BC, much closer to the time of Harappan cultural deterioration.

Giosan et al. (2012) showed that fluvial sedimentation was still active in the western part of the Thar Desert as late as *c.* 1000 BC, with fluvial courses joining the Nara valley. The question of whether these rivers flowed much further south, and reached the Arabian Sea in a different place from the modern Indus mouth, was addressed using the lead isotope approach to cores in the Nara valley (Figure 2.2) (Alizai, Clift et al., 2011). Figure 2.6B shows how sedimentary grains from river deposits in the Nara valley compare with the desert dune systems, as well as with sands in the modern Indus at nearby Sukkur. Within the uncertainties of the measurements and the moderate number of analyses, it is hard to distinguish any significant resolvable differences between the old channel sands in the Nara valley, the modern Indus and the desert. More recently, it has been argued that the desert may be providing significant sediment to the main Indus (East et al., 2015), especially as the western edge of the desert migrated further west during the latter parts of the Holocene, at least in the region of the Nara valley. In any case, the dunes appear to be supplied largely by sediment blown north from the delta plain by strong summer monsoon winds.

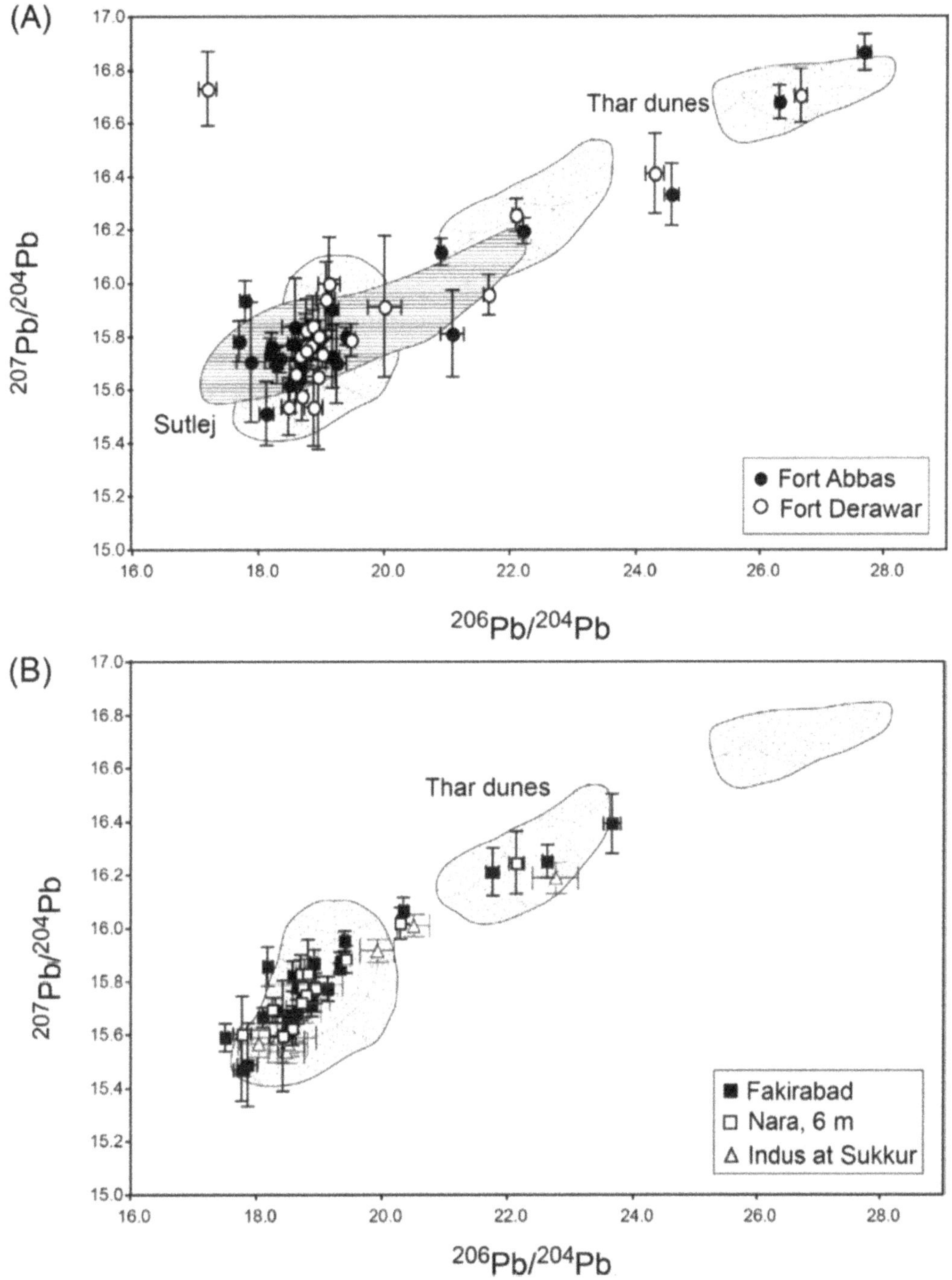

Figure 2.6 A. Pb isotopic plots of feldspars from Forts Derawar and Abbas, located on the edge of the Thar Desert, compared with compositions from the neighbouring River Sutlej and the dune sands of the Thar Desert. B. Pb isotopic ratio plots of feldspars from the core site at Nara and the nearby pit at Fakirabad (see Figure 2.4 for locations), compared with those from the Indus at Sukkur and with the range of values found in Thar Desert dunes. Modified from Alizai, Clift et al., 2011

The fluvial channel sands in the Nara region appear to be essentially indistinguishable from desert dune sands, as was true further north, which again favours an origin through significant reworking of material out of the desert. This does not preclude the presence of a major independent river in the southern parts of the floodplains north of the Rann of Kutch, but it does indicate that there is no easy way to tell whether this was a large river, or whether it originated in the Ghaggar catchment in the Himalayan foothills or within a High Himalayan glacially sourced region. Although a significant independent river might have been present in this area, it is essentially impossible to prove this as a result of sediment exchange between the rivers and the desert. In the case of the River Nara samples this means that we have evidence of a fluvial channel active as recently as 1200 BC (i.e., post-dating the Harappan civilisation), but the origin of this channel is unknown and could simply represent a bifurcation from the main Indus (Giosan et al., 2012). The large buffering effect of the Thar Desert is clear at the mouth of the River Nara in the Rann of Kutch, an infilled embayment of the Indian Ocean, where work has suggested that geochemical and isotopic fingerprinting cannot separate a distinct 'Sarasvati' source in the Holocene (Khonde et al., 2017).

River geometries can be further investigated by applying the detrital zircon U-Pb dating method to the sites discussed above. This method dates the age at which zircon crystals cool to below around 750°C, and has been used to effect in fingerprinting source terranes, allowing ancient drainage pans to be reconstructed, although it is somewhat susceptible to multiple cycles of recycling because of the durability of the crystals (Hodges, 2003). Nonetheless, its effectiveness in distinguishing different tributaries within the Indus system has been demonstrated (Alizai, Carter et al., 2011), and it provides an effective way of looking at ancient drainage patterns in the floodplains. Drilling in the Pakistani Punjab has provided samples at the Marot location (Figure 2.4) which have been dated roughly to between 47,000 and 8,000 years ago (Clift et al., 2012). In general, these sands provide zircon U-Pb ages consistent with an origin in the Himalayas, which is perhaps no great surprise. There is also some reworking of material out of the desert. In cases where the sediment has a more dominant Himalayan signature this requires a high-energy stream, with a strong flux of material from the mountains that overwhelms local reworking.

The oldest sample at Marot (47,000–8,000 years ago) yielded a particularly distinctive array of zircon grain ages, especially if we focus on those from between 1,700 and 2,000 million years ago (Ma). These represent the erosion of relatively old bedrock, probably in the Lesser

Himalayas (DeCelles et al., 2000). When this particular part of the age spectrum is analysed in detail it is noteworthy that the oldest sample has a peak in its age population slightly younger than 1,900 Ma, which contrasts with the generally younger ages closer to 1,850 Ma in the younger sediments. When these population ages are compared, using a kernel density estimation (KDE) plot, with the main Indus, the desert dunes and both the Sutlej and Yamuna rivers (Figure 2.7), it is seen that there is a good match between the Indus, the desert dunes and the younger fluvial sands. However, the oldest sand from Marot makes a good match with sediment from the River Yamuna. This would seem to corroborate the earlier hypothesis that the Yamuna used to supply the Ghaggar-Hakra around the northern edge of the desert. However, the depositional age constraints of this sediment clearly rule out the possibility that this combined river was young enough to represent the Sarasvati. This is because the earliest possible date of this flow would have been before 8000 BC, and more likely it is correlative with the 28,000–18,000 BC channels observed by Saini et al. (2009). Indeed, those channels seem to flow directly between the Yamuna course and the drill site at Marot, which is suggestive of a direct link. Nearby coring by A. Singh et al. (2016) was able to penetrate deeper into the section and reveal a system of older

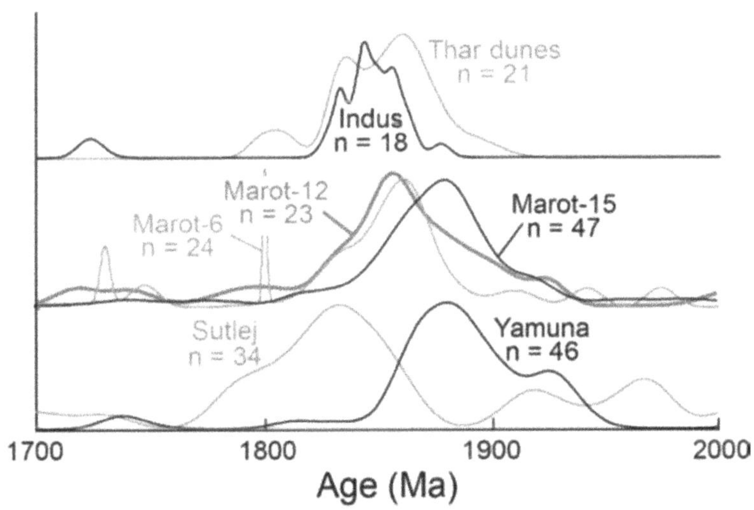

Figure 2.7 Kernel density plots of U-Pb ages of zircon grains from the Marot borehole compared with possible source rivers. The peak at ~1800 Ma matches well with the Yamuna in Marot-15 (> 49 ka), but is more like the Sutlej or Thar dunes in Marot-12 and -6 (6–49 ka). Reprinted from Clift et al., 2012, with permission from the Geological Society of America

channels, but they still show the cessation of most sandy channel sedimentation after *c.* 21,000 BC and *c.* 10,000 BC in their two cores and a total end of deposition shortly after 4400 BC. It appears that the zircon data supports the idea of stronger river flow in the early Holocene, but does not support the presence of any large river flowing around the west edge of the Thar Desert in the location of the Harappan sites at the time they were settled. This interpretation is in accord with the grain-size decrease in the fluvial sands reported by Giosan et al. (2012) and A. Singh et al. (2016).

Statistical analysis using a multidimensional scaling (MDS) plot, a form of factor analysis (Vermeesch, 2013), now allows us to compare river sediments from palaeo-channels in the Holocene floodplain both with modern rivers and with the desert sands. Figure 2.8 shows that the sediments in the southern part of the Thar Desert (samples UN1 and NM3) do closely resemble sediment in the river mouth shortly after the Last Glacial

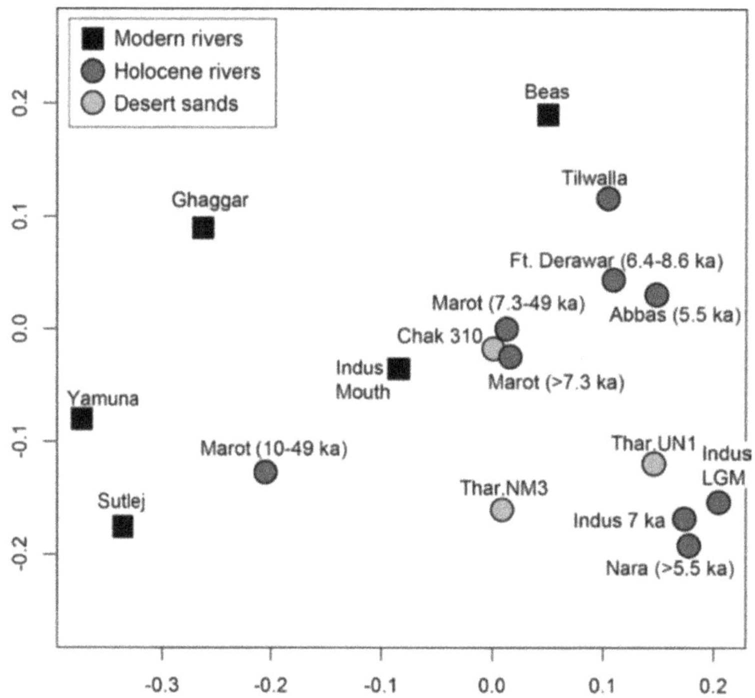

Figure 2.8 Multidimensional scalar (MDS) plot showing the similarities and differences between the zircon populations in a variety of analysed sands from the River Indus system. Modern river data is from Alizai, Carter et al., 2011 and Clift et al., 2004. Holocene floodplain sediments are from Clift et al., 2012 and the desert sands are from East et al., 2015. Image source: authors

Maximum (*c.* 13,000 BC), as well as the Mid-Holocene, around 5000 BC. Sands in the Nara valley are also extremely close to these samples and suggest large degrees of recycling in that system, so that the continuity of a Himalayan-fed river is essentially impossible to demonstrate.

As we can infer from the KDE diagrams shown in Figure 2.8, the MDS plot also shows that the oldest sediments at Marot show similarity to River Yamuna sand and contrast with the younger sediments at the same location. The MDS diagram shows the very close similarity between some of these Holocene fluvial sands and the desert sands in the northern part of the desert, particularly that at Chak 310 (Figure 2.4). It is particularly noteworthy that all the zircons from the Holocene floodplain sands do not show any similarity to the River Ghaggar and are also quite dissimilar to the Himalayan-fed Yamuna and Sutlej rivers, which argues against their derivation from any joint flow between the Ghaggar, the Yamuna and the Sutlej. The sand from Tilwalla hints that the River Beas may have been connected into a Ghaggar-Hakra river system during the Mid-Holocene (Clift et al., 2012), but any larger-scale river during the time around the Harappan settlements appears to be precluded by this analysis. The proposed Ghaggar–Beas connection is presently unsupported by any clear geomorphology evidence or by the U-Pb zircon dating.

Any connection between the abandoned Ghaggar-Hakra channels and large Himalayan rivers can be ruled out with certainty during the latter part of the Holocene. Thus, reorganisation of large rivers does not appear to have played a major role in the evolution of the floodplains of northern India along the edge of the desert where the Harappan sites clustered during the latter part of the Holocene. As a result such processes do not seem to have been significant in the growth or demise of the Harappan civilisation, implying that monsoon strength (i.e., regional drying) may instead be dominant.

Closer to the Himalayas, a number of studies have now targeted the channels feeding the upper reaches of the Ghaggar-Hakra between the desert and the mountains themselves. Electrical resistivity surveys have identified channels in the subsurface between the northern edge of the desert and the point at which the River Sutlej exits the mountains, just to the west of Chandigarh (Figure 2.4) (Sinha et al., 2013). Cores taken in this area now reveal the age of sedimentation, which extends back around 150,000 years in the region of Site GS-10 (A. Singh et al., 2016). Neodymium isotope ratios (ε_{Nd}) of the clastic sediment fraction have a long history of being used as a provenance proxy in South Asia (Clift and Blusztajn, 2005; Clift et al., 2002; Zhuang et al., 2015). Analysis of neodymium isotopes from these cores shows a gradual evolution of ε_{Nd}

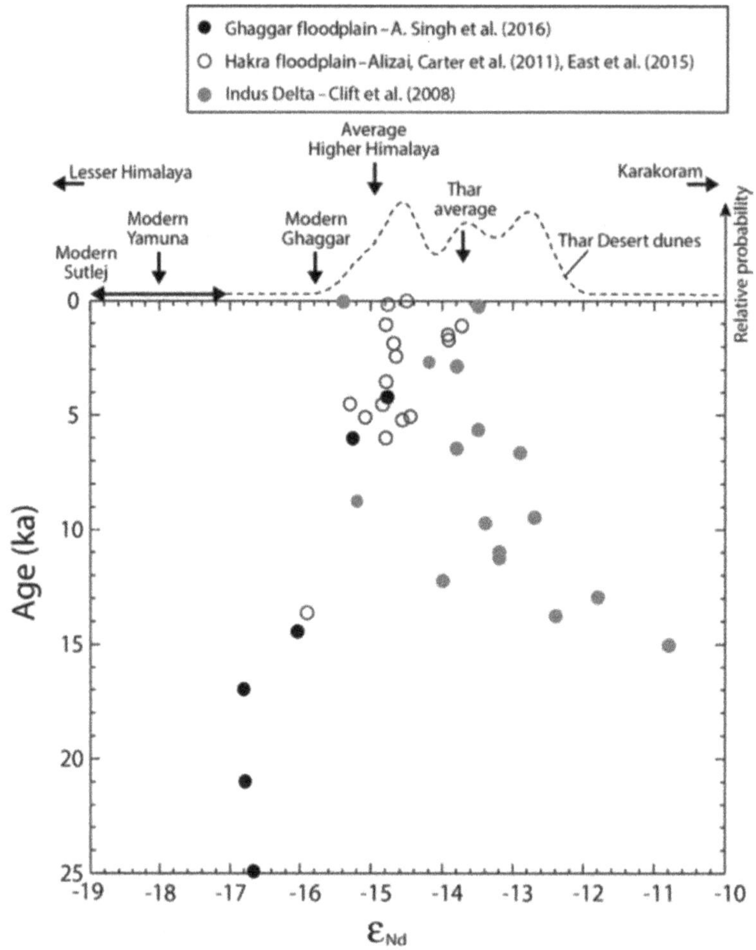

Figure 2.9 Evolution in the Nd isotopic character of the whole Indus basin as recorded at the delta, compared with the compositional evolution in the floodplains of the Ghaggar-Hakra. The results from A. Singh et al. (2016) are consistent with those dated by an earlier study from the floodplains in modern Pakistan that formed a continuous floodplain but from an ephemeral river that did not drain the Lesser or Higher Himalayas. The trend towards desert dune compositions reflects the progressive demise of this river and its choking by the expansion of the Thar Desert during the Holocene. Source: authors

values from relatively negative, around −17, around 23,000 BC, to −4 to −13 by 3000 BC (A. Singh et al., 2016) (Figure 2.9). These values were interpreted as indicating a source in the glacial High Himalayas. The isotopic trend in north-west India was interpreted as indicating a gradual shift of sediment supply away from the High Himalayas during times of glaciation, with more erosion from the Lesser Himalayas at those times (A. Singh et al., 2016). This interpretation, however, is inconsistent with an earlier model developed from the Indus Delta, which receives sediment from across the entire Indus basin (Clift et al., 2008). In the delta record, from around 13,000 BC the reverse trend was observed, ε_{Nd} values falling from −11 to around −13 to −14 in the present day. In contrast with the study of A. Singh et al. (2016), the delta record was used to suggest a shift towards Lesser Himalayan erosion during interglacial times when the monsoon was stronger, essentially the opposite of what was proposed for this channel connecting the Sutlej and the Ghaggar-Hakra. Figure 2.9 shows how the two contrasting trends relate to one another and also demonstrates that neodymium data from the Hakra floodplain itself in Pakistan is continuous with the data from northern India, as might be expected for a single continuous river system showing a steady fall in ε_{Nd} values during the latter part of the Holocene.

Clift et al. (2016) argued that the isotopic trend in northern India since 13,000 BC can be interpreted as indicating a steadily increasing mixing of river sediment with the Thar Desert rather than strong changes in the source areas themselves. Work in the northern part of the floodplains has already demonstrated that recycling of sediment during the incision of river channels is important during the Holocene (Giosan et al., 2012), and, as discussed above, many of the sands and the river channels have strong flavours of desert involvement in them. Consequently, it seems likely that this evolution of the channel connecting the Sutlej and the Ghaggar-Hakra reflects progressively greater involvement of desert material in the river channel as the climate dried during the Holocene, rather than changes in erosion patterns in the upstream part of the Sutlej. This model is in accord with observations that rivers tend to incise their floodplains when the climate starts to dry and as the sediment load from the upper reaches decreases, which allows the river to cannibalise its earlier deposits (Bookhagen et al., 2006; Jonell et al., 2017).

A connection via the Sutlej to the Ghaggar-Hakra seems likely (A. Singh et al., 2016), but could not have lasted after the incision of the Sutlej (Giosan et al., 2012). Clift et al. (2012) favoured a connection between the Sutlej and the Hakra somewhere near the location of Marot over the upstream location proposed by Sinha et al. (2013). Both models would have resulted in an increased discharge into the Ghaggar-Hakra. However, the evidence from the incision data summarised earlier in this study argues that a larger river should have pre-dated 8000 BC, since there is no incised Sutlej valley of this type connecting the two, and that regional incision of the river channels had started in earnest by this time. We argue instead that High Himalayan contributions to the

Ghaggar-Hakra from the Sutlej and Yamuna have been important in the past, but all connections significantly pre-date the Mature Harappan civilisation.

Such a conclusion does not mean that rivers were unimportant to the sustenance of that society. Giosan et al. (2012) argued that a smaller river, close to the settlements, with floods less dangerous than those of Himalayan rivers, was in many ways more useful to the maintenance of an urban society because it provided seasonal flooding for the fields that was relatively easy to control and allowed a reliable agricultural system to be established. In this view large rivers close to urban centres may have been as much a risk to the establishment of the settlements as a bonus. The tendency of natural large rivers to burst their banks and cause destruction during times of particularly strong monsoon would be good for agriculture. The flooding observed in Pakistan in 2010 shows the scale of hazard that exists today, not in the Bronze Age. Regular seasonal flooding would be more helpful to agriculture, but it is dependent on consistent rainfall, so that long-term drying of the climate, as reconstructed in South West Asia (Figure 2.5), would result in a steady deterioration of the situation, potentially to the point at which agriculture would no longer be productive.

Conclusions

Fluvial systems in South Asia that drain the Himalayas have undergone significant amounts of reorganisation in the relatively recent geological past. There is both geochemical and geophysical evidence that the River Yamuna used to flow west and probably joined a Ghaggar-Hakra stream flowing around the western edge of the Thar Desert. However, OSL ages indicate that this connection lasted at least 10,000 years, so it could not have affected the evolution of the Harappan culture. It is likely that the Sutlej, and probably the Beas, also supplied water to the Ghaggar-Hakra. The lack of channels incised into the floodplain that characterised the Himalayan river valleys indicates that these were probably lost before 8000 BC too. The Harappan settlement sites appear to have been sustained largely by the smaller volume flows of the Ghaggar-Hakra itself, which may have collected the waters of smaller sub-Himalayan tributaries. The Ghaggar-Hakra was relatively strong in the middle Holocene, but underwent reduction as the climate dried, particularly after 3000–2000 BC. Evidence of drying climates is strong in South Asia during the middle

Holocene, and they are probably the primary mechanism underlying the reduction of water supply to the region in which the densest settlements were located. Some adjustment to a changing climate is possible through water management once seasonal flows and inundation had reduced below a critical level, but it would no longer have been practical for larger concentrations of people to be sustained by the agriculture that could be developed in the immediate vicinity of the river.

References

Aklujkar, Ashok. 2014. 'Sarasvati drowned: Rescuing her from scholarly whirlpools.' In Nalini Rao (ed.), *Sindhu-Sarasvati Civilization: New Perspectives: A Volume in Memory of Dr. Shikaripur Ranganatha Rao*, 89–187. New Delhi: D. K. Print World Ltd.

Alizai, Anwar, Andrew Carter, Peter D. Clift, Sam VanLaningham, Jeremy C. Williams and Ravindra Kumar. 2011. 'Sediment provenance, reworking and transport processes in the Indus River by U-Pb dating of detrital zircon grains.' *Global and Planetary Change* 76 (1–2), 33–55.

Alizai, Anwar, Peter D. Clift, L. Giosan, Sam VanLaningham, R. Hinton, A. R. Tabrez, M. Danish and EIMF. 2011. 'Pb isotopic variability in the modern-Pleistocene Indus River system measured by ion microprobe in detrital K-feldspar grains.' *Geochimica et Cosmochimica Acta* 75 (17), 4771–95.

Bhadra, B. K., A. K. Gupta and J. R. Sharma. 2009. 'Saraswati Nadi in Haryana and its linkage with the Vedic Saraswati River: Integrated study based on satellite images and ground based information.' *Journal of the Geological Society of India* 73 (2), 273–88.

Bookhagen, B., D. Fleitmann, K. Nishiizumi, M. R. Strecker and R. C. Thiede. 2006. 'Holocene monsoonal dynamics and fluvial terrace formation in the northwest Himalaya, India.' *Geology* 34 (7), 601–4.

Clemens, Steven C., Warren L. Prell and Youbin Sun. 2010. 'Orbital-scale timing and mechanisms driving Late Pleistocene Indo-Asian summer monsoons: Reinterpreting cave speleothem $\partial^{18}O$.' *Paleoceanography and Paleoclimatology* 25 (4) (PA4207). doi:10.1029/2010PA001926.

Clift, Peter D. and Jerzy S. Blusztajn. 2005. 'Reorganization of the western Himalayan river system after five million years ago.' *Nature* 438, 1001–3.

Clift, Peter D., Ian H. Campbell, Malcolm S. Pringle, Andrew Carter, Xifan Zhang, Kip V. Hodges, Ali Athar Khan and Charlotte M. Allen. 2004. 'Thermochronology of the modern Indus River bedload: New insight into the control on the marine stratigraphic record.' *Tectonics* 23 (5), TC5013. doi:10.1029/2003TC001559.

Clift, Peter D., Andrew Carter, Liviu Giosan, Julie Durcan, Geoff A. T. Duller, Mark G. Macklin, Anwar Alizai, Ali R. Tabrez, Mohammed Danish, Sam VanLaningham and Dorian Q. Fuller. 2012. 'U-Pb zircon dating evidence for a Pleistocene Sarasvati River and capture of the Yamuna River.' *Geology* 40 (3), 211–14.

Clift, Peter D., Liviu Giosan, Jerzy Blusztajn, Ian H. Campbell, Charlotte Allen, Malcolm Pringle, Ali R. Tabrez, Mohammed Danish, M. M. Rabbani, Anwar Alizai, Andrew Carter and Andreas Lückge. 2008. 'Holocene erosion of the Lesser Himalaya triggered by intensified summer monsoon.' *Geology* 36 (1), 79–82.

Clift, Peter D., Liviu Giosan and Amy E. East. 2016. 'Comment on "Geochemistry of buried river sediments from Ghaggar Plains, NW India: Multi-proxy records of variations in provenance, paleoclimate, and paleovegetation patterns in the late quaternary" by Ajit Singh, Debajyoti Paul, Rajiv Sinha, Kristina J. Thomsen, Sanjeev Gupta.' *Palaeogeography, Palaeoclimatology, Palaeoecology* 455, 65–7.

Clift, Peter D., Jae Il Lee, Peter Hildebrand, Nobumichi Shimizu, Graham D. Layne, Jerzy Blusztajn, Joel D. Blum, Eduardo Garzanti and Athar Ali Khan. 2002. 'Nd and Pb isotope variability in the Indus River system: Implications for sediment provenance and crustal heterogeneity in the western Himalaya.' *Earth and Planetary Science Letters* 200 (1–2), 91–106.

DeCelles, P. G., G. E. Gehrels, J. Quade, B. LaReau and M. Spurlin. 2000. 'Tectonic implications of U-Pb zircon ages of the Himalayan orogenic belt in Nepal.' *Science* 288 (5465), 497–9.

Deotare, B. C., M. D. Kajale, S. N. Rajaguru, S. Kusumgar, A. J. T. Jull and J. D. Donahue. 2004. 'Palaeoenvironmental history of Bap-Malar and Kanod playas of western Rajasthan, Thar desert.' *Proceedings of the Indian Academy of Sciences (Earth and Planetary Sciences)*, 113 (3), 403–25.

Dixit, Yama, David A. Hodell and Cameron A. Petrie. 2014. 'Abrupt weakening of the summer monsoon in northwest India ~4100 yr ago.' *Geology* 42 (4), 339–42.

Durcan, J. A., H. M. Roberts, G. A. T. Duller and A. H. Alizai. 2010. 'Testing the use of range-finder OSL dating to inform field sampling and laboratory processing strategies.' *Quaternary Geochronology* 5 (2–3), 86–90.

Dykoski, Carolyn A., R. Lawrence Edwards, Hai Cheng, Daoxian Yuan, Yanjun Cai, Meiliang Zhang, Yushi Lin, Jiaming Qing, Zhisheng An and Justin Revenaugh. 2005. 'A high-resolution, absolute-dated Holocene and deglacial Asian monsoon record from Dongge Cave, China.' *Earth and Planetary Science Letters* 233 (1–2), 71–86.

East, Amy E., Peter D. Clift, Andrew Carter, Anwar Alizai and Sam VanLaningham (2015). 'Fluvial–aeolian interactions in sediment routing and sedimentary signal buffering: An example from the Indus Basin and Thar Desert.' *Journal of Sedimentary Research* 85 (6), 715–28.

Enzel, Y., L. L. Ely, S. Mishra, R. Ramesh, R. Amit, B. Lazar, S. N. Rajaguru, V. R. Baker and A. Sandle. 1999. 'High-resolution Holocene environmental changes in the Thar Desert, northwestern India.' *Science* 284 (5411), 125–8.

Fleitmann, D., S. J. Burns, M. Mudelsee, U. Neff, J. Kramers, A. Mangini and A. Matter. 2003. 'Holocene forcing of the Indian monsoon recorded in a stalagmite from southern Oman.' *Science* 300 (5626), 1737–9.

Ghose, Bimal, Amal Kar and Zahid Husain. 1979. 'The lost courses of the Saraswati River in the Great Indian Desert: New evidence from Landsat imagery.' *Geographical Journal* 145 (3), 446–51.

Giosan, Liviu, Peter D. Clift, Mark G. Macklin and Dorian Q. Fuller. 2013. 'Sarasvati II.' *Current Science* 105 (7), 888–90.

Giosan, Liviu, Peter D. Clift, Mark G. Macklin, Dorian Q. Fuller, Stefan Constantinescu, Julie A. Durcan, Thomas Stevens, Geoff A. T. Duller, Ali R. Tabrez, Kavita Gangal, Ronojoy Adhikari, Anwar Alizai, Florin Filip, Sam VanLaningham and James P. M. Syvitski. 2012. 'Fluvial Landscapes of the Harappan Civilization.' *Proceedings of the National Academy of Sciences* 109 (26), 1688–94.

Gupta, Anil K., David M. Anderson and Jonathan T. Overpeck. 2003. 'Abrupt changes in the Asian southwest monsoon during the Holocene and their links to the North Atlantic Ocean.' *Nature*, 421 (6921), 354–7.

Gupta, A. K., J. R. Sharma, G. Sreenivasan and K. S. Srivastava. 2004. 'New findings on the course of river Sarasvati.' *Journal of the Indian Society of Remote Sensing* 32 (1), 1–24.

Hodges, K. 2003. 'Geochronology and thermochronology in orogenic systems.' In Roberta L. Rudnick (ed.), *The Crust*, 263–92. Amsterdam: Elsevier Science.

Jonell, Tara N., Peter D. Clift, Long V. Hoang, Tina Hoang, Andrew Carter, Hella Wittmann, Philipp Böning, Katharina Pahnke and Tammy Rittenour. 2017. 'Controls on erosion patterns and sediment transport in a monsoonal, tectonically quiescent drainage, Song Gianh, central Vietnam.' *Basin Research* 29 (S1), 659–83.

Khonde, Nitesh, Sunil Kumar Singh, D. M. Maurya, Vinai K. Rai, L. S. Chamyal and Liviu Giosan. 2017. 'Tracing the Vedic Saraswati River in the Great Rann of Kachchh.' *Scientific Reports* 7, art. no. 5476. doi:10.1038/s41598-017-05745-8.

Lawler, Andrew. 2008. 'Indus collapse: The end or the beginning of an Asian culture?' *Science* 320 (5881), 1281–3.

Madella, Marco and Dorian Q. Fuller. 2006. 'Palaeoecology and the Harappan Civilisation of South Asia: A reconsideration.' *Quaternary Science Reviews* 25 (11–12), 1283–1301.

Maemoku, Hideaki, Yorinao Shitaoka, Tsuneto Nagatomo and Hiroshi Yagi. 2013. 'Geomorphological constraints on the Ghaggar River regime during the Mature Harappan period.' In Liviu Giosan, Dorian Q. Fuller, Kathleen Nicoll, Rowan K. Flad and Peter D. Clift (eds), *Climates, Landscapes, and Civilizations*, 97–106. Washington, DC: American Geophysical Union. doi: 10.1029/GM198.

Mehdi, Syed Muntazir, Naresh C. Pant, Hari Singh Saini, S. A. I. Mujtaba and Prabhas Pande. 2016. 'Identification of palaeochannel configuration in the Saraswati River basin in parts of Haryana and Rajasthan, India, through digital remote sensing and GIS.' *Episodes*, 39 (1), 29–38.

Orengo, Hector A. and Cameron A. Petrie. 2017. 'Large-scale, multi-temporal remote sensing of palaeo-river networks: A case study from northwest India and its implications for the Indus civilisation.' *Remote Sensing* 9 (7), 735. doi:10.3390/rs9070735.

Ponton, Camilo, Liviu Giosan, Tim I. Eglinton, Dorian Q. Fuller, Joel E. Johnson, Pushpendra Kumar and Tim S. Collett. 2012. 'Holocene aridification of India.' *Geophysical Research Letters* 39 (3), L03704. doi:10.1029/2011GL050722.

Possehl, Gregory L. 1999. *Indus Age: The Beginnings*. Philadelphia: University of Pennsylvania Press.

Prasad, Sushma and Yehouda Enzel. 2006. 'Holocene paleoclimates of India.' *Quaternary Research* 66 (3), 442–53.

Prell, W. L. and W. B. Curry. 1981. 'Faunal and isotopic indices of monsoonal upwelling: Western Arabian Sea.' *Oceanologica Acta* 4 (1), 91–8.

Saini, Hari Singh, Sampat K. Tandon, S. A. I. Mujtaba, N. C. Pant and R. K. Khorana. 2009. 'Reconstruction of buried channel-floodplain systems of the northwestern Haryana Plains and their relation to the "Vedic" Saraswati.' *Current Science* 97 (11), 1634–43.

Singh, Ajit, Debajyoti Paul, Rajiv Sinha, Kristina J. Thomsen and Sanjeev Gupta. 2016. 'Geochemistry of buried river sediments from Ghaggar Plains, NW India: Multi-proxy records of variations in provenance, paleoclimate, and paleovegetation patterns in the Late Quaternary.' *Palaeogeography, Palaeoclimatology, Palaeoecology* 449, 85–100.

Singh, G., R. D. Joshi, S. K. Chopra and A. B. Singh. 1974. 'Late quaternary history of vegetation and climate of the Rajasthan desert, India.' *Philosophical Transactions of the Royal Society of London, Series B, Biological Sciences* 267 (889), 467–501.

Sinha, Rajiv, G. S. Yadav, Sanjeev Gupta, Ajit Singh and S. K. Lahiri. 2013. 'Geo-electric resistivity evidence for subsurface palaeochannel systems adjacent to Harappan sites in northwest India.' *Quaternary International* 308–9, 66–75.

Staubwasser, M., F. Sirocko, P. M. Grootes and M. Segl. 2003. 'Climate change at the 4.2 ka BP termination of the Indus valley civilization and Holocene south Asian monsoon variability.' *Geophysical Research Letters* 30 (8), 1425. doi:10.1029/2002GL016822.

Swain, A. M., J. E. Kutzbach and S. Hastenrath. 1983. 'Estimates of Holocene precipitation for Rajasthan, India, based on pollen and lake-level data.' *Quaternary Research* 19 (1), 1–17.

Tyrrell, S., P. D. W. Haughton, J. S. Daly, T. F. Kokfelt and D. Gagnevin. 2006. 'The use of the common Pb isotope composition of detrital K-feldspar grains as a provenance tool and its application to Upper Carboniferous paleodrainage, northern England.' *Journal of Sedimentary Research* 76 (2), 324–45.

Valdiya, K. S. 2002. *Saraswati: The River that Disappeared*. Hyderabad, India: Universities Press.

Valdiya, K. S. 2013. 'The River Saraswati was a Himalayan-born river.' *Current Science* 104 (1), 42–54.

Vermeesch, Pieter 2013. 'Multi-sample comparison of detrital age distributions.' *Chemical Geology* 341, 140–6.

Wünnemann, Bernd, Dieter Demske, Pavel Tarasov, Bahadur S. Kotlia, Christian Reinhardt, Jan Bloemendal, Bernhard Diekmann, Kai Hartmann, Joachim Krois, Frank Riedel and Nidhi Arya. 2010. 'Hydrological evolution during the last 15 kyr in the Tso Kar lake basin (Ladakh, India), derived from geomorphological, sedimentological and palynological records.' *Quaternary Science Review* 29 (9–10), 1138–55.

Yashpal, Sahai B., R. K. Sood and D. P. Agrawal. 1980. 'Remote sensing of the "lost" Saraswati river.' *Proceedings of the Indian Academy of Sciences (Earth and Planetary Sciences)* 89 (3), 317–31.

Zhuang Guangsheng, Yani Najman, Stéphane Guillot, Martin Roddaz, Pierre-Olivier Antoine, Grégoire Métais, Andrew Carter, Laurent Marivaux and Sarfraz H. Solangi. 2015. 'Constraints on the collision and the pre-collision tectonic configuration between India and Asia from detrital geochronology, thermochronology, and geochemistry studies in the lower Indus basin, Pakistan.' *Earth and Planetary Science Letters* 432, 363–73.

3
Habitat hysteresis in ancient Egypt

Judith Bunbury

Abstract

Habitats in Egypt are strongly cyclic, on the annual and decadal scales and on the scale of a century, a millennium or more. State recording of and intervention in the annual flood is well documented, but in this chapter I examine the evidence for longer-term cyclicity in ancient Egyptian literature and practice. Some evidence, in grain storage at temples and from building strategies on islands, suggests that local knowledge of decadal and centurial cycling was also common knowledge. Beyond this, some wider climatic excursions, for example the period of drying associated with the First Intermediate Period (2160–2055 BC), clearly came as a surprise and were lamented as such. Later, during New Kingdom and Roman times, it seems that scholarship and the recognition of past occupants in newly habitable areas were a guide to locations of water, for example the wells of the Kharga basin. A practical understanding of longer-term Nile behaviour during these later periods also led to successful schemes of Nile management and hence the extension of existing habitats.

Introduction

Egypt has been inhabited since the Palaeolithic (700,000 BC and perhaps before). During this long period, climate cycling on a whole range of timescales has continued, mediated in some cases by human activity and at other times wholly as a result of natural climate change. Some habitats are short-lived, for example those formed when a playa lake receives rainfall and vegetation germinates, attracting occasional visitors, or when the Nile floods and recedes, leaving cultivable land. In other cases, the kinder climate can persist for centuries, which means that permanent lakes and forests develop, supporting established, settled communities. Evidence of the expansion of populations across the Saharan region takes the form of abandoned settlements far out into the Sahara Desert that are being rediscovered now through satellite imaging and other forms of exploration. This chapter considers the ways in which annual and longer-term changes of climate can create new habitat and how humans interact with these ecological niches to survive and thrive; they are locked into a dialogue with their environment. When humans recognise the cyclicity of a habitat, they may plan to harvest, store or redirect water in order to form new habitats or extend the stability of that habitat, so creating new landscapes. Equally, adopting new technology, for example domesticating animals or plants, as well as water-harvesting devices, can also extend the viability of a habitat.

Landscape history of Egypt

Egypt today is a tract of desert that has the sparse vegetation of the Sahel to the south and the Mediterranean Sea to the north. Between these the green thread of the Nile valley travels from south to north, providing a rich habitat for humans. The annual cycle of the Nile, with its summer floods fed by the Ethiopian monsoon, is long studied and well understood. However, over longer climate cycles, the monsoon waxes and wanes, and during the last ice age it even ceased. The same climate cycles may extend the Sahel and savanna habitats northwards, bringing winter and sometimes summer rains into the Egyptian deserts. In the early Holocene (between around 8000 BC and 5000 BC), episodic climate amelioration meant that much of the Saharan region became green and annually formed playa lakes that provided a rich, if seasonal, habitat for settlers. An annual cycle of rainfall, flooding, desiccation and harvest followed by a dry patch was established. Late Pleistocene lakes, extensive before the last ice age (Maxwell et al., 2010; Haynes et al., 1979; Haynes, 1980), had already deposited fertile sediments throughout the Saharan region, and only the addition of water was required for bounty.

Researchers in the Sahara Desert find evidence of many ancient colonies living partially sedentary lives around ephemeral (and sometimes longer-lived) lakes. Garcea (2006) and Bubenzer and Riemer (2007) describe how, during the North African Neolithic, fish, fowl, pastoralism and the harvesting of wild plants made a sedentary life possible without the effort of agriculture. Settlements in the Kharga (under investigation by Salima Ikram and colleagues) and Farafra oases (Barich et al., 2014) provide evidence of numerous groups of this type during the sixth millennium BC. Ethnobotanical evidence from Farafra even suggests that, at times, there were both winter and summer rains. Knowledge of the varying resources associated with playa lakes through the year fostered the development of permanent settlements. Although we do not have written records of activity during the Saharan Neolithic, we do have abundant finds and associated rock art.

Climate cycles in the Saharan region

Beyond the cycle of annual or twice-annual rains, archaeological evidence shows that there were periods of more persistent wetness. A survey of finds around the playa lakes, for example Dolfin Playa in the Kharga Oasis (Figure 3.1), shows that large settlements flourished on the shores as well as within the dry beds of the playas. The shoreline settlements are mostly distinguished by flint knapping and hut rings on high ground, while activity in the lakebeds includes well-digging near to sources of water from the local sandstone aquifer and grinding of grains on large (scarcely portable) grindstones. After observation of one such playa lake that formed during winter rains in January 2013, we saw, within days of the rainfall, six types of plants that had germinated around the fringes of the lake. There was also evidence of gazelles and birds visiting the newly created oasis. Sinkholes and underground channels showed that water was draining from the basin into the local sandstone aquifer (the surface water sandstone), and after ten days or so the pool was reduced to a briny puddle from which gypsum and other evaporites crystallised (Figure 3.2). No plants could grow in the salt-crusted muds, but, one year later, tamarisk and acacia were still flourishing around the former shoreline. Again there was evidence of gazelles visiting the area, presumably from the permanent population at the persistent spring, Ain Amur, and birds were nesting in the shrubs around the basin.

Rock art from panels around numerous small lake basins in the area suggest that fish, crocodiles, elephants and ostriches, among a range of as yet unidentified fowl and quadrupeds, might have been on the menu in the area. The sites are also associated with images of giraffes led by men. Nearby, abundant large grinders, found mainly within the lakebeds, suggest that processing grain was an important part of the daily round and

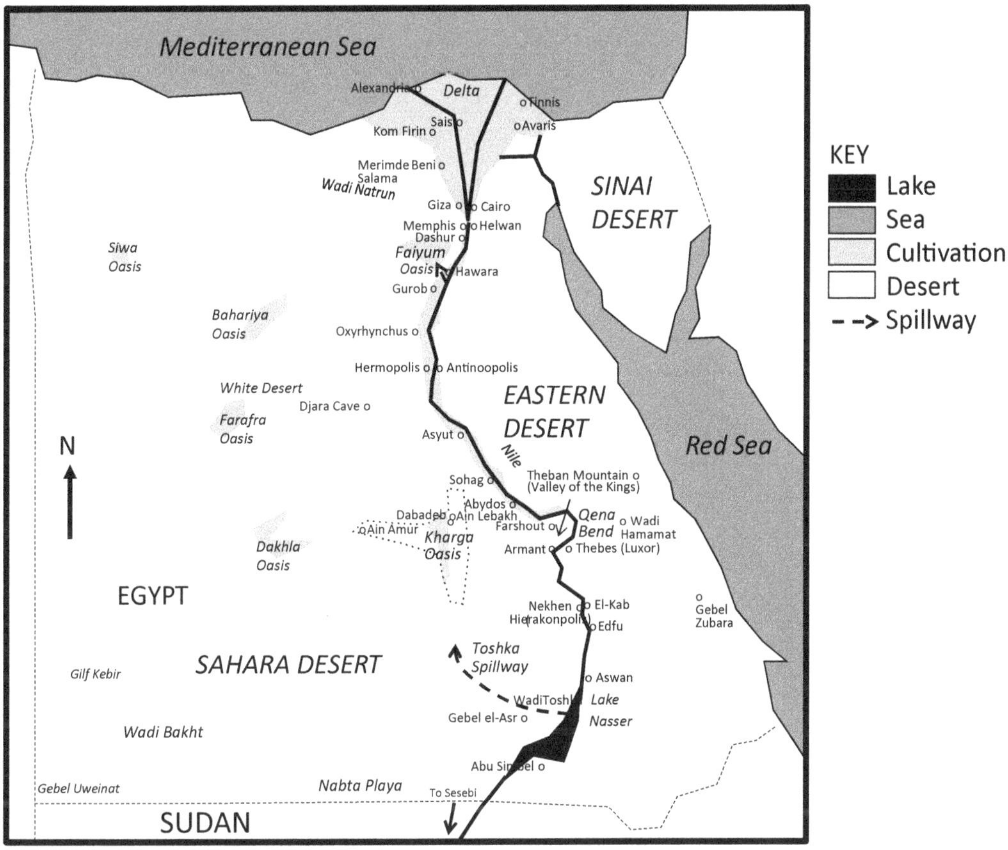

Figure 3.1 Map of Egypt showing the Nile traversing Egypt and the main places mentioned in the text. Source: author

therefore that grain was an important component of the diet. Pottery, although known from this period, is sparse in both the Farafra and Khargan areas, while ostrich eggshell was abundant, which led the team at Farafra to suggest that ostrich eggshell, being both lighter and stronger than pottery, was preferred. In Kharga, abundant ostrich eggshell fragments around structures interpreted as wells demonstrate that eggshells were used for collecting water during the Neolithic. In many cases, these spots were revisited by New Kingdom and Roman travellers, of which more below.

We infer annual cyclicity in the habitation of the Khargan playa lakes. When the rains came and the lakeshore sites were occupied the main activities were hunting, fishing and the production of rock art. As the lakes retreated, incidentally filling the surface water sandstone aquifer, plants grew in the lakebeds and could be thereafter harvested, and ground if required, for consumption. Ironically, the best-preserved sites from this period are from Dolfin Playa (Figure 3.3), where many artefacts were abandoned during an incursion of water and sediment.

Figure 3.2 The small playa, Lake Ephemera, one year after rainfall. Source: author

The same deposit also captured the remains of trees that were growing in the basin at that time. Analysis of sediments from nearby Treewater shows that they are typically thin (around 3 cm) and that the upper part of the deposit was mud that had dried completely and cracked. Some layers even contained gypsum crystals of the type described from Lake Ephemera, which indicates that the water was briny and evaporated completely. At times small sand ripples formed upon the mud-cracked surface of the playa before the next downpour brought more mud and sand. Although some sequences show the wet and dry pattern described above, at Wadi Bershama and Apot much thicker beds, up to 1.5 m thick, have formed, and the trunks and roots of trees growing in the lakebed sediments are preserved.

The observations from Kharga and Farafra correspond with others from across the Egyptian Sahara, from the Faiyum in the north to Nabta Playa (Wendorf and Schild, 1998) in the south. At Nabta Playa the longest periods of habitation are around 500 years, while stone tools typical of the Djarra B type from the White Desert further to the north (Karin Kindermann, personal communication) are typical of the period 5800 BC to 5300 BC. Further north again, contemporary settlements in

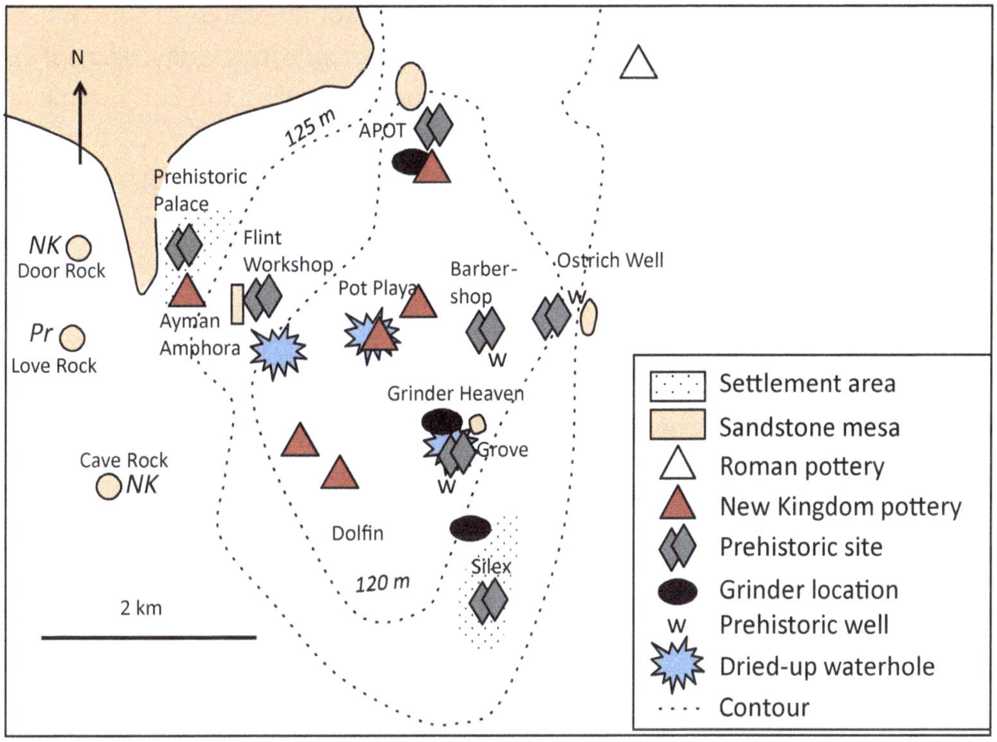

Figure 3.3 Map of a small playa, Dolfin Playa, in the Kharga basin, showing the distribution of sites around the playa including rock art sites of Prehistoric (Pr) and New Kingdom (NK) age. Source: author

the Faiyum focus on hearths (rather than huts) on high ground near the lakebed (Holdaway et al., 2018); the remains suggest the staple diet was fish enhanced by grains, with occasional game and limited pastoralism. Well-preserved baskets sunk into the ground show that already, at this period, grain was being stored.

During periods of global warming, rainfall in the deserts and an enhanced flood in the Nile valley may tip the balance of habitat for humans from the Nile valley to the Saharan lakes. Conversely, during times of global cooling, for example at the end of the Saharan Neolithic, the collapse of the playa habitats with failure of springs and wells forced populations back into the Nile valley. Temperature excursions may be a brief amelioration of climate, little more than a blooming of the desert or a few years of rainier weather, but when cycles extend over hundreds or thousands of years, whole new habitats and ecosystems arise.

As the Holocene warm period ended and populations moved into the Nile valley, the river became an essential partner to life. Fed by the heavy summer rains of the Ethiopian monsoon, the annual ebb and flow of the Nile's flood created a fertile floodplain, an excellent habitat for humans. There was increased reliance on domesticated grains, which were fortunately available from neighbours in Mesopotamia to the north-east. Patterns of river behaviour that emerged during the Early Dynastic (*c.* 3000–2686 BC) persisted until the construction of the Aswan High Dam in the 1960s. These patterns are now overlain by those of urban expansion and year-round irrigated agriculture, which have expanded across the whole Nile floodplain. Evidence of desert habitation is also being overwritten, as water pumped up into the desert has promoted re-greening of considerable areas, including large tracts of the New Valley Governorate of which Kharga is a part. Between the patterns of millennial cycling of the greened deserts and the annual flooding of the Nile there have also been decadal and century-long cycles of climate and landscape variation.

Annual cycles in the Nile valley

In the annual cycle as well as in longer cycles there emerge a number of phases: flooding, greening, harvesting and desiccation. A microcosm of the Egyptian response to climate cycling is the Egyptian management of the annual flood. From the earliest unification of Egypt, the annual cycle was well understood. This knowledge formed the basis of the Coptic calendar, and the fluctuations were closely monitored by the Egyptian state, using nilometers. The calendar was divided into three seasons: Akhet (June to September), Peret (October to February) and Shemu (March to May). In the Akhet phase, the river started to rise in the south, and the flood travelled north over the following weeks. As the river rose, levees were strategically breached to irrigate flood-basins and the inhabitants were forced to retreat to high ground, such as the flanking terraces of the Nile valley, river levees and, in the delta, ancient sandbanks known as *gezireh* (islands). A flood too low was disastrous and resulted in famine, but one that was too high caused widespread disruption of irrigation networks and dwellings. Prayers asked for a rise of 13 cubits, neither too little nor too much, and taxes were set accordingly, on the assumption that an ideal flood was productive of the richest crop. Ancient accounts tell of the flooding of Karnak temple, and Roman papyri from Oxyrhynchus (Parsons, 2007) describe the emergency measures used when an embankment was accidentally breached, including men pressing furniture into the gap while earth could be brought to stem the flow.

Where settlers were confined to the tops of the river levees and sand mounds, cultural accumulations known as tells or koms formed and became so large (Bunbury et al., 2017) that they also provided refuge from the flood in all but the most extreme events. Also during the flood season, large ships transferred goods and grain barges travelled the length of Egypt, collecting the taxes. After the flood, there was a wait of a month or more until it began to recede and the land could be prepared for agriculture, which took place in the season of Peret. The Ottoman traveller Evliya Çelebi, during his visit to Egypt in the late seventeenth century (AD 1672–1680), described the festivities that took place when the flood began to recede, and the frenzied activity that ensued as fresh mud emerged from the water. People were so busy making bricks and preparing for agriculture that he described there being 'no place to drop a needle' (Çelebi, 1938).

As Kemet (the Black Land), as the Egyptians knew it, emerged once again from the floodwaters, the agricultural season, Shemu, began. Although in the oases a local supply of artesian water meant that there could be year-round agriculture, in general in the Nile valley there was a relatively short growing and harvesting season; as the land dried out, the season of preparation began. Again we know from Roman accounts (Parsons, 2007) of duties shirked that a labour tax was exacted from each able-bodied member of the community so that the cleaning and digging of ditches and the repair of embankments could be effected before the floodwaters rose and the cycle began again. Indeed, to garner sufficient food in anticipation of the inactivity of the flood season, it was crucial to improve irrigation and productivity as far as possible. Since grain was the main currency (bread and beer being the standard payment in the Old Kingdom (2686–2160 BC)), and was taxed, the state invested a good deal in the expansion of irrigation. Juan Caros Garcea Moreno (personal communication) suggests that the records indicate that marginal marshes and swamps were an important source of fish, fowl and game, an alternative larder upon which the population drew more heavily when state intervention was weaker.

Above, I have outlined two types of annual cycle and the ways in which, by anticipating them, the ancient populations were able to survive, in either the Saharan region or the Nile valley. It is clear, from the ancient measurements and the images of intervention, that the people understood and responded to these cycles. Examples include the construction of wells and nilometers, the recording of flood levels and the construction of irrigation systems. Even from the Early Dynastic (*c.* 3000–2686 BC), a mace-head found at Hierakonpolis (Nekhen) shows one of the earliest attested kings, Narmer (*c.* 3200 BC) brandishing a hoe ready for the ceremonial breaching of the dykes as the inundation rose.

The extent and control of irrigation basins generally increased over time until, by the New Kingdom, there were extensive systems, and new spur channels were dug to take branches of the Nile to all parts of the floodplain, and earlier marginal marshes were re-irrigated to attract waterfowl to, for example, the lake of Abusir (Earl, 2010).

Longer-term climate cycling

I would like to explore whether, and to what extent, longer-term climate cycles were understood and anticipated, or whether the changes were experienced as the arbitrary actions of the gods. As we shall see, these longer cycles tended to tip the balance of activity from the Nile valley to the deserts and back. Two dramatic changes are known: the first at the end of the Pre-Dynastic around 3000 BC, and a second known as the First Intermediate Period (FIP), said to be between 2160 and 2055 BC.

Late Pre-Dynastic desertification

By the Late Pre-Dynastic, a pattern of settlements in the mouths of dried river valleys (wadis) flanking the Nile valley had developed. These locations offered access to the still-green wadi ranges for pastoralism as well as for the rich flora and somewhat hostile fauna of the Nile valley. Evidence from the shores of the Faiyum and the Nile valley, and from Merimde Beni Salama, shows that hippopotamuses, hartebeests and aurochs (Van Neer and Linseele, 2016) roamed the valley and its swamps. Indeed, Shirai (2016) even postulates that fields had to be made further from Lake Faiyum to discourage local hippos from devastating the crops.

However, as the climate cooled during the late Pre-Dynastic, the vegetation in the wadis died during periods of drought. In the by then rare rainstorms, flash floods moved large quantities of sandy sediment into the Nile floodplain and swept away wadi-mouth settlements like that near Giza at Maadi (*c.* 4000–3200 BC, El-Sanussi and Jones, 1997; Dufton and Branton, 2009). In the wadi around Hierakonpolis, there is evidence of considerable vegetation in the Pre-Dynastic. There is also evidence of a great deal of use of wood in brewing, baking and architecture (Adams, 1995). Unfortunately, the ancient wood-cutters almost certainly did not anticipate that the denudation of vegetation would make living conditions unsustainable in the new climate regime and soon afterwards their wadis started to wash out.

The First Intermediate Period

Another example of a well-studied and relatively long-range change of climate is the First Intermediate Period (FIP), an episode of increasing drought in the late Old Kingdom that is credited with devastating ancient Egypt and inspiring the Middle Kingdom tradition of lamentation literature. Typically, this period is given as 2160–2055 BC, a time of dynastic change and uncertainty in Egypt. Desiccation is evident not only from the literature referring to the period, but also from the sediments and settlement locations.

However, the evidence from sand incursion, somewhat in opposition to the ancient reportage, suggests that sand, long captured by the grasslands of the Saharan area, was already being released some 500 years earlier. In the run-up to the FIP, the sand flowed into the Nile valley, mantling the cliffs that bound the floodplain and accumulating dune fields in some places, for example Memphis (Bunbury et al., 2017) and Middle Egypt (Verstraeten et al., 2017). At Dashur around 5 m of sand accumulated during the latter part of the Old Kingdom and into the Middle Kingdom (Alexanian et al., 2010), with a further 2 m to follow since the beginning of the New Kingdom.

The sand flux affected the behaviour of the Nile, creating many islands and sandbars and maturing the floodplain and the delta (Pennington et al., 2016), reducing the number of Nile channels and making the remaining ones more stable with well-formed levees. Abandoning desert areas, the population settled in the now tamer Nile valley, particularly along its levees (Jeffreys and Tavares, 1994) and upon the tops of islands in the delta (Bunbury et al., 2017; Tristant, 2004). There is reference to this type of change in the literature of the north, for example the (retrospective) 'Prophecy of Neferty', set in the time of Sneferu (*c.* 2614–2579 BC) although it was written in the reign of Amenemhat I (*c.* 1976–1947 BC).

> Egypt's river shall run dry so that one may walk dry-shod across it; They shall seek water for ships to sail on – the River's course is now but dusty land. Riverbank shall turn to flood, and water's home shall be the place for shore.
>
> 'Prophecy of Neferty', trans. John L. Foster, 2001

The changes were noted and lamented but they were probably not predictable enough to inspire a coordinated response. The power of the king over the unified Upper and Lower Egypt diminished, and the number of local nomarchs (rulers of the regional divisions of land, the *nomes*) increased. Although the FIP was experienced as an environmental disaster around Giza and Memphis, life in the south of Egypt seemed to continue much as before. In the Semna Dispatches (British Museum) we read of small bands of travellers that come through the desert to trade, with only the occasional murmur in Dispatch 5 about a period of hunger. Whether it is regarded as a disaster or not, the FIP ushered in a period of increased regionalisation and devolved power. Literacy became widespread among nomarchs and wealthy private citizens; it was no longer a royal prerogative. We can only debate the extent to which the emergence of a reading class helped to inform the longer-term planning that is evident from the Middle Kingdom onwards.

Landscape design in the Middle Kingdom (2055–1650 BC)

A general reduction in the flood after the end of the Old Kingdom meant that the Nile no longer naturally overflowed the Hawara gap to refresh Lake Faiyum, which consequently subsided. To increase the agricultural area during the Middle Kingdom (MK), Amenemhat III (1831–1786 BC) enacted a scheme to redirect the Bahr Yusuf (a western and minor channel of the Nile) into a reservoir and the Faiyum basin and thus irrigate the former lakebed for agriculture. This is probably one of the earliest pieces of evidence of the redirection of the Bahr Yusuf, an art that continued to develop through the New Kingdom (see below).

We can conclude that the experience of climate change at the end of the Pre-Dynastic, as well as during the FIP, was mixed, but that the response was reactive rather than anticipatory. Events associated with the FIP climate change were recorded in somewhat apocalyptic form and the response was relatively experimental; different regions explored different ways of living in the new landscapes during the FIP and the MK. Note that the appearance of a new landscape suitable for the cultivation of grain went without mention in the literature, but adoption of domesticated species of grain from Mesopotamia was enthusiastic. Perhaps, as today, the ancient populace was quicker to hear the good news than the bad.

Folk wisdom

However, not all types of knowledge are reflected in the surviving texts, for example the behaviour of islands, even though these landscape features are prominent in the mythology. Observations of modern islands in the Nile show how they form, develop and are eventually bonded to the mainland within about a century. New Kingdom temple layouts suggest that these patterns were already familiar to the constructors of the temple. Just as, today, in the Nile valley, old men and women remember the topography of the changing islands and channels of their youth, similar topographical information was available in antiquity. By comparing the development of contemporary islands with the changes in those that they recall from their youth, they develop a broad-brush predictive tool.

The use of scholarship

In the New Kingdom and Roman periods, the role of ancient scholarship in the recognition of strategies for survival becomes explicit. For example, we know from the reports of the priests to whom Herodotus spoke (*c.* 430 BC) that:

> One fact which I learnt of the priests is to me a strong evidence of the origin of the country. They said that when Moeris was king, the Nile overflowed all Egypt below Memphis, as soon as it rose so little as eight cubits. Now Moeris had not been dead 900 years at the time when I heard this of the priests; yet at the present day, unless the river rise sixteen, or, at the very least, fifteen cubits, it does not overflow the lands.
>
> Herodotus (1885): Book 2, Chapter 13

Our observations of sediment in the area show that the priests' account is probably an accurate description of the changing geography, which demonstrates that they had passed historical knowledge down through almost a millennium.

Work on diverting the Bahr Yusuf, which had begun during the Middle Kingdom, continued in the New Kingdom. Thutmosis III (1479–1425 BC) had a channel dug at Gurob which was controlled by the associated palace. By the Ptolemaic period (332–30 BC), the art of channel management had become a science and the Bahr Yusuf was diverted across the floodplain further south of the Faiyum to supply the new town, safely constructed on the desert edge, of Oxyrhynchus (Figure 3.4). By the time of the Roman emperor Hadrian, channel management had extended from the Bahr Yusuf to the main Nile, which has been trapped at Antinoupolis by Roman revetments since the foundation of the city in AD 130 (see the blog

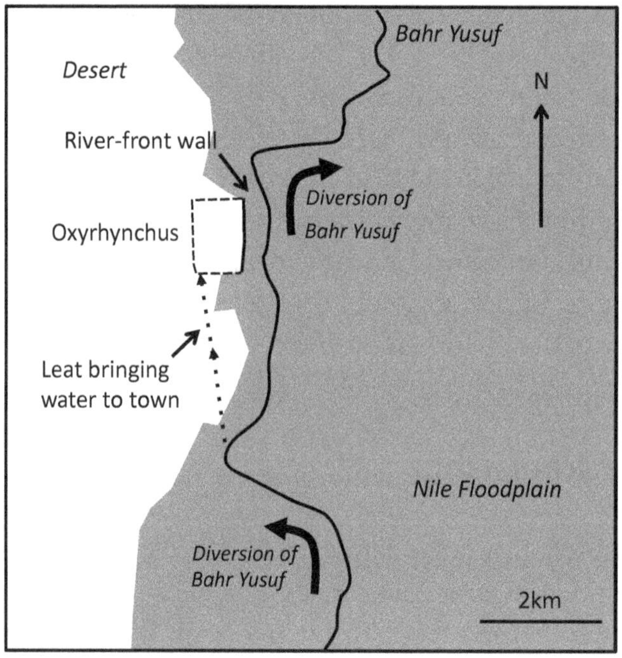

Figure 3.4 Map of the area around Oxyrhynchus showing how the Bahr Yusuf is diverted westwards to the desert edge in order to serve the town. Source: author

of the Istituto Papirologico 'G. Vitelli' and the Antinoupolis Foundation, by Jay Heidel, at http://antinoupolis.net/blog).

We might expect that climate cycles of a century or less would most easily be interpreted with the benefit of either folk wisdom or scholarship. Indeed, we know that Amenhotep, Son of Hapu, who served under Amenhotep III (1390–1352 BC), was famous in the New Kingdom for his erudition, which included both ancient and foreign languages. The reign of Amenhotep III lasted for around 39 years and the record of events during this time is curiously episodic. While we have widely dispersed ceremonial scarabs from the first decade trumpeting the achievements of the king, the second decade goes unmarked. The third decade sees a change of location to the western desert edge of the Theban region and an extraordinary investment in the excavation of large lakes or reservoirs, of which Birket Habu alone is 2.3 km^2 in area. At the same time, new family tombs are excavated 6 km into the desert beyond the palace. Devotions to the goddess Sekhmet (responsible for pestilence) are prominent; there are many Sekhmet statues commissioned for the Temple of Millions of Years at Kom el-Hetan. By regnal year 30 we have reports of exceptional harvests and the beginning of a number of festivals at the palace of Malqata. Sedimentary evidence from the surrounding desert (Litherland, 2015) indicates that the period of construction of the tombs was uncharacteristically wet, and the desert became quite widely explored and occupied. Given the royal distaste for recording negative events it is tempting to think that the silent second decade of the reign was a time of adverse conditions. Can we speculate that the magnificent project of Birket Habu, coupled with fervent petitions to Sekhmet, was intended to ensure that they were never caught out again?

Thought to be from a similar period, the biblical narrative of Genesis, the history of a group of expatriates living in the north of Egypt, contains the hero story of Joseph who, anticipating seven years of famine after seven years of plenty, stores surplus grain against the bad times. Although this narrative cannot be dated precisely, it is known from the New Kingdom onwards and is often placed in the Eighteenth Dynasty, a time when 'many-coloured cloth' was a popular import from trade associates in the eastern Mediterranean. A later reference to Joseph in the account of Moses, which can be linked to the Nineteenth Dynasty, is consistent with this interpretation.

From independent climate records of this period trapped in ice cores from the Greenland and Kilimanjaro ice sheets (Thompson et al., 2002), we can see that in two periods, the New Kingdom and the Roman, there were climate-warming excursions that matched the magnitude but not the duration of the Saharan Neolithic. Perhaps it is therefore no surprise that during these reigns there is much evidence of the revisiting of Neolithic waterholes and wells in the Kharga basin as well as of the revival of cross-desert routes like that between Thebes (modern Luxor) and Farshout. Indeed, the frequent juxtaposition of Neolithic ostrich eggshell beads and fragments with New Kingdom and Roman pottery and wells in the Kharga area may even imply a systematic approach.

During the New Kingdom period of amelioration of the desert, areas of water collection were arranged and a large jar was left in place so that subsequent travellers could harvest the water. However, by Roman times, there was a more 'academic' approach. To find more water, they determined where signs of earlier activity suggested there had previously been water, and then dug deeper. This recognition included identification of Neolithic pools that were associated with snail shells and ostrich eggshell beads. The Romans also brought or developed the Persian technology of the *qanat* (known in Egypt as *manawir*) to collect water more efficiently. *Qanats* around the abandoned settlement of Debadeb used many kilometres of underground channel to supply water to the town. Even where the tunnels have collapsed, their traces can still be seen where the frequent maintenance shafts puncture the ground surface. At Debadeb the relict *qanat* system still supplies sufficient water for an acacia grove to flourish, exactly the sort of refuge from which the desert can be re-inseminated when the climate moderates.

Combining the Greenland ice core and the records from glaciers from Mount Kilimanjaro (Alley, 2000; Thompson et al., 2002), we can explore further the effect New Kingdom climate cycling had on the Egyptian Empire and on Egyptian behaviour. Although the calendar for the kings (both men and women) of the New Kingdom is well understood (Figure 3.5), the ice core records cannot be closely correlated with the political upheavals. The main reason for the lack of a tight chronology is the way in which the air incorporated in the glacier remains in contact with the atmosphere for up to 50 years after the snowfall with which it is associated. To determine historical global temperatures, isotopic analysis of air and water is used to provide a proxy for temperature and a date (Thompson et al., 2002). Only as snow accumulates above it and the ice is compacted into firn and then into glacial ice are the air bubbles isolated, capturing the ambient air. The results necessarily present a somewhat blurred average of the conditions pertaining over several decades.

What evidence we have suggests that the first king of the Eighteenth Dynasty (New Kingdom), Ahmose I (1550–1525 BC) was afflicted by an

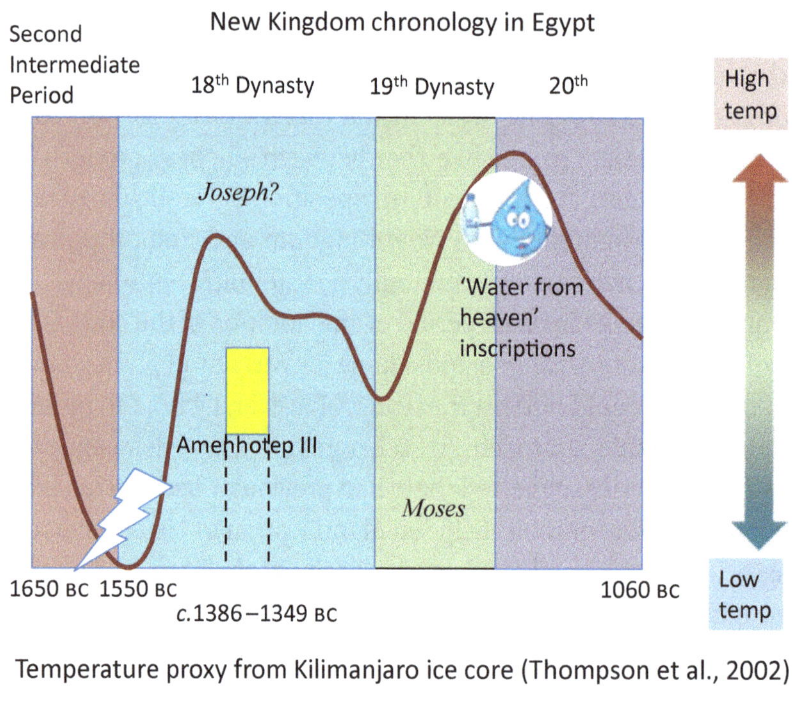

Figure 3.5 Diagram of climate changes from the Mount Kilimanjaro ice core superposed on the political events in Egypt discussed in the text. The ice core records are means and are not closely calibrated with the regnal years, being up to 50 years earlier or later. Source: author

extreme weather event, described in the Tempest Stele and including such adverse conditions that there were '[... corpses(?)] floating on the water like skiffs of papyrus' (Ritner and Moeller, 2014: 6). Some authors (Polinger et al., 1996) also link the events described to the eruption of Thera, which caused temporary global cooling. Ice core records suggest that, having reached something of a low, global temperatures started to rise. Subsequently, the rising temperature increased local rainfall and intensified the monsoon in Ethiopia, raising the level of the annual summer flood. With higher floods, fertile Nile sediments extended over the flanks of the Nile valley and impinged upon the mouths of the wadis that meet it (Macklin et al., 2015). Direct rainfall in the Saharan hinterland fostered the development of vegetation in the wadis and playas, creating additional habitat for game. Although high floods may damage canals, sluices and other irrigation infrastructure, during the agricultural season that follows a high flood large areas of fresh, wet soil are available. Incidentally, the expansion of the river into old channel beds and the refilling of marshes in low ground may, at this stage, lead to a higher risk of waterborne disease, of which bilharzia and malaria are known to be common in Egypt and whose traces are found on ancient burials.

As the climate cycle returns to drier conditions, lakes and marshes dry, exposing rich sediments that are relatively wet and easily irrigable. Local aquifers are full and can release water back into the sediment. The risk of disease lessens as marshes start to dry, and the dry, sunny conditions ensure that a bumper harvest, such as the one described from the thirtieth year of the reign of Amenhotep III (1390–1352 BC), can be obtained. However, if drier conditions persist, the flood intensity is reduced, aquifers become depleted, desert wells and vegetation are exhausted and productivity falls again, with resultant pressure on the now expanded population. Salts dissolved in the water of playa lakes and the residual waterholes become concentrated, and the water source may become bitter and sterile.

Within this cycle of greening and drying, the phase most advantageous to human habitation is that during which drying begins. If desiccation continues, however, what follows is insufficient flooding for irrigation and fertilisation, coupled with the eventual collapse of the wadi beds, which narrows the floodplain further. The sediments around Luxor provide an excellent example of the magnitude of these changes. Relicts of early Holocene Nile silts are 30 m above the current level of the floodplain and overlie wind-blown sand thought to have accumulated during the arid period of the last ice age. At the time of deposition water must have penetrated far into the wadi mouths of the Theban Mountain and combined with additional rainfall in the massif, forming an excellent habitat with easy access to the Nile valley. Like the Khargan area, the Theban Mountain contains many examples of rock art, Pre-Dynastic settlements as well as New Kingdom and Roman shelters, springs and wells. For example, rock art in Wadi F in the Theban Mountain near Luxor (Litherland, 2015) demonstrates the use of rock shelters during this period.

Our observations from sediments, settlements and graffiti in the Theban Mountain suggest that the reign of Amenhotep III saw the greatest extent of New Kingdom activity there. During recent excavations from 2014 onwards of the shaft tombs of a group of women associated with Amenhotep III in the Wadi Bariya, we noted that the sediments that pre-dated the excavation of the tombs indicated increasingly wet conditions. Towards the Eighteenth Dynasty surface, the rounding of clasts increased and they were topped by a layer of mud that bore New Kingdom pottery and the imprints of hands, feet and knees where the mud had been extracted and mixed with temper to plaster the interior of the tomb. Later, after the abandonment of the tombs, the sediments suggest a return to arid conditions, with large diurnal temperature changes leading to frost-cracking of clasts between episodes of re-deposition during rainfall. These wetter periods were followed by a return to current desert conditions and the onset of deflation by the dry trade winds.

Some 2.5 km to the east of Wadi Bariya in Wadi F, sedimentary observations show a similar pattern of climatic variation. Some 30 cm of angular clasts were deposited, followed by beds with rounder clasts, here interpreted as the product of a milder, wetter climate, and finally by a layer of coarse, well-rounded sand. The upper surface of the sand dipped steeply down the wadi (*c.* 15°), which further suggests that vegetation had helped to stabilise the steep wadi floor as it descended through the canyon. At eye level to this surface are New Kingdom (NK) inscriptions, which are consistent with it being the same wet period as that captured by the evidence at Wadi Bariya. The smooth wadi floor probably rendered it more accessible than it is today, and there are many inscriptions in a small side wadi, suggesting many visits. Round stone structures nearby may have been NK huts, but later excavation of them makes the context difficult to interpret.

The same area was intensively investigated during the Third Intermediate Period (TIP) by a state-sponsored team hoping to recover gold and other valuables from the earlier tombs in the area. The absence of pottery or other material culture, which was presumably simply discarded by them at Wadi Bariya, suggests that this ancient team did not discover tombs in Wadi F. Carter recorded that he found no tombs, and some of the excavations may be attributable to him. The TIP visits are recorded in numerous inscriptions, some of which are below the postulated NK surface; this suggests that the erosion had already occurred by the TIP, since the erosion of the sediment would otherwise have erased the fragile inscriptions.

Although the Wadi Bariya tombs seem to be far out in the desert, more than 6 km from the floodplain and the palace at Malqata, they are associated with other infrastructure: a winding chariot road and a cross-desert donkey route. In Pierre Anus's description of blocks reused at Karnak (Anus, 1971; Gräzer Ohara, 2012) from the Amenhotep III temple at Kom el-Hetan, all the elements of the Wadi Bariya, as we understand it to have been during the New Kingdom, are present. We see images of wadis with jackals (still present today) and gazelles, representations of scrub vegetation, chariots being put through their paces, copper targets being taken out for archery practice and two people sporting in a pool. The blocks also paint a lively picture of the palace at Malqata, identifiable from the neighbouring mounds of Birket Habu (a large reservoir) that impinge upon it, and showing the palace, a menagerie, vineyards, pools, gardens and a fenced stock enclosure.

However, the climate change curve suggests that Amenhotep III's reign was characterised by a decrease in temperature which may have been the part of the cycle in which the flood started to diminish and the rains to cease. The additional sunshine falling upon the fresh muds may have been partly responsible for the bumper wheat harvest of regnal year 30. Conversely the move to Malqata, near to the Temple of Millions of Years at Kom el-Hetan, and the construction of the Birket Habu, a very large reservoir of 2.3 km^2, with sources of water from the wadis and spurs from the Nile channel, may be an indication of increasing anxiety about water security. Almost as well known as Amenhotep III is his contemporary Amenhotep-son-of-Hapu, a renowned scholar able to read foreign and ancient languages. The development, during this period, of water-management measures is likely to arise from deliberate scholarship in an attempt to solve an immediate problem.

Additional evidence of erosion of the wadi system after the New Kingdom arises from the chariot track that surrounds the Wadi Bariya site. Since the chariots had fairly narrow wooden wheels, they could not cross the normal desert surface, which is littered with stones. Obstructions were therefore scraped to the sides to clear a track a couple of metres wide and consistent with use by chariotry of the type preserved in the tomb of Tutankhamun (1336–1327 BC), the grandson of Amenhotep III. Some of the terraces along which the old road passed have now been dissected by up to two metres of erosion, and a trail of rounded sherds washed 8 km across the desert from a New Kingdom and Roman donkey station (WB8) on the plateau is further testament to the amounts of rainfall and erosion since the New Kingdom, when wadi sediments seem to have reached a peak. In addition to the road, texts containing detailed requests for horses, equipment and horse-trainers to be sent from the Near East are common in the contemporary Amarna archive (Moran, 2000), and there are a number of strands of evidence to suggest that use of the newly arrived horse and chariot was an important activity, which would tie the use of the track to the Eighteenth Dynasty. In addition to the construction of the track, the banks of Birket Habu were adapted to further their use for training and exercising horses.

After the Eighteenth Dynasty, rainfall was recorded during the Nineteenth-Dynasty inscriptions near waterfalls in the Theban Mountain (Dorn, 2016). Rainfall around this time also caused flash floods in the Valley of the Kings, overwhelming the entrance to Tutankhamun's tomb and concealing it. Back in Wadi F, inscribed blocks in the side wadi described above were also overlain by shattered rock apparently brought down by rainstorms. While the ice core records cannot reveal the decadal climate variations inferred from the archaeological evidence, they can reveal centurial cycles, which may have affected the stability and extent of the ancient Egyptian regimes. However, evidence of changing activity in the desert supports the view that during the New Kingdom and Roman warming episodes, people were quick to return to the deserts and exploit the resources, including water, game and minerals, available there. Some evidence, both archaeological and textual, suggests that scholarship played a role in mitigating the negative effects of the climate cycles as well as in finding scarce resources as the environment improved.

Conclusions

There is abundant evidence of measurement and recording of annual cycles of habitat caused by the seasonality of the flood in the Nile valley. Earlier Neolithic sites echo this pattern of interaction with an annual cycle, since there are many sites associated with the lakes and playas of the Saharan region. On a decadal timescale there is evidence from myth and from the storage of grain by temples and the construction of large reservoirs to indicate a general appreciation of the cyclicity of climate change. On timescales of up to a century, there is some practical evidence of anticipation of landscape change, from a consistent pattern of choice of sites for settlements and temples on islands in preference to levees of the Nile. Surviving texts mention islands as abundant in the afterlife, and they are prominent in the foundation myth of Egypt, which is said to have emerged as an island from the waters of chaos. Practical advice about the use and patterns of geographical development of islands is unknown in the texts but seems to have been transmitted orally, much as it is in the Nile valley today.

Repetition of climate warming over several centuries during the Neolithic, New Kingdom and Roman periods seems to have been noted by practical scholars, who, faced with a shortage of water, examined past precedents. They learned to find water in the environment by excavating ancient pools, but again written records of this knowledge are unknown. Adoption of developments such as more productive seed stock for grains and irrigation strategies from the Levant to provide food security in a cyclical habitat, as well as the adoption of water-moving and -harvesting devices such as the *qanat* from Persia, shows that ideas were borrowed from neighbours as required. The dispersion of lapis lazuli from Pakistan and Afghanistan to Egypt shows that trade percolated through the whole region from the Old Kingdom onwards.

Acknowledgements

I would like to thank the Supreme Council of Antiquities for their permission to work on many sites in Egypt, and my many colleagues who have generously allowed me to join them on their excavations. In this work I have drawn particularly on the work of the North Kharga Oasis Survey directed by Salima Ikram and the work of the New Kingdom Research Foundation at Thebes directed by Piers Litherland and Geoffrey Martin.

References

Adams, B. 1995. *Ancient Nekhen: Garstang in the City of Hierakonpolis*. Egyptian Studies Association Publication No 3. New Malden: SIA Publishing.

Alexanian, Nicole, Wiebke Bebermeier, Dirk Blaschta, Arne Ramisch, Britta Schütt and Stephan Johannes Seidlmayer. 2010. 'The necropolis of Dahshur: Seventh excavation report autumn 2009 and spring 2010.' German Archaeological Institute/Free University of Berlin.

Alley, Richard, 2000. 'The Younger Dryas cold interval as viewed from central Greenland.' *Quaternary Science Reviews* 19, 213–26.

Anus, P., 1971. 'Un domaine thébain d'époque « amarnienne ». Sur quelques blocs de remploi trouvés à Karnak [avec 4 planches]' (A Theban realm of the 'Amarna' period: On several reused blocks at Karnak (with 4 plates)). *Bulletin de l'Institut français d'archéologie orientale* 69, 69–88.

Barich, Barbara E., Giulio Lucarini, Mohamed A. Hamdan and Fekri A. Hassan (eds). 2014. *From Lake to Sand: The Archaeology of Farafra Oasis Western Desert, Egypt*. Florence: All'Insegna del Giglio.

Bubenzer, Olaf and Heiko Riemer. 2007. 'Holocene climatic change and human settlement between the central Sahara and the Nile Valley: Archaeological and geomorphological results.' *Geoarchaeology* 22 (6), 607–20.

Bunbury, Judith, Ana Tavares, Benjamin Pennington and Pedro Gonçalves. 2017. 'Development of the Memphite floodplain.' In Harco Willems and Jan-Michael Dahms (eds), *The Nile: Natural and Cultural Landscape in Egypt. Proceedings of the International Symposium held at the Johannes Gutenberg-Universität Mainz, 22–23 February 2013*, 71–96. Mainz Historical Cultural Sciences. Mainz: transcript Verlag.

Çelebi, Evliya.1938. *Seyahatname. Volume 10: Egypt and the Sudan*. Transliteration to Turkish by Ahmet Cevdet. Istanbul: Ikdam Matbaası. (First published 1672.)

Dorn, Andreas. 2016. 'The hydrology of the Valley of the Kings: Weather, rainfall, drainage patterns, and flood protection in antiquity.' In Richard H. Wilkinson and Kent R. Weeks (eds), *The Oxford Handbook of the Valley of the Kings*, 30–40. Oxford: Oxford University Press.

Dufton, D. and T. Branton. 2009. 'Climate change in early Egypt.' *Egyptian Archaeology* 36, 2–3.

Earl, E. 2010. 'The Lake of Abusir, northern Egypt.' Unpublished dissertation, Cambridge University.

El-Sanussi, Ashraf and Michael Jones. 1997. 'A site of the Maadi culture near the Giza pyramids.' *Mitteilungen des Deutschen Archäologischen Instituts, Abteilung Kairo* 53, 241–53.

Foster, J. L. (trans.). 2001. 'The Prophecy of Neferty'. In *Ancient Egyptian Literature: An Anthology*, 76–84. Austin: University of Texas Press.

Garcea, Elena A. A. 2006. 'Semi-permanent foragers in semi-arid environments of North Africa.' *World Archaeology* 38 (2), 197–219.

Gräzer Ohara, Aude, 2012. 'Le palais des monts sur un bloc de remploi de Karnak: Marou d'Amon et/ou complexe jubilaire d'Amenhotep III à Malqata?' [The palace of the mountains on a re-used block at Karnak: Amon's *Marou* and/or a jubilee complex of Amenhotep III at Malqata?]. *Bulletin de l'Institut français d'archéologie orientale* 112, 191–213.

Haynes, C. Vance. 1980. 'Geochronology of Wadi Tushka: Lost tributary of the Nile.' *Science* 210 (4465), 68–71.

Haynes, C. Vance, Peter J. Mehringer and El Sayed Abbas Zaghloul. 1979. 'Pluvial lakes of north-western Sudan.' *Geographical Journal* 145 (3), 437–45.

Herodotus (ed. and trans. George Rawlinson). 1885. *The History of Herodotus*, vol. 2. New York: D. Appleton and Company.

Holdaway, Simon, Rebecca Phillipps, Joshua Emmitt, Veerle Linseele and Willeke Wendrich. 2018. '*The Desert Fayum* in the twenty-first century.' *Antiquity* 92 (361), 233–8.

Jeffreys, David and Ana Tavares. 1994. 'The historic landscape of Early Dynastic Memphis.' *Mitteilungen des Deutschen Archäologischen Instituts, Abteilung Kairo* 50, 143–73.

Litherland, Piers. 2015. *The Western Wadis of the Theban Necropolis: A Re-examination of the Western Wadis of the Theban Necropolis*. London: New Kingdom Research Foundation.

Macklin, Mark G., Willem H. J. Toonen, Jamie C. Woodward, Martin A. J. Williams, Clément Flaux, Nick Marriner, Kathleen Nicoll, Gert Verstraeten, Neal Spencer, Derek Welsby. 2015. 'A new model of river dynamics, hydroclimatic change and human settlement in the Nile Valley derived from meta-analysis of the Holocene fluvial archive.' *Quaternary Science Reviews* 130, 109–23.

Maxwell, Ted A., Bahay Issawi, C. Vance Haynes, Jr. 2010. 'Evidence for Pleistocene lakes in the Tushka region, south Egypt.' *Geology* 38 (12), 1135–38.

Moran, William L. (ed. and trans.). 2000. *The Amarna Letters*. Baltimore, MD: Johns Hopkins University Press.

Parsons, Peter. 2007. *City of the Sharp-Nosed Fish: Greek Lives in Roman Egypt*. London: Weidenfeld & Nicolson.

Pennington, B.T., J. M. Bunbury and N. Hovius. 2016. 'Emergence of civilization, changes in fluvio-deltaic style, and nutrient redistribution forced by Holocene sea-level rise.' *Geoarchaeology* 31 (3), 194–210.

Polinger Foster, Karen, Robert K. Ritner and Benjamin Foster. 1996. 'Texts, storms, and the Thera eruption.' *Journal of Near Eastern Studies* 55 (1), 1–14.

Ritner, Robert K. and Nadine Moeller. 2014. 'The Ahmose "Tempest Stela", Thera and comparative chronology.' *Journal of Near Eastern Studies* 73 (1), 1–19.

Shirai, N. 2016. '*The Desert Fayum* at 80: Revisiting a Neolithic farming community in Egypt.' *Antiquity* 90 (353), 1181–95.

Thompson, Lonnie G., Ellen Mosley-Thompson, Mary E. Davis, Keith A. Henderson, Henry H. Brecher, Victor S. Zagorodnov, Tracy A. Mashiotta, Ping-Nan Lin, Vladimir N. Mikhalenko, Douglas R. Hardy and Jürg Beer. 2002. 'Kilimanjaro ice core records: Evidence of Holocene climate change in tropical Africa.' *Science* 298, 589–93.

Tristant, Y. 2004. *L'Habitat Pré-Dynastique de la Vallée du Nil: Vivre sur le rives du Nil aux V et IV Millénaires*. Oxford: Archaeopress.

Van Neer, Wim and Veerle Linseele. 2016. 'Interaction between man and animals in the prehistoric Nile Valley.' In Sonia R. Zakrzewski, A. J. Shortland and Joanne Rowland, *Science in the Study of Ancient Egypt*, 110. New York & Abingdon: Routledge.

Verstraeten, Gert, Ihab Mohamed, Bastiaan Notebaert and Harco Willems. 2017. 'The dynamic nature of the transition from the Nile floodplain to the desert in central Egypt since the mid-Holocene.' In Harco Willems and Jan-Michael Dahms (eds), *The Nile: Natural and Cultural Landscape in Egypt. Proceedings of the International Symposium Held at the Johannes Gutenberg-Universität Mainz, 22–23 February 2013*, 239–54. Mainz Historical Cultural Sciences, 36. Mainz: transcript Verlag.

Wendorf, Fred and Romuald Schild. 1998. 'Nabta Playa and its role in Northeastern African prehistory.' *Journal of Anthropological Archaeology* 17, 97–123.

4

Geoarchaeology of prehistoric moated sites and water management in the Middle River Yangtze, China

Duowen Mo and Yijie Zhuang

Abstract

Moated sites appeared very early in the Middle River Yangtze, and their numbers continued to grow throughout Neolithic times, indicating the increasing investment in hydraulic engineering undertaken by prehistoric societies. These hydraulic features formed a robust water-management system that sustained continuous socio-economic growth in the region. We present evidence from our regional geoarchaeological surveys of prehistoric sites in the Middle River Yangtze, focusing particularly on the moats at these sites. We demonstrate how the construction and maintenance of these moated sites were related to and constrained by the changing environmental and socio-economic conditions. Moats became increasingly important in water management, especially flood prevention, and particularly when the settlements occupied large areas of local landscapes, often encroaching upon low-lying areas, because of population growth and migration. By the Qujialing-Shijiahe period (3000–2100 BC), regional competition had become one of the main driving forces that stimulated large-scale infrastructure construction and production of various consumables among large to medium-sized settlements. This not only changed patterns of economic development in the region, a region that was driven and accelerated by ritual activities and associated social gatherings, but also profoundly transformed the local landscape into one that was densely occupied by moats, ditches and other water-management infrastructures. It is this unique entanglement between social competition, intensification of economic production, and infrastructure construction that needs to be unpacked in future research efforts to demonstrate the long-term water-management practices in this region and their relationship with environmental changes and the corresponding social evolution.

Introduction

Genetic studies and computational simulations have predicted that both the Middle and Lower Yangtze rivers were important places for rice domestication (Silva et al., 2015). While recent research has illustrated the cultivation, domestication and intensification processes of a rice economy in the Lower River Yangtze (Fuller et al., 2009; Gross and Zhao, 2014), the subsequent developmental trajectory of a rice economy in the prehistoric Middle River Yangtze remains unclear. Rice consumption began very early in this region (C. Zhang, 2000; Z. Zhao, 1998), which raises questions about how the transition from possible early cultivation to established rice farming took place and how rice domestication occurred.

In contrast to the uncertainty surrounding the development of the prehistoric rice economy is the early appearance of large moated settlements and the continuous development of hydraulic systems throughout prehistoric times in the Middle River Yangtze. These hydraulic systems were on a large scale compared with earlier periods, consisting of moats, earthen walls and levees, with associated features such as steps, bridges and piers located along the moats and nearby rivers. Together, these features, arguably, formed a robust water-management system that sustained continuous socio-economic growth in the region (see Pei, 2004a, 2004b). Further questions may be asked about the relationship between the changing palaeo-environmental and climatic conditions, the prehistoric settlements, and the associated hydraulic systems. To what extent did these hydraulic systems function as irrigation facilities supplying water to rice farming or as flood protection infrastructures? And how were the construction and management of these systems related to social evolution and economic development in the region? To address these questions we present evidence from our regional geoarchaeological surveys of prehistoric sites and their surrounding environments in the Middle River Yangtze. We focused particularly on the moats and demonstrated how their construction and maintenance was related to, and constrained by, the changing environmental and socio-economic conditions.

Regional environmental settings

The Middle River Yangtze region consists of two large alluvial plains, the Jianghan Plain and the Dongting Plain (Figure 4.1), both of which are encircled by mountains. These plains are low-lying, around 20–40 m above sea level. The low-lying areas are subject to active alluvial activities and prone to flooding, especially during the monsoon season. Many systematic coring surveys have demonstrated that the Holocene stratigraphies are dominated by fine alluvial and lacustrine sediments that have rapidly built up since the beginning of the Holocene (H. Deng et al., 2009). These cores sometimes contain sandy or coarser particle deposits derived from periodic flooding events. Scattered in the middle or along the edges of the alluvial plains are places that have been overlain by Middle to Late Pleistocene loess and red clay, often several metres higher than their surroundings. These places are called by local people 'slopes' or 'tablelands' (Figure 4.2A). They would have been ideal places for prehistoric occupation, as would the areas transitional between foothills and alluvial fans, which were generally higher than their surroundings.

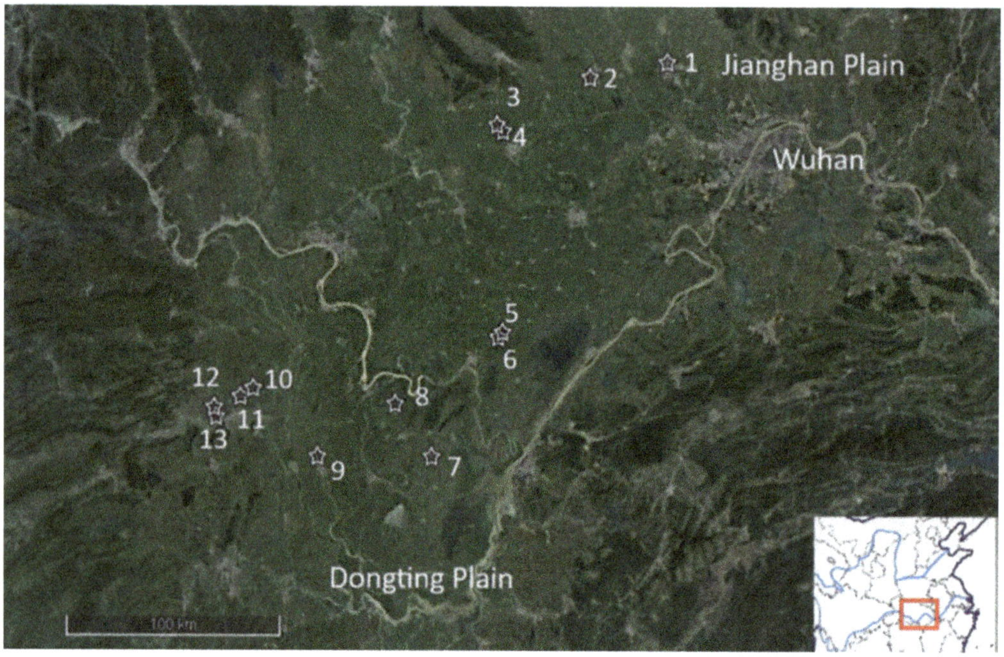

Figure 4.1 The Jianghan Plain and the Dongting Plain in the Middle Yangtze, and the locations of major archaeological sites mentioned in the text. 1 Yejiamiao; 2 Menbanwan; 3 Shijiahe; 4 Sanfangwan; 5 Futian; 6 Liuguan; 7 Fenshanbao; 8 Zoumaling; 9 Huachenggang; 10 Bashidang; 11 Jijiaocheng; 12 Chengtoushan; 13 Pengtoushan. Source: authors

Figure 4.2 A. Typical loess slope at Shiligang, Hunan. B. The Pengtoushan site: the moat, which is still functional, and the site (where the person stands). Source: authors

The studied region has humid, rainy summers and cold, relatively dry winters. The present average annual temperature is 15–18°C, while annual precipitation ranges from 800 to 2,000 mm. Its eastern part, closer to the monsoon front, receives significantly more rainfall than its western areas. As documented in high-resolution stalagmite records, the Holocene climate has slowly but steadily deteriorated from the middle to late Holocene onwards (Y. Wang et al., 2005), meaning that the climate was warmer and more humid during the middle Holocene than during the latter half of the Holocene. It is estimated that the middle Holocene temperatures were 1–2°C higher and precipitation was 200–400 mm greater than at present. However, it is important to note that many decadal and centennial oscillations can also be observed from the reconstructed curves of moisture changes. While this general trend of Holocene precipitation is clear, its impact on the Holocene alluvial hydrology is complicated by the fact that the hydrology is influenced by other factors, such as sea-level fluctuations and sedimentation rates. Generally speaking, in the short term, the peak charge of river water is indeed directly linked to concentrated summer rainfall. However, in the longer term, river flooding regimes are more closely related to landforms and changes in sedimentation rates. This means that it is currently difficult to assess the impact of Holocene climate on flooding regimes in this region.

Prehistoric societies and subsistence

Parallel with the long tradition of rice consumption in the Lower Yangtze River, the earliest rice remains discovered at the Yuchanyan and Xianrendong cave sites in the Middle River Yangtze can be dated to before 8000 cal. BC (C. Zhang, 2000; Z. Zhao, 1998). Together with, or earlier than, this important development was the invention of pottery. The earliest pottery is dated to around 18,000 cal. BC (X. H. Wu et al., 2012). A popular theory is that this pottery was for food processing and thus represents a significant technological leap driven by subsistence change (see Kuzmin, 2013; Shelach, 2012). However, because of the lack of archaeological finds for this chronological gap between 18,000 cal. BC ago, when pottery first appeared, and the early Holocene, when rice first began to be consumed, it is still difficult to understand fully the relationship between the technological advancements, cultural adaptation and the environment. Given that most of these early sites were located in the mountainous areas and were primarily reliant for food on hunting and gathering, the role of rice, as a grassy plant grown in wet habitats, in the local diet was probably minor at this time.

The earliest Neolithic culture found in the Middle River Yangtze is the Pengtoushan, dating to 7100–5600 cal. BC (Hunan Provincial Institute of Cultural Relics and Archaeology, 2006: 617). According to C. Zhang (2015: 207) and Pei (2013), the majority of the Early Neolithic sites were located on river terraces and tablelands, ranging in size from several thousand to three hectares, including some shell midden sites which were situated closer to rivers. The Pengtoushan culture sites were already quite 'mature' in terms of their settlement structures. The Pengtoushan site was encircled by a moat (Figure 4.2B). Similarly, the Bashidang site was built next to a river. One side of the site was connected to a lake, the other three sides being enclosed by a moat and earthen walls (Hunan

Provincial Institute of Cultural Relics and Archaeology, 2006; Pei, 2013; C. Zhang, 2015). At Bashidang, single-roomed houses appeared in rows in the north-west and north-east corners of the enclosed area, each containing a fireplace or hearth. Abundant artefacts, and organic remains such as animal bones and seeds, were found next to the residential area. Rice was also present. Some scholars have claimed that the rice was already cultivated (W. X. Zhang, 2006; W. X. Zhang and Pei, 1997; D. L. Zhao and Pei, 2000), with gathered wild plants probably still an important component of the diet. Along with mammal bones, aquatic species including fish and shells were also abundant. This evidence suggests that hunting and gathering were still the main sources of food. However, there is also the suggestion that granaries were built and used to store agricultural surpluses at the site (C. Zhang, 2015: 209). If this was true, it would mean that these granaries were most likely used to store durable grains such as rice or other cereals. Other contemporary or slightly later period settlements, such as Fenshanbao (C. Zhang, 2015: 209), have houses with similar structures but moats are absent at these sites.

The following, Chengtoushan period (5600–3300 BC; we use 'the Chengtoushan period' here as a chronological term: the cultural affinities of the archaeological remains of this period are more complicated than for other periods) saw the boom of settlement sites, with expanding residential features in the Dongting and Jianghan plains. Around 300 sites have been found (C. Zhang, 2013: 510). Most sites are 3–5 ha in size, while several reach 5–10 ha (Table 4.1). Inside the enclosed and walled area at Chengtoushan (Figure 4.3), not only have houses been excavated, but a ceremonial feature has also been brought to light through recent excavations. This ceremonial feature is a round-shaped altar built of pure yellow earth, around one metre in height and 200 m^2 in size. Shallow pits, urn burials and graves have been discovered, along with abundant archaeological remains, including cooking vessels, rice and animal bones. Equally significant was the discovery of paddy fields at this site (He and Yasuda, 2007; Toyama, 2007; Yasuda, 2013). The fields appeared to have already reached a large scale, 20 × 2.7 m, compared to the small field units discovered elsewhere (see Chapter 5 in this volume). This series of new archaeological phenomena indicates that rice farming was further developed and that the farming regimes themselves underwent fundamental changes, such as the planting of rice stands in managed fields. This developed rice economy explains why abundant rice remains were discovered at the aforementioned ceremonial altar. Also worth mentioning are the millet remains discovered at Chengtoushan, along with weedy plants indicative of dry habitats (Guedes, 2011; Nasu et al., 2007). This suggests that millets were being cultivated locally, after being introduced to the area, most

Table 4.1 Dimensions of the moats and walls at prehistoric sites in the studied area

Site	Cultural phase	Enclosed area (ha), dimensions	Moat dimension (m)	Wall dimension (m)	Source(s)
Bashidang	Pengtoushan	3.7, 270 m × 180 m	Top width 4, bottom width 2, and depth 2	Bottom width 5, top width 2, height 2	Guo, 2010
Shanlonggang	Pengtoushan	3.6			Guo, 2010
Pengtoushan	Pengtoushan	3,160 m × 190 m			Guo, 2010; Hunan Provincial Institute of Cultural Relics and Archaeology, 2006
Chengtoushan	Mid- to late Daxi	9	Top width 10, bottom width 6.15 and depth 4	Top width 5, bottom width 8, height 1.6	Z. H. Zhang, 1998; Guo, 2007, 2010
Yinxiangcheng (only partially surveyed)	Daxi/Qujialing?	*c.* 19, 350 m × 580 m	Width 30–40 and depth 4	Width 10–25, height 1–2	Jia, 1998a; H. X. Wang, 2003; Yuan, 1997
Longzui	Daxi	8.2, 305 m × 270 m	Surrounded by lakes on three sides; one side has moat: width 18 and depth 1.5–2.7	Bottom width 17, height 1–3	Hubei Provincial Institute of Cultural Relics and Archaeology, 2008.
Huachenggang	Daxi-Qujialing	*c.* 4, 200 m × 200 m	Width 10–15	Very low	He, 1983; Yin, 2005
Shijiahe	Qujialing-Shijiahe	*c.* 120	Width 60–100, depth 2	Bottom width 50, top width 15, height 6	Liu and Yu, 2016; Joint archaeological team of Shijiahe, 1992

Table 4.1 (Continued)

Site	Cultural phase	Enclosed area (ha), dimensions	Moat dimension (m)	Wall dimension (m)	Source(s)
Menbanwan	Qujialing-Shijiahe	20, 550 m × 400 m	Top width 59, depth 1.8–2.5	Top width 13.5–14.7, bottom width *c.*40, height 6	H. X. Wang, 2003
Taojiahu	Qujialing-Shijiahe	67, 1000 m × 850 m	Top width 20–45	Bottom width 30, height 1.5–3	Li and Xia, 2001; H. X. Wang, 2003
Tunzishan	Qujialing-Shijiahe	8		Width *c.* 30, height 1–5	Jia and Yang, 2005
Xiaocheng	Qujialing-Shijiahe	9.8, 156–305 m × 250–360 m	Moat connected to lakes encircling three sides of the settlement	Bottom width 23, height 0.6–1.6	W. X. Huang et al., 2007
Chenghe	Qujialing-Shijiahe	71.8, 600–800 m × 550–650 m	Width 15–50, depth 1–3	Width 8–50, height 1–5	Hubei Provincial Institute of Cultural Relics and Archaeology, 2014; Jingmen Institute of Cultural Relics and Archaeology, 2008
Wangguliu	Qujialing-Shijiahe	88, 1000 m × 880 m			Zeng, 1990
Qinghe	Qujialing-Shijiahe	6, 300 m × 200–240 m	Width 30–50	Width 30, height 0.5–2	Jia and Yang, 2005
Yejiamiao	Qujialing-Shijiahe	15, 870 m × 650 m	Top width 40, depth 2	Top width 27–30, height 2–6	Liu, Hu et al., 2012

(*Continued*)

Table 4.1 (Continued)

Site	Cultural phase	Enclosed area (ha), dimensions	Moat dimension (m)	Wall dimension (m)	Source(s)
Zoumaling	Qujialing-Shijiahe	7.8	Width *c*. 35, depth 2	Bottom width *c*. 28 and height 5	H. X. Wang, 2003; Z. H. Zhang, 1998
Jijiaocheng	Qujialing-Shijiahe	14, 400 m × 470 m	Multiple rings of moats and ditches, the innermost one 40–70 in width; other moats and ditches *c*. 20 in width	Width 40–60 and height 2–5	Guo, 2010; H. X. Wang, 2003; Yin, 2002
Jimingcheng	Qujialing-Shijiahe	15, 500 m × 400 m	Width 20–30 and depth 1–2	Top 15 m in width, bottom 30 m in width and 2–3m in height	Jia, 1998b
Majiayuan	Qujialing-Shijiahe	24	Width 30–50 and depth 4–6	Bottom width 30–35, top width 6–8, and height 1.5–6	C. F. Wang and Tang, 1997; Z. H. Zhang, 1998
Shijiahe phase at Chengtoushan	Qujialing-Shijiahe	8	Width 40–50	Width 20 and height 2–4	Guo, 2007, 2010; Joint Archaeological Team of Shijiahe, 1992
Zhangxiwan	Qujialing-Shijiahe	9.8, 295 m × 335 m	Width 25–35, depth 3–3.5	Width 24, height 1.5	Liu, Guo et al., 2012

Figure 4.3 The Chengtoushan site. A. Aerial photograph showing the plan of the site, with the excavated trenches located in the southern part of the round enclosed area. Source: Courtesy of Hunan Provincial Institute B. The moat and the earthen wall. Source: authors

likely from the north (Guedes, 2011). Some smaller sites of this period, such as Huachenggang, also had moats and walls. However, these sites either were severely damaged or lacked large-scale excavations, so the full scope of their structures remains unclear.

Dramatic changes in settlement structures can be seen during the Qujialing-Shijiahe period (5000–4100 cal. BP). The number of settlements reached more than a thousand (Table 4.2; C. Zhang, 2015: 227). These settlements can be divided into at least four tiers according to their sizes (C. Zhang, 2013: 514, 2015: 228). The Shijiahe site complex was composed of a large rectangular walled area, with numerous archaeological features inside and outside this enclosed area (Figure 4.4).

Table 4.2 Number of sites from different cultural periods

Cultural period	Number	Source
Pengtoushan	*c.* 30	Pei, 2013
Daxi-Yangshao	*c.* 300	C. Zhang, 2015: 210
Qujialing-Shijiahe	>1000	C. Zhang, 2015: 227

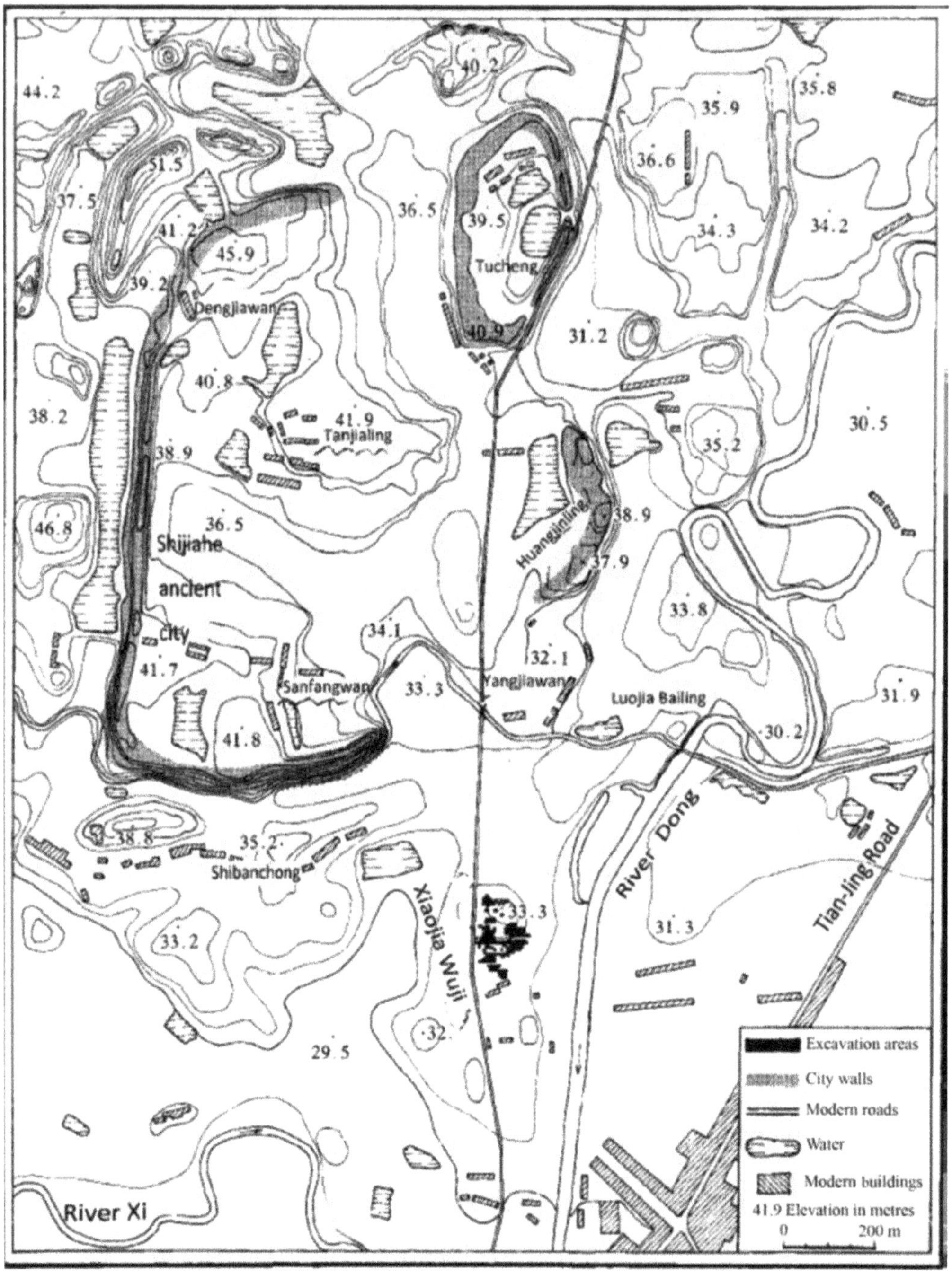

Figure 4.4 Plan of the Shijiahe site complex (after C. Zhang, 2013)

Excavations at several locations inside the walled area have revealed houses, large graves with rich burial goods, and most significantly, ritual objects and features. The latter include hundreds of thousands of ceramic cups, concentrated in a small area and covered by earth, two pits in the cemetery containing several thousand terracotta figures of animals and humans, and many jars lying horizontally, with one attached to another in row after row (Figure 4.5). Outside the moated area were more than 20 residential areas, each containing houses and living facilities. This spatial relationship between ritual activities and residential areas leads some scholars to think that Shijiahe was a regional centre where important ceremonial events were held and participated in by the populace. Such a central role would have meant that its settlement structure became a model for neighbouring local communities to follow. This may have triggered regional competition, resulting in the construction of large-scale walls and moats by contemporaries. Indeed, settlements

Figure 4.5 An excavated ceremonial feature at the Dengjiawan location inside the walled area (after Yang, 2003)

belonging to the second and third tiers all possessed moats and walls. Several of them also formed into clusters, but their structures are poorly understood because of the scarcity of archaeological data. Behind this emerging regional phenomenon, of intensifying engineering taking place on the local landscape, was a fundamental shift in the agricultural economy and unprecedented population growth. During this phase, not only were the large, central sites able to organise the construction of large-scale infrastructures, but medium or smaller settlements also had the means to build walls and moats.

Archaeobotanical flotation was conducted at a number of Qujialing-Shijiahe-period sites. The predominant plant recovered at both Yejiamiao and Sanfangwan was rice. These results suggest that rice farming had become very important, although millets continued to be cultivated (Deng et al., 2013; C. R. Wu et al., 2010). Though no paddy fields from this period have been found (they are likely to be buried underneath late Holocene alluvial deposits), a kind of dyke has been commonly found 'all over the Liyang Plain' and is thought to represent the type of 'dyke and reservoir works' used in irrigation practices. This implies that rice farming was practised on a much larger scale to feed the increasing 'rural and urban population' (C. Zhang, 2013: 523).

The construction of the prehistoric moats and walls

Unlike the Late Neolithic walled sites in North China, constructed by well-developed earth-pounding techniques and consolidated by standing wooden logs inside the walls (H. Zhao and Wei, 2002: 18), the moats and walls in the Middle River Yangtze were constructed with rather simple technologies that remained largely unchanged during this period. During the early stage, earth dug up from the moats was simply piled up on the same spot to build the walls, without any further treatment. This simple process was sustainable because the construction remained on a small scale, and the size of the moats and walls was also fairly small, and therefore it did not require sophisticated technology to build them. By the Qujialing-Shijiahe period, as the size of the moats and walls grew exponentially, pounding and trampling technology, using wooden sticks or other tools, was occasionally applied at some sites when earth was piled up. This enabled the construction of higher walls, which were more solid and resistant to surface erosion. However, the majority of the walls in the prehistoric Middle River Yangtze were still being built in a very simple manner as the heavy, clayey earth did not require heavy pounding

(Figure 4.6). Indeed, resistance to surface erosion was a critical factor in this rainy and humid region with abundant surface run-off. Hence, many Neolithic walls have a prismatoid cross-section with a very wide base and gentle slopes to reduce surface erosion (Table 4.1).

The construction of the moats and walls was closely related to changes in social structure and agricultural growth throughout Neolithic times. Table 4.1 shows the size increases in moats and walls from the Pengtoushan cultural period to the Qujialing-Shijiahe period. None of the early-period moats enclosed an area larger than 10 ha. Most of these moats were rounded or irregular in shape. This spatial arrangement was to connect parts of the existing natural water bodies with the artificial moats to save labour costs. Assuming low productivity among the earth workers, Pei suggests that 100 labourers would have needed to work for 20 days to dig the moat at Bashidang (Pei, 2004a). As mentioned above, by the Qujialing-Shijiahe period a clearly ranked system of walled and moated sites had developed, which is inferred from site size estimates.

Figure 4.6 The earthen wall at Shijiahe (Meng et al., 2012)

The largest one was the Shijiahe site, whose enclosed area reached 120 ha. The second-tier sites fell into a size range of 50–100 ha and the third-tier ones measured 10–20 ha (C. Zhang, 2015: 228). In addition, at and around large regional centres such as Shijiahe smaller sites were often concentrated to form one huge site complex, revealing an unprecedented effort by a dense population to transform a landscape into a heavily engineered environment. Completion of these increasingly large moats and walls would have consumed substantial resources. Zhang calculates that at least half a million m^3 of earth would have been dug from the moats to build the walls at Shijiahe. This would mean about 1,000 workers working for about 10 years, who 'had to be fed by probably 20,000 to 40,000 persons' (Nakamura 1997, cited in C. Zhang, 2013: 519). Labour investment and organisation on such a large scale would have only been possible in a society with a considerable agricultural surplus and advanced organisational skills. Also by the Qujialing-Shijiahe period, some sites, such as Jijiaocheng and Zoumaling, had multiple rings of moats encircling a large area (Chen, 1998; Yin, 2002). This further suggests that sophisticated planning and logistics existed not only at large sites, but also at smaller sites which had economic and cultural connections with them.

Along with these structural changes and the increase in size of the moats and walls came an increasing emphasis on the construction along or next to the moats of associated facilities, which served a key role in their functioning. At Bashidang, from the early period, an initial attempt to connect the residential area with its surrounding water bodies was evidenced by the discovery of pebble stairs descending towards the river (Figure 4.7). In the following period, at the Chengtoushan site, not only were reeds and wooden planks used to consolidate the banks along the moat, but wooden paddles and parts of boats were also discovered inside the moat (He and Yasuda, 2007). The latter suggest that the moat was also used for transport. By the Qujialing-Shijiahe period, in addition to the continuous building of stairs along the moats to facilitate water usage, which can be seen at sites such as Zhangxiwan (Liu, Guo et al., 2012), wooden bridges had been built between the moats and the walled areas. Bridges of this type have been found at the Shijiahe site (Meng et al., 2012; Figure 4.8). Only a few wooden stakes are left, but the construction of such bridges demonstrates a greater investment in the moats. The most significant development was the construction of 'water gates' at many Qujialing-Shijiahe walled sites. These gates were often built in the low-lying areas surrounding the moats and connected to rivers and lakes nearby. C. Zhang (2013) speculates that an existing

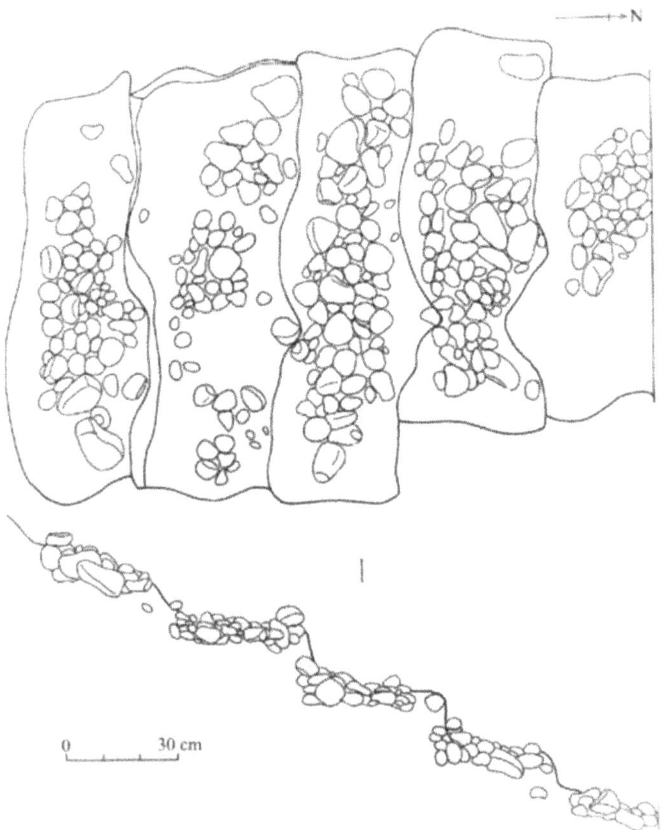

Figure 4.7 The stone steps discovered at the Bashidang site (after Hunan Provincial Institute, 2006)

Figure 4.8 The excavated bridge remains at the Tanjialing location of the Shijiahe site complex (after Meng et al., 2015)

gap (*c*. 450 m in width) in the south-eastern corner of the Shijiahe walls was the place where one of those gates was situated (but the precise dimension will remain unknown until the area is excavated), probably fenced with wooden logs. Such gaps have also been found at other sites, such as Menbanwan (H. X. Wang, 2003).

Water management and landscape engineering

The functions of these moats and their associated facilities has been much debated. Were they mainly used for flood protection? To what extent did they function as a defensive system? What specific roles did they play in irrigation, in the supply of water for daily use, and in water transport? And did these functions evolve through time? Despite continuing disagreement on these questions, there is a growing consensus that these issues need to be examined within the wider socio-economic and environmental contexts.

As mentioned above, the Holocene landscape in much of the Jianghan and Dongting plains was crisscrossed by natural rivers. In our survey, we have found that the majority of the prehistoric sites were located 35 m above sea level and at least 2 m higher than the plains that were often flooded by the rivers (see also Zhu et al., 2007). These settlements were encircled by moats, which were situated on lower ground and connected to natural water bodies through artificial ditches. The ample rainfall and its uneven temporal and spatial distribution have been a constant challenge to daily life and agriculture in the region in the past and the present. When the rainfall reaches its peak, it quickly gathers in ditches and low-lying areas in the field, becoming a threat to communities living there. On the other hand, however, it is necessary for villages to be connected to natural water bodies to guarantee water availability during dry seasons. Small modern villages in the region indeed rely entirely on small rivers for water for daily use and agricultural irrigation.

The extent to which these prehistoric sites were affected by Holocene floods remains unclear. Evidence derived from regional environmental surveys suggests that, even though the volume of water flow per second of the Yangtze and its major tributaries and associated lakes was higher during the middle Holocene than at present, the water levels of major rivers were actually much lower than now; the rivers benefited from the low silt-laden water and deep and wide river channels with larger volumes (J. Li et al., 2011; Zhou, 1994). Thus, these rivers and lakes would have served to absorb the low- to medium-magnitude floodwater, mitigating the danger of flooding at the prehistoric sites located on higher ground and not too close to the major rivers. Indeed, there is so far no convincing evidence that archaeological remains were directly overlain by flooding deposits. Rather, excavations of many moats have revealed that they were filled up predominantly by organic-rich, dark, clayey sediments (Hui Liu, personal communication), without the pebbles and sands which are normally deposited in standing water conditions. Such depositional processes suggest either that the local water systems were well maintained and regulated and the flow and velocity of water were well controlled and stable, or that these moats were not flooded on a regular basis, as floods would have resulted in the deposition of much coarser particles than clay.

By the Chengtoushan and Qujialing-Shijiahe periods, though most of the sites were still located on higher ground, settlements were expanding more extensively over the landscape, presumably because of the pronounced increase in population. Associated with this increase was an intensification of the rice economy, which means that substantial areas of arable fields, including the land situated in the low-lying areas, would have been reclaimed for rice farming. The elevations of the Liuguan and Futian sites, for instance, decreased to lower than 20 m above sea level, which put these sites at greater risk of being flooded. Because of increasing deforestation and land clearance (J. Li et al., 2011; Shi et al., 2010; see also Y. Li et al., 2010), the rivers were silting up more quickly than before. These changes in settlement patterns and river hydrological regimes posed a significant challenge to water-management practices at these sites, especially when the growth of rice plants coincided with the peaks of summer rainfall, meaning that rice fields needed to be drained quickly (see Chapter 5 in this volume). One of the most significant efforts to mitigate this potentially detrimental effect was further engineering of the local landscapes to enhance the function of large-scale moats and their associated water-management systems. Surveys at the Jijiaocheng site have revealed a well-developed water-management system (Figure 4.9). This walled area is surrounded by a network of ditches and weirs, some of which have been excavated recently. These ditches and weirs are located in low topographic areas and many of them are still in use for growing aquatic plants such as lotus and for routine activities such as washing. Among the ditches are agrarian fields. These ditches and weirs formed a highly integrated water-management system, which played a crucial role in water drainage and flood mitigation by receiving overflow from the farming fields. Floodwater was most likely diverted before it caused damage to agricultural yields, and quickly discharged into the rivers. The key

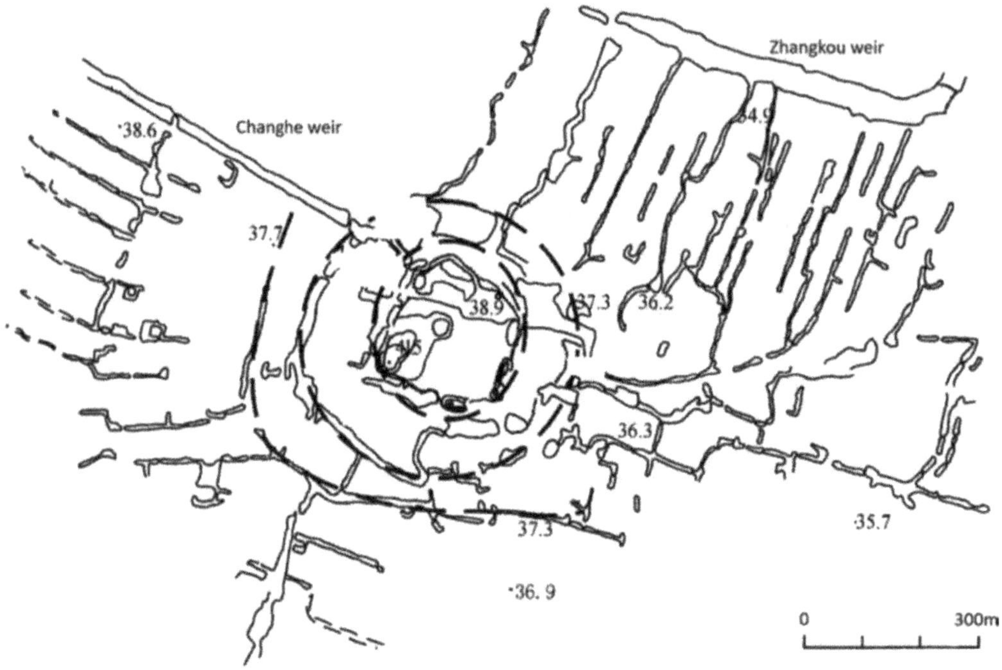

Figure 4.9 Plan of the Jijiaocheng site. The bold single lines are the moat and the two circles of earthen walls. The double-lined features are ditches and weirs. Some of them were excavated and their chronology confirmed to be of the Qujialing-Shijiahe period, but reports are not published yet. The figures show the elevations of different places at the site. Redrawn after Guo, 2010

to the normal functioning of these combined water-management systems of moats and ditches was ensuring that these ditches were not silted up. This would have involved careful management and maintenance of the ditches by enforcing frequent clearance and related activities, although evidence of such activities has not been found yet. Between the ditches and the moat, sluices, water outlets and other small facilities would have been built to control water flow.

Discussion

Environmental constraints and opportunities in the prehistoric Middle River Yangtze

The moated and walled sites in the Middle River Yangtze were among the earliest and most developed in prehistoric China. The scale of these early moated sites was large and their number continued to increase, with a pronounced leap during the Qujialing-Shijiahe period. Moats and walls became integral parts of Neolithic sites in this region and played a central role in the socio-economic activities that took place there. The early and consistent emphasis on the construction of moats and walls was an unprecedented phenomenon that was closely related to its environmental context.

High-resolution climate change data, especially precipitation patterns, for the Middle River Yangtze, is lacking. However, the whole region has been very humid throughout the Holocene, characterised by heavy summer rainfall. In this sense, we suggest that rainfall (as a constant variable) was not the paramount parameter for the evolution of prehistoric water-management systems. Rather, as discussed above, it was the combination of geomorphological processes and characteristics of local landforms that played a decisive factor in the planning, expansion and maintenance of these systems. Compared with other regions, the landforms in the Middle River Yangtze are dominated by relatively flat alluvial plains. Variations in the elevations are often less than 5 m. These places were ideal for the construction of mega-scale walled and moated sites such as the Shijiahe site. These sites took advantage of the developed natural water systems in their large-scale engineering works of local landscapes for various purposes. The water system and engineered landscape at Jijiaocheng reached a mature level, characterised by the construction of well-designed moats and ditches for robust channelling ditches with the construction of moats for water management and flood prevention. This mature water-management system shows a sophisticated understanding of the local ecology and an ability to manage micro-topography in the field. Indeed, these moated sites gradually developed multiple functional water-management systems. Irrigation, supplying water for daily use, drainage and flood prevention, water transportation, and other functions which, despite fluctuating water needs (depending on local economic and environmental conditions), make these moats and walls an intrinsic part of both the seasonal and the temporal functioning of these sites. However, the construction and maintenance of these increasingly large water-management systems required large economic resources and a large labour force, as well as managerial skills. This is one reason why moats and walls were only built at medium-to-large sites. This pattern prompts one to ask what the relationship is between the intensification of water management and pathways of social complexity in the prehistoric Middle River Yangtze.

Intensification of water management and social complexity

Variables that determine trajectories of intensification include population and labour, demand and supply patterns, and related key factors such as environmental conditions and availability of natural resources. In many classic studies of intensification (e.g., Boserup, 1965), the importance of environmental conditions has largely been overlooked. As we have discussed above, the construction and especially the expansion of large moated and walled sites were fundamentally constrained by local landforms. On the one hand, the alluvial plains made the construction and expansion possible; on the other, there was always the necessity to choose the optimal space on the plains to avoid floods.

From about the fifth millennium BC onwards, the increasingly large moated and walled sites, by means of their socio-economic impact accumulated sufficient economic resources and knowledge for the dramatic development witnessed during the succeeding Qujialing-Shijiahe period. Indeed, the scale of some of the walled and moated sites during that period was enormous, and the density of cultural and economic activities was so high that it caused a fundamental transformation, both qualitative and quantitative, of prehistoric landscapes. We see a positive correlation between the size of the walled and moated sites and their productivity in terms of water management and other economic functions. When the size of the moats and walls increased n times, the size of the enclosed area increased n^2 times. This meant that more space and resources became available for the growing population at these large sites. In turn, this improved economic efficiency accelerated social and economic division, gearing the societies up for further intensification and specialisation of economic production.

An example that illustrates this point is the Shijiahe site complex, which is dominated by a wide variety of functional areas inside and around the enclosed area, as mentioned briefly above (Figure 4.4). The hundreds of thousands of ceramic cups and pottery figurines appear to represent large-scale social gatherings or similar events, which would have been a crucial arena for enhancement of social identity and acquisition of social power. An excavation in 2017 recovered even more ceramic cups and evidence of their production (Hui Liu, personal communication). What was taking place at Shijiahe was imitated by many sites in the region (Table 4.1). This process stimulated regional competition and considerable resources were invested in building walls and moats as well as in the production of consumables such as the ceramic cups. Such competition for space and resources resulted in conflict. This made defence by means of the walls and moats increasingly vital, a situation that would further encourage regional competition in the building of larger walls and moats and the consumption of more cups as a way of enhancing and increasing social power. We can therefore see another distinctive pathway to intensification, initiated by social-cultural actions, becoming closely related to the acquisition of social power in the society. This pathway might also have provoked a fundamental shift in the use of moats towards multiple functions, in which defence played a key role.

Conclusion

The early appearance of walled and moated sites in the Middle River Yangtze is unparalleled among contemporary Neolithic cultures in different regions of China. Moated and walled sites in this region continued to see quantitative and qualitative growth throughout the Neolithic period. Because the development of a rice economy is not clearly understood at present, it is hard to assess the role of moats in irrigation. Nonetheless, these moats did become crucial in water management, especially in flood prevention, when the settlements expanded into larger areas of local landscapes, including the low-relief areas, because of population growth and migration. In addition, the ability of the ditches to absorb surface run-off quickly during the rainy season makes them especially efficient in water management. The frequent water flow in turn clears away sediments in the ditches, guaranteeing their normal function in flood prevention. These moats and ditches form an extremely effective and unique water management system in the Middle River Yangtze, which is dominated by flat lands and receives heavy summer rainfall. By the Qujialing-Shijiahe period, regional competition had become one of the main driving forces that stimulated large-scale infrastructure construction and the production of various consumables. This competition not only changed the pattern of economic development in the region, one that was driven and accelerated by ritual activities and social gatherings, but also profoundly transformed the local landscape into one that was densely occupied by moats, ditches and other water-management facilities. It is this unique entanglement between social competition and intensification of economic production and infrastructure construction that needs to be further unpacked in future research to demonstrate the long-term water-management practices of this region and their relationship with environmental changes and social evolution.

References

Boserup, E. 1993 [1965]. *The Conditions of Agricultural Growth: The Economics of Agrarian Change under Population Pressure*. Oxford: Earthscan.

Chen, G. T. 1998. 'Hubei shishoushi zoumaling xinshiqi shidai yizhi fajue jianbao' (A preliminary report of the excavation at the Zoumaling Site in Shishou City, Hubei). *Kaogu/Archaeology* (4), 16–38.

Deng, H., Y. Y. Cheng, J. Y. Jia, D. W. Mo and K. S. Zhou. 2009. '8500 a BP yilai changjiang zhongyou pingyuan diqu guwenhua yizhi fenbu de yanbian' (Distributional patterns of archaeological sites in the Middle Yangtze River since 8,500 years ago). *Acta Geographica Sinica* 64, 1113–25.

Deng, Z. H., H. Liu and H. P. Meng. 2013. 'Hubei tianmenshi shijiahe gucheng sanfangwan he tanjialing yizhi chutu zhiwu yicun fenxi' (Archaeobotanic remains from Sanfangwan and Tanjialing sites of the Shijiahe City site, Tianmen City, Hubei). *Kaogu/Archaeology* (1), 91–9.

Fuller, Dorian Q., Ling Qin, Yunfei Zheng, Zhijun Zhao, Xuguo Chen, Leo-Aoi Hosoya and Guo-Pin Sun. 2009. 'The domestication process and domestication rate in rice: Spikelet bases from the lower Yangtze.' *Science* 323 (5921), 1607.

Gross, Briana L. and Zhijun Zhao. 2014. 'Archaeological and genetic insights into the origins of domesticated rice.' *Proceedings of the National Academy of Sciences* 111 (17), 6190–7.

Guedes, Jade d'Alpoim. 2011. 'Millets, rice, social complexity, and the spread of agriculture to the Chengdu plain and southwest China.' *Rice* 4 (3), 104–13.

Guo, W. M. 2007. 'Chengtoushan chengqiang haogou de yingzao jiqi suo fanying de juluo bianqian' (Construction of the walls and moats at Chengtoushan and settlement pattern changes). *Cultural Relics of Southern China* (2), 70–6.

Guo, Weimin. 2010. *Xinshiqi Shidai Liyang Pingyuan yu Handong Diqu de Wenhua he Shehui* (Culture and society in the Liyang plain and the Handong area during the Neolithic period). Beijing: Cultural Relics Press.

He, J. J. 1983. 'Anxiang huachenggang xinshiqi shidai yizhi' (The Neolithic site at Huachenggang in Anxiang county, Hunan province). *Acta Archaeologica Sinica* (4), 427–70.

He, J. J. 2007. *Lixian Chengtoushan* (Chengtoushan in Lixian: Excavation report of the Neolithic site). Beijing: Cultural Relics Press.

He, J. J. and Yasuda, Y. (eds). 2007. *Lixian Chengtoushan: Zhongri Hezuo Liyang Pingyuan Huanjing Kaogu yu Youguan Zonghe Yanjiu* (Chengtoushan in Lixian: Sino-Japanese environmental archaeology project on the Liyang Plain). Beijing: Cultural Relics Press.

Huang, W. X., W. Zhou and Y. M. Zhang. 2007. 'Hubei tianmen xiaocheng chengzhi fajue jianbao' (Excavation of the Xiaocheng walled site in Tianmen city, Hubei province). *Kaoguxuebao/Acta Archaeologica Sinica* (4), 469–88.

Hubei Provincial Institute of Cultural Relics and Archaeology. 2008. 'Hubeisheng tianmenshi longzui yizhi 2005 nian fajue jianbao' (A preliminary report of the excavation at the Longzui site in Tianmen city, Hubei in 2005). *Jianghan Archaeology* (4), 3–13.

Hubei Provincial Institute of Cultural Relics and Archaeology. 2014. '2013 nian hubeisheng wenwu kaogu yanjiusuo kaogu gongzuo zhuyao shouhuo' (Major achievement of archaeological works in 2013). *Jianghan Archaeology* (1), 7–20.

Hunan Provincial Institute of Cultural Relics and Archaeology (ed.). 2006. *Pengtoushan and Bashidang*. Beijing: Science Press.

Jia, H. Q. 1998a. 'Hubei jingzhoushi yinxiangcheng yizhi 1995 nian fajue jianbao' (A preliminary report of the excavation at the Yinxiangcheng site in 1995, in Jingzhou City, Hubei). *Kaogu/Archaeology* (1), 17–28.

Jia, H. Q. 1998b. 'Hubei gongan jimingcheng yizhi de diaocha' (Survey of the walled site at Jimingcheng in Gong'an county). *Cultural Relics* 6, 25–30.

Jia, H.Q. and K. Y. Yang. 2005. 'Hubei gongan shishou sanzuo gucheng kancha baogao' (Survey of the three prehistoric walled sites in Gongan County and Shishou City, Hubei). *Ancient Civilizations* (5), 391–412.

Jingmen Institute of Cultural Relics and Archaeology. 2008. 'Hubei jingmenshi hougang chenghe chengzhi diaocha baogao' (Report of the survey at the Hougang Chenghe walled site in Jingmen, Hubei). *Jianghan Archaeology* (2), 27–34.

Joint archaeological team of Shijiahe. 1992. 'Shijiahe yizhiqun diaocha baogao' (Report of the surface survey at the Shijiahe site complex). *Southern Ethnic Archaeology* (5), 213–94.

Kuzmin, Yaroslav V. 2013. 'Origin of old world pottery as viewed from the early 2010s: When, where and why?' *World Archaeology* 45 (4), 539–56.

Li, J., S. Y. Wang and D. Mo. 2011. '6000 a BP yilai dongtinghu chenji jilu de huanjing yanbian jiqi tong renlei huodong de guanxi' (Environmental changes and relationships with human activities of the Dongtinghu Plain since 6,000 years ago). *Journal of Peking University (Science edition/Scientiarum Naturalium Universitatis Pekinensis)* 47, 1041–8.

Li, Taoyuan and Xia Fang. 2001. 'Hubei yingcheng taojiahe guchengzhi diaocha' (Survey of the Taojiahu walled site, Yingcheng, Hubei). *Cultural Relics* (4), 71–6.

Li, Yiyin, Jing Wu, Shufang Hou, Chenxi Shi, Duowen Mo, Bin Liu and Liping Zhou. 2010. 'Palaeoecological records of environmental change and cultural development from the Liangzhu and Qujialing archaeological sites in the middle and lower reaches of the Yangtze river.' *Quaternary International* 227 (1), 29–37.

Liu, H., C. J. Guo, J. Zhang and Y. X. Xie. 2012. 'Wuhanshi huangpiqu zhangxiwan xinshiqi shidai chengzhi fajue jianbao' (A preliminary report of the excavation at Zhangxiwan City in the Huangpi Area in Wuhan City, Hubei). *Kaogu/Archaeology* (8), 42–68.

Liu, H., J. J. Hu, N. Tang and C. J. Guo. 2012. 'Hubei xiaogan yejiamiao xinshiqi shidai chengzhi fajue jianbao' (A preliminary report of the excavation at the Yejiamiao walled site in Xiaogan City, Hubei). *Kaogu/Archaeology* 8, 3–28.

Liu, H. and L. Yu. 2016. 'Shijiahe yizhi 2015 nian fajue de zhuyao shouhuo' (The 2015 excavation season at the Shijiahe site complex). *Jianghan Archaeology* (1), 36–41.

Meng, H. P., H. Liu, Z. H. Deng and Q. F. Xiang. 2012. 'Hubei tianmenshi shijiahe gucheng sanfangwan yizhi 2011 nian fajue jianbao' (A preliminary report of the excavation at the Sanfangwan site of Shijiahe City in 2011, in Tianmen City, Hubei). *Kaogu/Archaeology* (8), 29–41.

Meng, H. P., H. Liu, Q. Xiang and Z. H. Deng. 2015. 'Hubei tianmenshi shijiahe gucheng tanjialing yizhi 2011 nian de fajue' (Excavation of the Tanjialing site of Shijiahe City in 2011). *Kaogu/Archaeology* (3), 51–73.

Nasu, Hiroo, Arata Momohara, Yoshinori Yasuda and Jiejun He. 2007. 'The occurrence and identification of *Setaria italica* (L.) P. Beauv. (foxtail millet) grains from the Chengtoushan site (ca. 5800 cal B.P.) in central China, with reference to the domestication centre in Asia.' *Vegetation History and Archaeobotany* 16 (6), 481–94.

Pei, Anping. 2004a. 'Liyang pingyuan shiqian juluo xingtai de tedian yu yanbian' (Characteristics and development of the prehistoric settlement patterns in the Liyang Plain). *Kaogu/Archaeology* (11), 63–7.

Pei, A. P. 2004b. 'Zhongguo shiqian de juluo weigou' (Prehistoric moats in China). *Southeast Culture* (6), 21–30.

Pei, A. P. 2013. 'The Pengtoushan culture in the middle Yangzi river valley.' In Anne P. Underhill (ed.), *A Companion to Chinese Archaeology*, 497–509. Chichester: Wiley-Blackwell.

Shelach, Gideon. 2012. 'On the invention of pottery.' *Science* 336 (6089), 1644–5.

Shi, C. X., D. W. Mo, H. Liu and L. J. Mao. 2010. 'Jianghan pingyuan beibu hanshui yidong diqu xinshiqi wanqi wenhua xingshuai yu huanjing de guanxi' (Development of Neolithic cultures in the eastern River Hanshu and northern Jianghan Plain and its relationship with the environment). *Quaternary Research* 30, 335–43.

Silva, Fabio, Chris J. Stevens, Alison Weisskopf, Christina Castillo, Ling Qin, Andrew Bevan and Dorian Q. Fuller. 2015. 'Modelling the geographical origin of rice cultivation in Asia using the rice archaeological database.' *PloS One* 10 (9), e0137024.

Toyama, S. 2007. 'Cong dixing fenxi he zhiguishi fenxi kan chengtoushan yizhi de huanjing ji daozuo' (Environment and rice cultivation in the Chengtoushan site based on the geomorphological and phytolith analysis). In J. He and Y. Yasuda (eds), *Chengtoushan in Lixian: Sino-Japanese Environmental Archaeology Project on the Liyang Plain*, 44–66. Beijing: Cultural Relics Press.

Wang, C. F. and X. F. Tang. 1997. 'Jingmen majiayuan qujialing wenhua chengzhi diaocha' (Survey of the Qujialing culture Majiayuan walled site in Jingmen). *Cultural Relics* (7), 49–53.

Wang, H. X. 2003. 'Cong menbanwan chenghao juluo kan changjiang zhongyou diqu chenghao juluo de qiyuan yu gongyong' (On the origins and functions of the moated settlements in prehistoric Middle River Yangtze: Evidence from the Menbanwan site). *Kaogu/Archaeology* (9), 61–75.

Wang, Y., H. Cheng, R. L. Edwards, Y. He, X. Kong, Z. An, J. Y. Wu, M. J. Kelly, C. A. Dykoski and X. D. Li. 2005. 'The Holocene Asian monsoon: Links to solar changes and North Atlantic climate.' *Science* 308 (5723), 854–6.

Wu, Chuanren, Hui Liu and Zhijin Zhao. 2010. 'Cong xiaogan yejiamiao yizhi fuxuan jieguo tan jianghan pingyuan shiqian nongye' (Prehistoric agriculture of the Jianghan Plain: Evidence from flotation results of the Yejiamiao site in Xiaogan). *Cultural Relics in Southern China* (4), 65–9.

Wu, Xiaohong, Chi Zhang, Paul Goldberg, David Cohen, Yan Pan, Trina Arpin and Ofer Bar-Yosef. 2012. 'Early pottery at 20,000 years ago in Xianrendong cave, China.' *Science* 336 (6089), 1696–1700.

Yang, Q. 2003. *Dengjiawan: Tianmen Shijiahe Kaogu Baogao Zhier* (Dengjiawan: Second excavation report of the Shijiahe site complex). Beijing: Cultural Relics Press.

Yasuda, Yoshinori. 2013. 'Discovery of the Yangtze River Civilization in China.' In Yoshinori Yasuda (ed.), *Water Civilization: From Yangtze to Khmer Civilizations*, 3–46. Tokyo: Springer.

Yin, J. S. 2002. 'Lixian jijiaocheng guchengzhi shijue jianbao' (Excavation of the ancient walled site of Jijiaocheng in Lixian County, Hunan). *Cultural Relics* (5), 58–68.

Yin, J. S. 2005. 'Hunan anxiang huachenggang yizhi dierci fajue baogao' (Report of the second excavation at the Huachenggang site in Anxiang County, Hunan). *Acta Archaeologica Sinica* 1, 55–108.

Yuan, W. Q. 1997. 'Hubei jingzhoushi yinxiangcheng yizhi dongchengqiang fajue jianbao' (A preliminary report of the excavation of the east wall at the Yinxiangcheng site in Jingzhou City, Hubei). *Kaogu/Archaeology* (5), 1–10.

Zeng, C. L. 1990. 'Hubei xiaogan diqu xinshiqi shidai yizhi diaocha shijue' (Survey and excavation of Neolithic sites in Xiaogan City, Hubei). *Kaogu/Archaeology* (11), 981–8.

Zhang, C. 2000. 'Jiangxi wannian zaoqi taoqi he daoshu zhiguishi yicun' (Pottery and the rice phytoliths discovered at the Xianrendong site, Wannian County, Jiangxi province). In W. M. Yan (ed.), *The Origins of Rice Agriculture, Pottery and Cities*, 43–50. Beijing: Cultural Relics Press.

Zhang, C. 2013. 'The Qujialing–Shijiahe culture in the Middle Yangzi river valley.' In A. P. Underhill (ed.), *A Companion to Chinese Archaeology*, 510–34. Chichester: Wiley-Blackwell.

Zhang, C. 2015. 'Changjiang zhongyou diqu de shiqian juluo yanbian yu zaoqi wenming' (The prehistoric settlements and early civilisations of the prehistoric Middle Yangtze River). In Centre for the Study of Chinese Archaeology (ed.), *Settlement Change and Formation of Early Civilizations*, 204–65. Beijing: Cultural Relics Press.

Zhang, W. X. 2006. 'Bashidang gu zaipeidao' (The ancient cultivated rice from the Bashidang site). In Hunan Provincial Institute of Cultural Relics and Archaeology (ed.), *Pengtoushan and Bashidang*, 544–61. Beijing: Science Press.

Zhang, W. X. and A. P. Pei. 1997. 'Lixiang mengxi bashidang chutu daogu de yanjiu' (The study of the ancient rice remains from Bashidang, Lixian County). *Cultural Relics* (1), 36–41.

Zhang, Z. H. 1998. 'Changjiang liuyu shiqian gucheng de chubu yanjiu' (A preliminary study of the prehistoric walled sites of the Yangtze River valleys). *Southeast Culture* (2), 5–13.

Zhao, D. L. and A. P. Pei. 2000. 'Hunan lixian bashidang yizhi gu zaipeidao de zaiyanjiu' (Re-examining the ancient cultivated rice from the Bashidang Site in Lixian County, Hunan province). *Chinese Journal of Rice Science* 14, 139–43.

Zhao, H. and J. Wei. 2002. 'Zhongguo xinshiqi shidai chengzhi de faxian yu yanjiu' (Discoveries and research of prehistoric walled sites in China). In Research Centre for Chinese Archaeology, Peking University (ed.), *Ancient Civilization*, 1–34. Beijing: Cultural Relics Press.

Zhao, Zhijun. 1998. 'The Middle Yangtze region in China is one place where rice was domesticated: Phytolith evidence from the Diaotonghuan Cave, Northern Jiangxi.' *Antiquity* 72 (278), 885–97.

Zhou, F. Q. 1994. 'Yunmengze yu jingjiang sanjiaozhou de lishi bianqian' (Evolution of the Yunmeng marsh and the Jingjiang Delta). *Journal of Lacustrine Science* 6, 22–32.

Zhu, C., Y. S. Zhong, C. G. Zheng, C. M. Ma and L. Li 2007. 'Hubei jiushiqi zhi zhanguo shiqi renlei yizhi fenbu yu huanjing de guanxi' (Relationship of archaeological sites distribution and environment from the Palaeolithic age to the Warring States time in Hubei province). *Acta Geographica Sinica* 62, 227–42.

5
Rice fields, water management and agricultural development in the prehistoric Lake Taihu region and the Ningshao Plain

Yijie Zhuang

Abstract

The Lake Taihu region and the Ningshao Plain are important in the development of prehistoric rice farming in China. Because of differences in hydrological regimes, soil conditions, availability of natural resources and socio-economic traditions, these two regions saw different developmental trajectories for rice farming. In the Ningshao Plain, even though rice was consumed at a very early date, the cultivation and domestication process was slow, and hunting and gathering remained important to the local subsistence economies. In the Lake Taihu region, a tendency towards rice farming intensification can be witnessed during the transition from the early (*c.* 5000–4000 BC) to late (*c.* 4000–3500 BC) Majiabang periods. Archaeological evidence for this process includes the construction of rice paddy fields of increasing size, and more complex in-field facilities for water management, along with greater control of wet–dry alternations in the fields for irrigation and drainage. The latter, according to some research, induced 'water stress' to force higher yields of rice. In these two regions, we can see that rice farming, water management and social development were becoming more intricately intertwined, but such relationships varied, depending on local environmental and socio-economic conditions.

Introduction

The landscapes of the rural areas around the present Lake Taihu are dominated by the so-called poldered or dyked fields. These are paddy fields that are demarcated by artificial dykes inside large water bodies to prevent floods in the fields (Figure 5.1). Poldered fields were constructed during the medieval period (*c.* seventh century AD), if not earlier, and have been continuously maintained for rice farming (Bray 1984: 114) (Figure 5.2). Taking the local hydrological and soil conditions into account, this simple yet efficient means of water management has played a crucial role in the successful operation of these poldered fields (Lu et al., 2005) and has greatly promoted the growth of rice farming. Unfortunately, traditional knowledge of water management, soil ecology and sustainable farming is fast disappearing because of

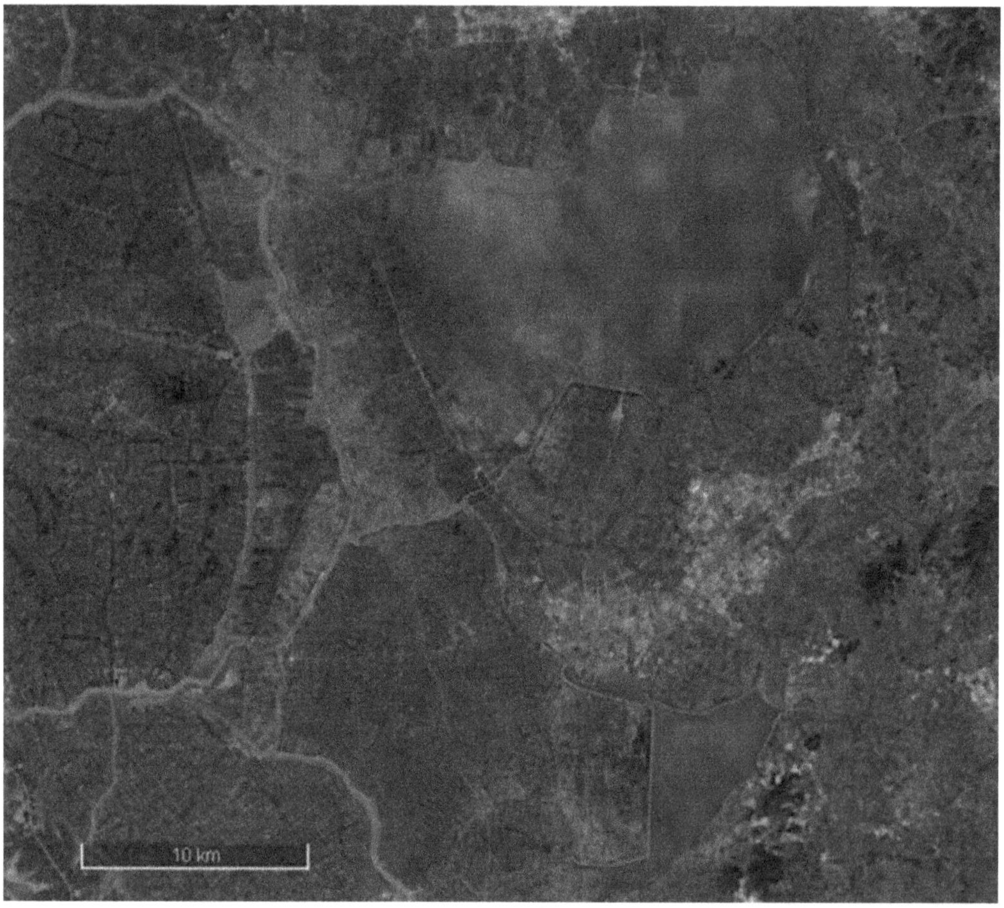

Figure 5.1 Lake Shijiu and the poldered fields and small lakes surrounding it. Note the poldered fields of various shapes on the left and the small lake to the south of the Shijiu Lake, half of it already disappeared. Source: authors

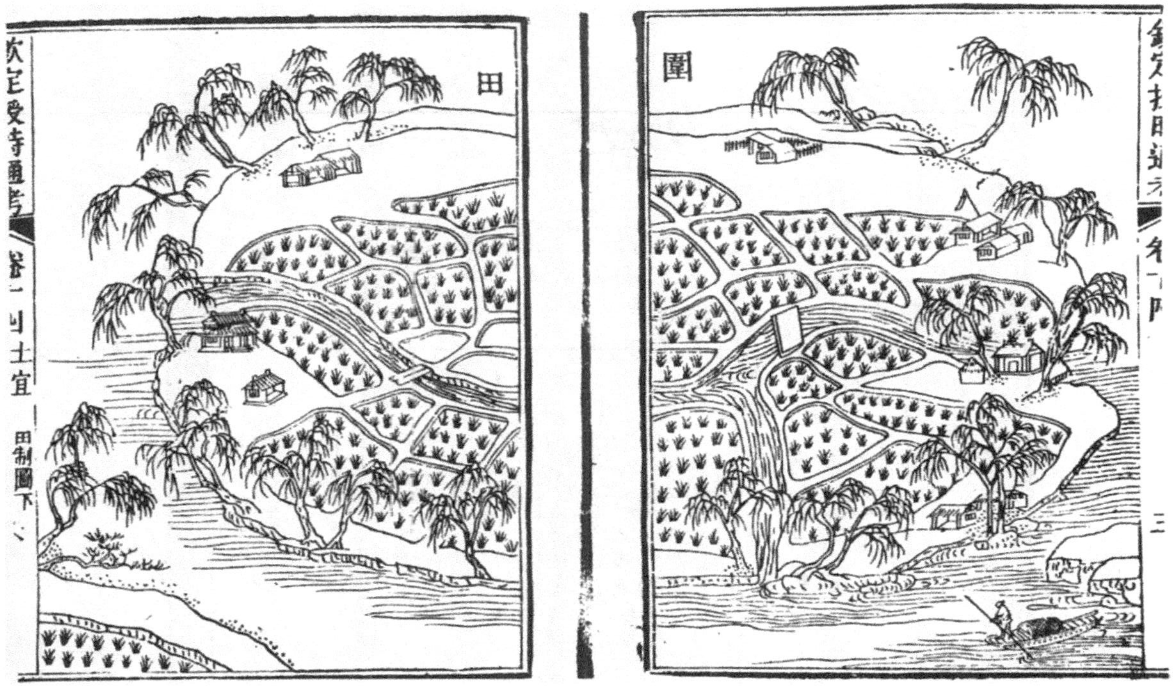

Figure 5.2 Historic poldered fields in the Lower River Yangtze recorded by Wang Zhen (AD 1271–1333) in his book Wang Zheng *Nong Shu* (after Bray, 1984: 115)

the increasing obsession with modern, advanced technologies in the context of rapid urbanisation. These technologies, often derived from or invented in entirely different environmental and cultural contexts, have indeed been accelerating urbanisation and economic growth on an unprecedented scale in the past few decades. Their ecological and environmental outcomes, however, have been largely neglected. Although this chapter does not directly examine the ecology and water-management strategies of these poldered fields, the questions raised above are relevant to the issues facing the study of both historical and contemporary poldered fields. The studied regions, the Lake Taihu region and the Ningshao Plain (Figure 5.3), have unique geomorphology, soil conditions and hydrological regimes, which were shaped by Holocene marine and alluvial activities. These conditions in turn fundamentally define the developmental patterns of water management and rice farming in these regions. I investigate the evolution of rice-farming systems and water-management practices during the Majiabang period (*c.* 5000–*c.*

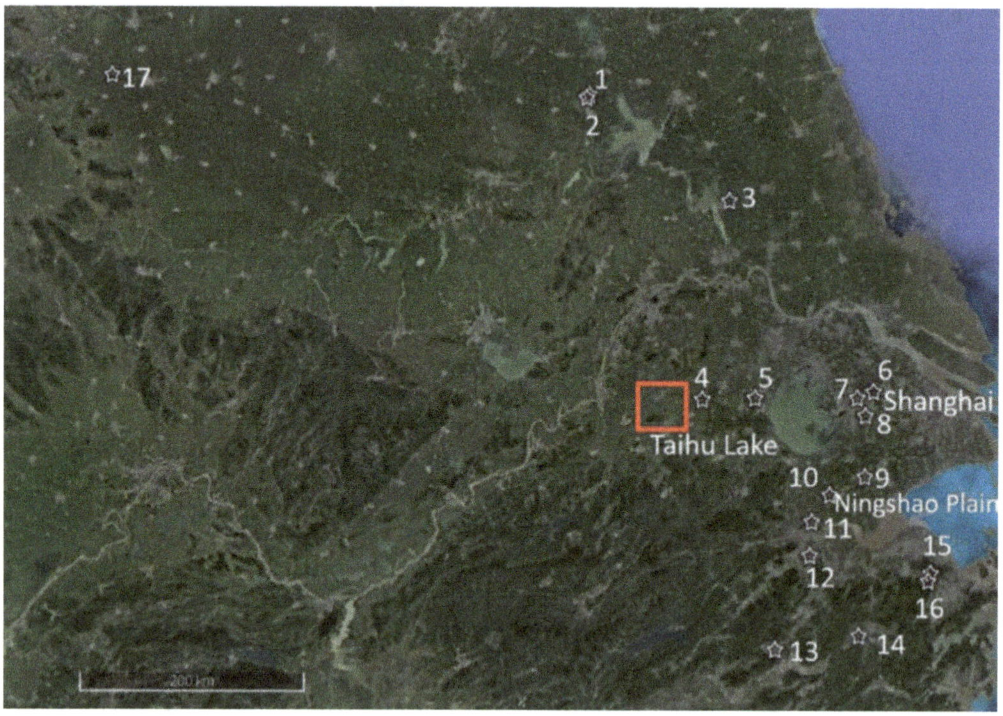

Figure 5.3 The Lake Taihu region and the Ningshao Plain region and locations of sites mentioned in the text. 1 Hanjing; 2 Shunshanji; 3 Longqiuzhuang; 4 Shendun; 5 Luotuodun; 6 Chuodun; 7 Caoxieshan; 8 Chenghu; 9 Majiabang; 10 Luojiajiao; 11 Maoshan; 12 Kuahuqiao; 13 Shangshan; 14 Xiaohuangshan; 15 Hemudu; 16 Tianluoshan; 17 Jiahu. The red rectangle is the area shown in Figure 5.1. Source: authors

3500 BC) of the Lake Taihu region, and the contemporary Hemudu culture (*c*. 5050–*c*. 3050 BC) in the Ningshao Plain, when and where settlement size and rice farming were expanding and the construction of rice fields was transforming the landscapes. Despite a pronounced lack of systematic investigation, the Majiabang and Hemudu cultures, especially in their later phases, can be seen to represent a period of considerable investment in rice farming and water management. I examine and compare how water management related to local hydrological and geological conditions, how the technologies of water management evolved, and how prehistoric societies gradually modified and transformed their landscapes for and with rice farming, and the associated water-management strategies in these two regions.

Geomorphology, soils and local environments

Despite it being primarily deposited on the Loess Plateau of North China, mosaic patches of loess can be found in the Lake Taihu region of South China. The so-called Xiashu loess was carried by strong westerly winds that reached this region during the Last Glacial Maximum (X. S. Li et al., 2001; Wu, 1985). This loess deposition resulted in the formation of small and sometimes isolated loess hills or tablelands. These small landform units became the key pre-lacustrine landscape that was then further transformed by active Holocene marine and alluvial geomorphological processes. The Yangtze Delta and many regions along the eastern coast experienced steady sea-level rise after the early Holocene, which reached their peak around 5000–4500 BC (Zong, 2004). As well as bringing marine sediments to the inland regions, and thereby significantly contributing to the formation of the Yangtze Delta, this high sea-level stand also caused long-standing inundation of low-lying areas. The evolution of Lake Taihu was a direct outcome of this geomorphological process. The western part of the lake, around 9000 BC, was a typical estuarine environment, fed by brackish water and covered by saltmarsh plants, while the eastern part only became submerged when the regional water level reached its highest *c* 4500 BC. Around 3000 BC, both the western and eastern sections of this area were inundated by freshwater, mainly from rivers and surface run-off (Han, 1998; Hong, 1991; Jing, 1985; Q. Wang and Chen, 1999; Zong et al., 2012) (Figure 5.4).

The Lake Taihu region was dominated by mosaics of wetlands during the middle to late Holocene. Pollen assemblages, derived from

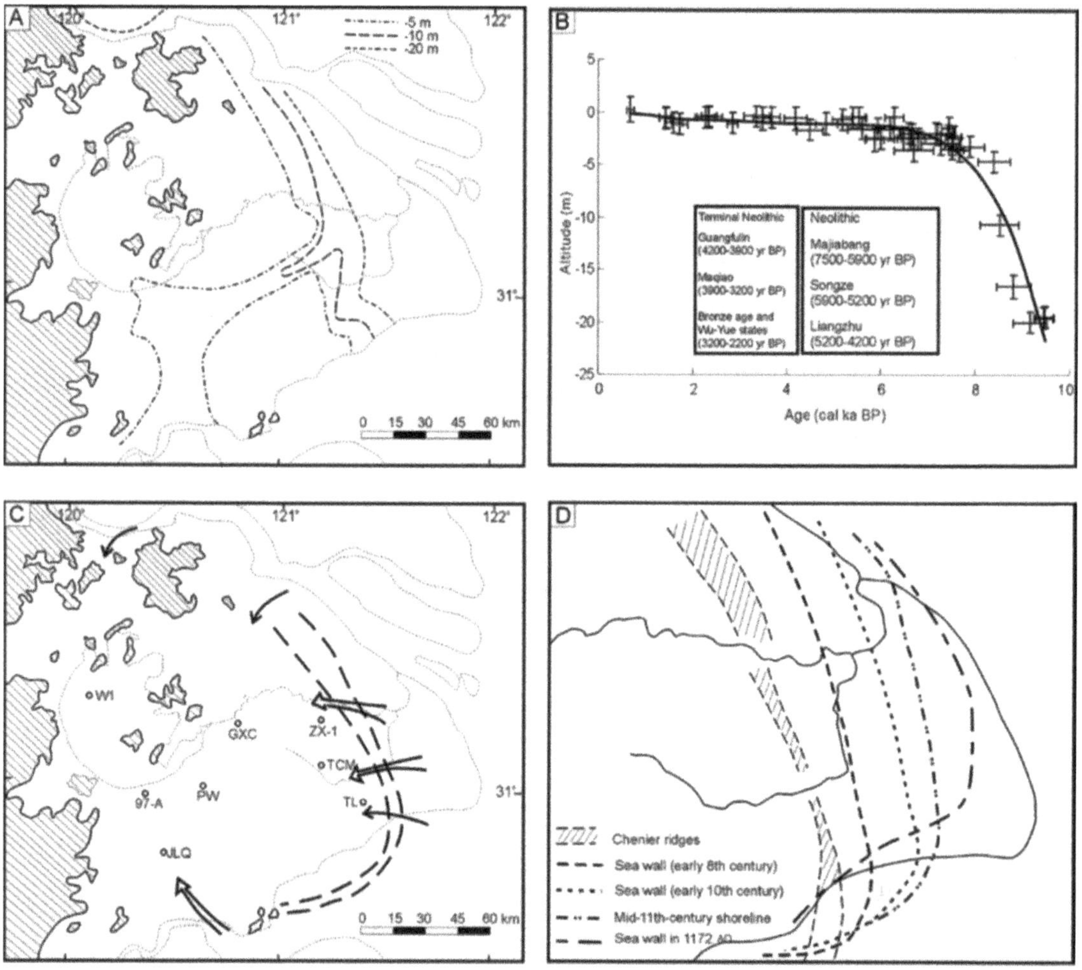

Figure 5.4 The Holocene geomorphological history of the Lake Taihu region and the Yangtze Delta. A. Pre-Holocene landscape. B. Relative sea-level fluctuations. C. The Chenier ridges and the Mid-Holocene shoreline. D. Sea wall constructions and land reclamation in the past 2,000 years. (Captions and figure after Y. Zong et al., 2012.)

palynological studies at a number of Majiabang culture sites, were dominated by non-arboreal species, especially aquatic plants that grow in or near the edge of natural water bodies and wetlands (Ding, 1999; K. F. Wang et al., 1980; K. F. Wang et al., 1984). The presence of arboreal pollens indicates that there were forests or woodlands growing not far from these natural water bodies and wetlands. Though many sites look like mounds, as they are higher than their immediate surroundings, some scholars have pointed out that the palaeo-surfaces during the occupation of these sites would have been much lower than their surroundings (L. Sun and Gao, 2006). The Majiabang-period archaeological sediments are often found at the bottom of these mounds, up to 11 m deep (e.g., at the Caoxieshan site). Thus, it is important to take into account the palaeo-environmental conditions at the time of occupation. To date, there are between 60 and 70 Majiabang culture sites found in the Lake Taihu region. They are mainly located in three kinds of local environments: slopes next to natural water bodies, alluvial plains immediately adjacent to rivers which are buried underneath the current surface or submerged in the water, and foothills (L. Sun and Gao, 2006). Affected by the high sea level, vast areas of the Ningshao Plain were inundated during the early to middle Holocene. From 5000 BC onwards the seawater gradually retreated, but the regional groundwater table remained high, which led to the formation of a widespread peat horizon (Shi et al., 2008; L. Wu et al., 2012). Palynological studies suggest a similar pattern of Holocene vegetation in the Lake Taihu region, with sub-tropical species predominating (Qin, 2006; Tan, 2015). Of the 40–50 Hemudu culture sites that have been found (Liu and Chen, 2012: 204), around 10 have been excavated (G. P. Sun, 2013). The majority of these sites are buried very deep, indicating that they were located on low-lying areas during their occupation.

Two distinctive types of late Pleistocene/Holocene sedimentary sequences are present in the two studied regions. In places with late Pleistocene loess deposition of the Lake Taihu region, the loess formed the base of the late Pleistocene/Holocene sequences (X. S. Li et al., 2001; Yang et al., 2007). As can be seen in the stratigraphy at the Chuodun site, what sits on top of the loess deposits are loess or reworked loess and alluvial sediments (Chen and Zhang, 1994; Shen et al., 2006; Yang et al., 2007) (Figure 5.5). Contrary to such loess–alluvium sequences, the late Pleistocene/Holocene sediments in the low-lying areas of the Ningshao Plain are characterised by clayey, often organic-rich, alluviums and marine sediments (C. M. Lin, 1997; Shi et al., 2008). Loess is well known for its water drainage ability, which is due to its porous properties (Y. Zhuang et al., 2016), while clayey deposits are normally good at water retention. Such a contrast in soil properties would have had a significant impact on prehistoric rice farming and water management, which is discussed below.

Holocene rice cultivation, domestication and expansion

The consumption of rice in the Lake Taihu, Ningshao Plain and surrounding regions can be traced back as far as the beginning of the Holocene. At some Shangshan culture sites (*c*. 9000–6500 cal. BC; Jiang, 2013,

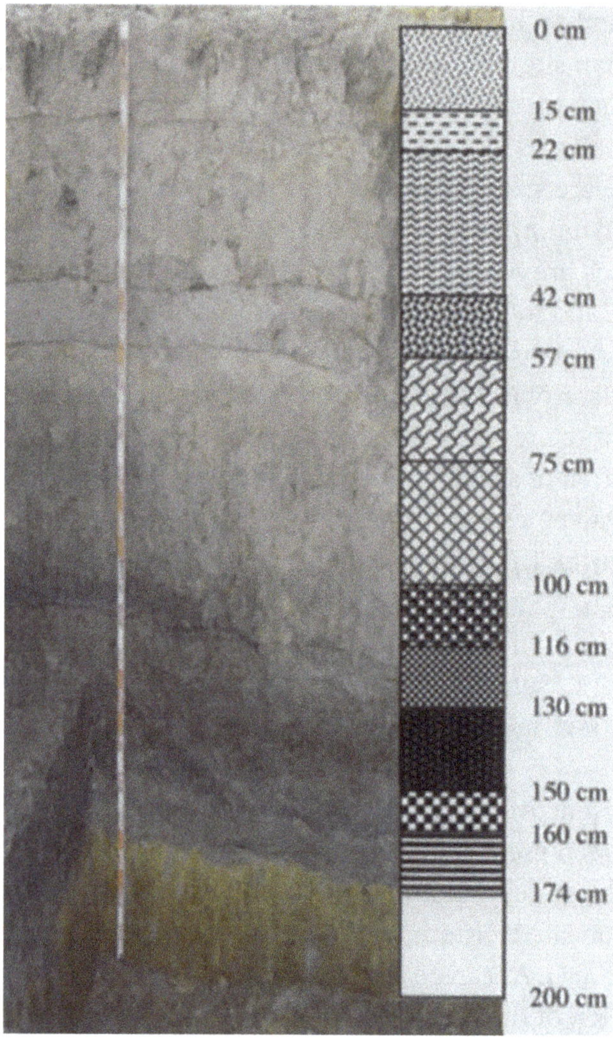

Figure 5.5 A typical Late Pleistocene–Holocene sequence at Chuodun. At the bottom is the yellowish loess and reworked loess of the Late Pleistocene age, overlain by multiple layers of cultivated soils from 174 cm upwards (after Shen et al., 2006)

but cf. Zuo et al., 2017 for a more conservative determination of the date, 7450–6500 cal. BC), not only was rice consumed but rice chaff was often used as tempering material in pottery production (Jiang, 2007). In the following period, at Xiaohuangshan (7000–5700 cal. BC) and Kuahuqiao (6000–5000 cal. BC), rice grains were consumed and rice chaff was used in ceramic temper (Y. F. Zheng et al., 2004; Y. F. Zheng et al., 2007). However, as many scholars have pointed out, subsistence strategies at these early Holocene sites were rather diverse: they included the consumption of a wide range of wild plants, notably aquatic plants, and wild animals. Rice accounted for only a small proportion of people's diet at this time (L. Qin, 2012). Also, in the light of recent discoveries, until 8,000 years ago the River Huai to the north of the Lake Taihu region was another important region for the early development of rice economy. At the Hanjing and Shunshanji sites (L. G. Lin et al., 2014), for instance, not only has the discovery of carbonised rice remains pushed the date of the beginning of rice consumption in the region to before 6000 cal. BC, almost as early as the dates of some other important discoveries (e.g., Jiahu located further north), but the presence of domesticated-type rice spikelet bases (L. N. Zhuang et al., 2017) also clearly indicates cultivation and/or intentional field management.

While careful measurement of the carbonised rice grains has revealed an analytical distinction between wild and possibly cultivated rice, the process of cultivation and domestication remains controversial. Debates on how to disentangle the domestication syndrome have become heated among scholars who examine rice remains from the Hemudu culture sites (Fuller et al., 2007; Liu et al., 2007; Zuo et al., 2017). The debate on when and how rice was domesticated is ongoing, and much of this debate concerns the lack of understanding of domestication as a lengthy evolutionary process. For the two to three thousand years of continuous cultivation experiments, there would have been plants under cultivation that had some morphological traits that characterise domestication but cannot be regarded as domesticated, and that are hard, if not impossible, to determine through archaeobotanical research (C. Stevens, personal communication). If this viewpoint is adopted, either Shangshan or Hemudu rice is fully domesticated. In spite of these controversies, it is certain that rice was becoming increasingly important in the diet of the Hemudu communities. This trend can best be demonstrated by the detailed archaeobotanical research conducted by Fuller and colleagues at the Tianluoshan site (Fuller et al., 2009). They identified rice spikelet bases with rough rachises, indicative of domestication, which appeared early (*c.* 4900 cal. BC). However, the overall percentage of rice in the total archaeobotanical assemblage was low at this time. The percentage of rice in the archaeobotanical samples continued to increase through time, while there was a corresponding decrease in other plant remains such as water chestnuts (Fuller and Qin, 2010). Thus, the Tianluoshan case in the Ningshao Plain provides a convincing illustration that rice domestication was a slow process.

In the Lake Taihu region, rice farming expanded rapidly during the Majiabang period. Luojiajiao (5300–4900 cal. BC) is one of the earliest Majiabang culture sites. Abundant pottery sherds, animal bones, plant remains and wooden architectural structures were discovered during the excavation. Rice remains include several hundred (*c.* 500) carbonised rice grains and rice chaff found inside pottery sherds as tempering material (Zheng et al., 2007; Zhu, 2004). Some of the carbonised seeds still had awns attached to the surface of the seeds (Zhu, 2004). Zheng and colleagues recently examined 100 rice spikelet bases from Luojiajiao and suggested that 50 per cent of them were from the cultivated *japonica* species (*O. sativa L.* subsp.), while the rest belonged to wild rice (Zheng et al., 2007). But this identification is not without a problem as it does not include immature spikelet bases, which might explain why the site generated so much higher a percentage of domesticated rice than Tianluoshan. At a site contemporary with Longqiuzhuang (which lasted into much later periods, 5000–4300 and 4300–3500 BC), located to the north of the Lake Taihu region, rice remains have been recovered, via flotation, from many excavated features (Longqiuzhuang Archaeological Team, 1999). In line with the trend observed in many contemporary or earlier sites (e.g., Tianluoshan), rice remains increased while the importance of gathered wild plant foods such as water chestnuts gradually diminished.

As suggested by L. Qin (2012), the tipping point between the opposing trends of rice and wild plant foods in the prehistoric diet occurred around 4000 BC, when rice farming experienced a leap forward in terms of technological development. By the late Majiabang period rice farming was significantly intensified. This take-off of rice farming benefited from improved farming practices and intensified water management. These processes stimulated pronounced social-economic changes, as illustrated by the increasing number of settlements, the peopling of the lowland areas and the considerably larger scale of rice farming. The number of late Majiabang culture sites increased to more than 30, much larger than the 10 and 17 of the early and middle phases respectively. Given that some late Majiabang culture sites may be buried underneath late Holocene alluviums, the actual number may be higher. By the late Majiabang period, with more lands emerging around Lake Taihu because of the combination of a falling lake level and increasing sediment supplies, settlements were located all over this area, with a particularly noticeable expansion towards the eastern alluvial plain. In the Ningzhen Plain, an even more pronounced increase in the number of settlements can be seen from the early (5000–3900 BC, about 10 sites) to late (3900–2900 BC, nearly 40 sites) Hemudu culture periods. These late Hemudu culture sites were distributed over a much larger region (Liu and Chen, 2012: 204).

Rather like the classic settlement structure excavated at the Hemudu site in the 1970s (Liu and Chen, 2012), the many contemporary Hemudu and Majiabang settlements were characterised by the so-called *ganlan* (raised-floor) wooden structure above the ground (Ji, 1983; G. P. Sun, 2013) as well as above-ground houses. Such above-ground houses can be found, for instance, at the Shendun and Luotuodun sites, where some houses measure almost 60 m^2 and have large storage jars located immediately outside (L. G. Lin et al., 2009; Tian et al., 2009). Because of a lack of systematic archaeobotanical investigation, only limited rice remains have been discovered at a few late Majiabang and Hemudu sites to date (Stevens and Fuller, 2017). However, the initial intensification of rice farming around 4000 BC is supported by the discovery of several paddy fields.

Construction of Neolithic rice fields

By examining microfossil plant remains and charcoal discovered at the Kuahuqiao site, Zong et al. (2007) were able to suggest that the Neolithic community was practising cultivation of rice *c.* 5700 cal. BC. Although controversy regarding the sampling location and the interpretation of the data persists (Shu et al., 2013), the modification of wild habitats of rice plants in the swampy wetlands and the management of blackish water formed a significant step in rice farming for the Kuahuqiao community (also see J. W. Sun et al., 2010). Domesticated-type spikelet bases are present (<10 per cent), which points to significant morphological changes in some rice plants, even though the numbers are very low compared with wild rice (Zheng et al., 2016; Zuo et al., 2016; Zuo et al., 2017). The rice fields at Tianluoshan are among the earliest rice fields to have been excavated. Similarly to those at Kuahuqiao, the Tianluoshan rice fields were located on or along a swamp, as confirmed by a systematic coring survey, from which phytolith samples were analysed (Zheng et al., 2009). They were located near the raised-floor houses (Figure 5.6). These fields can be dated to two periods, *c.* 4650–4490 cal. BC and *c.* 3340–3090 cal. BC, and were buried *c.* 2 m and *c.* 0.95–1.7 m below the surface respectively. Some wooden and bone tools were found in the fields, including 'dibble sticks' and spades (G. P. Sun et al., 2007). The soil is characterised by dark clayey particles (Figure 5.7) and is rich in organic matter, and, as noted by the excavators, 'no evidence of an irrigation system, which should include ditches, field ridges/bunds for controlling drainage and water retention, was found' (Zheng et al., 2009: 2613). This may indicate that the ecology

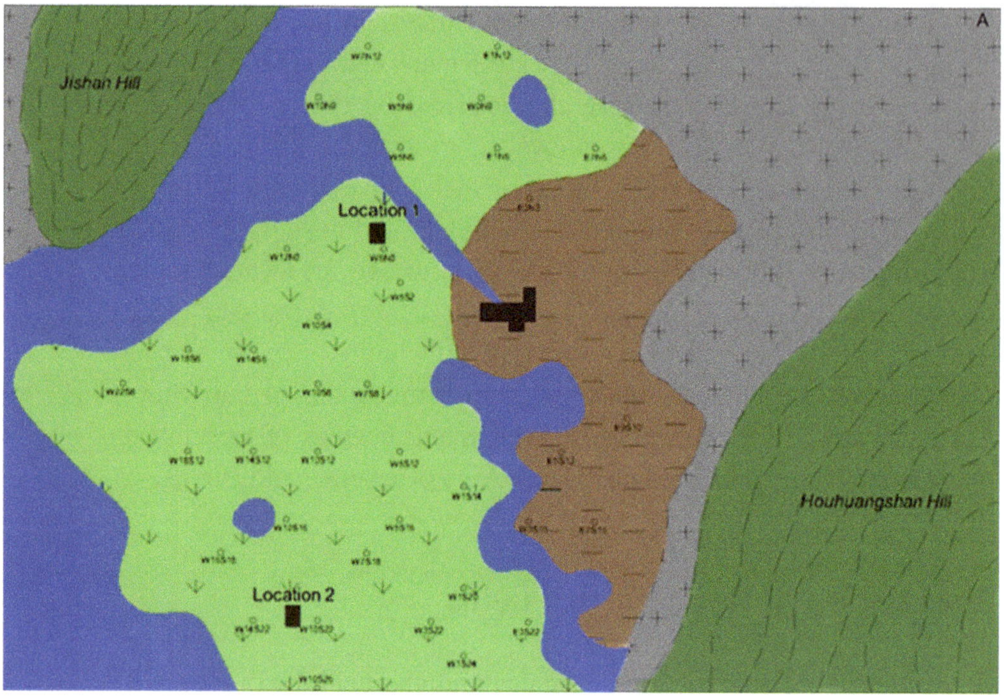

Figure 5.6 Distribution and excavation of the rice fields around the settlement at Tianluoshan, based on phytolith analysis of soil samples collected from many locations. The dark blue represents reconstructed water bodies, the dashed lines the settlement area, and the crosses rice fields (after Zheng et al., 2009, redrawn by C. Stevens)

of rice farming at Tianluoshan remained 'crude', without intensive modification of the local environment, resembling that found at Kuahuqiao. We should therefore call these 'rice fields' rather than 'paddy fields'. The main difference between these two categories is that there is a lack of investment in the labour of building bunds and clearing irrigation ditches in early rice fields. This situation is quite different from that in the early paddy fields found in the Lake Taihu region, where one can see considerably more effort being invested in building and maintaining the fields.

Caoxieshan is situated next to Lake Yangcheng and is surrounded by wetlands. The site measures *c.* 45 ha, according to the excavator. However, it is important to note that it is a multiple-period site and the extent of the Majiabang period deposits may be smaller than is estimated. The site is poorly preserved, except for the central part, where several mounds still stand on the ground. An area of more than 1,000 m^2 of the site was excavated in the 1970s, while the paddy fields were excavated in the 1990s by a joint Sino-Japanese team. A total of 44 (1,400 m^2) Majiabang-period paddy fields were unearthed through the excavation (Figure 5.8). These

Figure 5.7 Early-phase rice fields at Tianluoshan. No clear archaeological features related to the field system are preserved. Note the dark clayey deposit (after Zheng et al., 2009)

fields, which were generally small and of irregular shape, can be divided into three phases. The early-phase fields were situated in the low-lying area. They were slightly larger than later-phase fields, but the lack of associated facilities for water management, such as water outlets, seemed to suggest that they were built without considerable labour investment. The middle-phase fields were built directly over early-phase fields after they had become flooded and silted up. For this middle phase, more effort was invested in building water outlets, sluices, wells and pathways across the fields. After the middle-phase fields were abandoned, the surface became higher. The sizes of the three phase fields ranged greatly, from several square metres to more than 10–20 m^2. They are mostly of single units, but some are comprised of multiple units. As well as digging into the surface to build the fields, labourers dug ponds and wells to facilitate water management. Steps can be found inside some of the wells at Caoxieshan (e.g., well no. J34); similarly, pots were found at the bottom of the wells that were used as water containers. In addition, a small pier structure was excavated. A wooden plank (1.1 × 0.7 m) found next to an area with steps led towards one of the ponds (Gu et al., 1998). All these discoveries

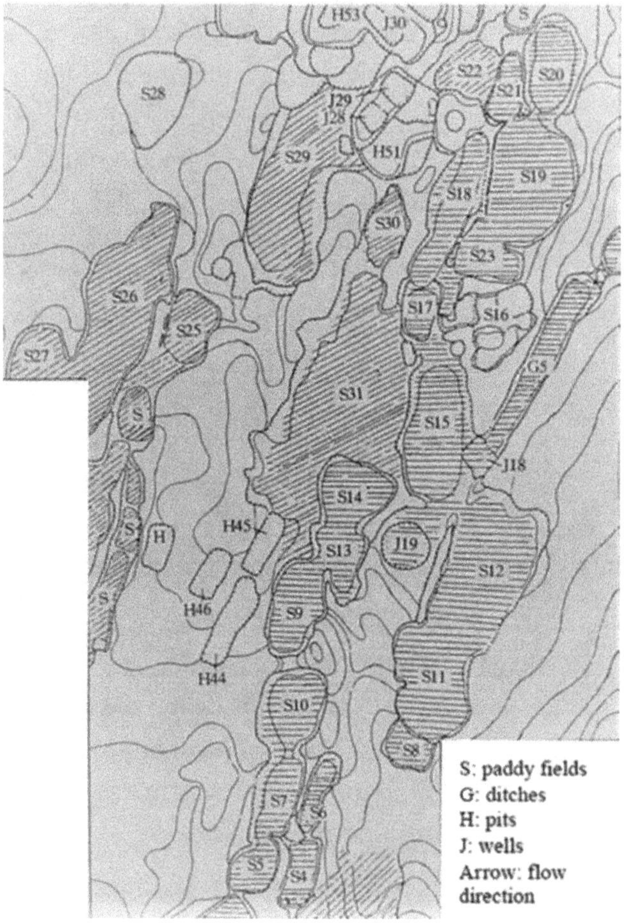

Figure 5.8 Distribution of paddy fields at the east excavation area, Caoxieshan. Note that the fields in this plan are divided into two phases (after Gu et al., 1998)

indicate increasing investment in the fields and in related facilities for water management.

Chuodun is another important site at which Majiabang-period paddy fields have been excavated. It is located on a narrow strip of land between Lakes Yangcheng and Kuilei. It is a site with multiple cultural-period occupations, measuring *c* 40 ha, but the size of the Majiabang-period settlement remains unclear. A total of 46 paddy fields were unearthed in two excavation seasons (Figure 5.9). From a recent geoarchaeological survey it is clear that the fields were built on the low-lying areas of the site (Cao et al., 2007). These fields were small (averaging 1 to 10 m^2) and of irregular shapes. Simple structures such as small water outlets, ditches and small ponds were built for water management (Gu, 2003).

Figure 5.9 Plans of the paddy fields at the Chuodun site, redrawn by C. Stevens

While the Caoxieshan and Chuodun paddy fields have similarities in their distributional patterns and structures, at a paddy field site from a later period (the Songze culture period, 3520–3260 cal. BC), Chenghu, more changes to the field structure can be seen, indicative of the further intensification of rice farming. The Chenghu site is situated next to Lake Chenghu. It is divided into a residential zone and a production zone. A ditch (no. G2) was dug between the residential and production zones. This ditch on the one hand served as a sewer for the residential area, and on the other received water from the River Xixupu, which was fed by the River Wusong before releasing water to the fields. The paddy fields were located around an artificial pond. These irregularly shaped fields vary greatly in size, from only about 1 m^2 to more than 100 m^2 (Ding, 2004). The latter demonstrates an unprecedented increase in size of the paddy fields in the region.

Water management and intensification of rice farming

While irrigation is always key to rice growth, on many occasions drainage is more vital to rice farming, especially during the harvesting season as lengthy saturation or waterlogging is extremely detrimental to rice plants in the growing stage of grain milking and filling. Indeed, effective drainage remained one of the biggest challenges in early rice farming.

The fields in the eastern part of Caoxieshan can be divided into several groups. Each group was surrounded by wells and connected with ditches and water outlets. Although the topographical differences between these features were minor (often less than a metre), they played a key role in channelling, diverting and draining water in the fields. When monsoonal rainfall is heavy, as it often is today in the summer (July–August especially), it accumulates very rapidly in the fields. A good understanding and efficient micro-management of the local topography by these early rice farmers would have enabled them to build and cultivate the paddy fields more effectively. The excavations have shown that at the low ends of the fields there were several shallow pits, into which water flowed via ditches and related features. In the western part of the sites, instead of being connected to wells the fields were connected to several ponds via water outlets. These ponds have a much larger volume than wells for holding and absorbing water. Thus, these features represented a significant development in water management, in that drainage remained crucial yet challenging. A phytolith study conducted by Weisskopf and colleagues provides supportive evidence for this emphasis on drainage at Caoxieshan. By grouping phytoliths into different categories based on comparisons with modern rice fields with known farming practices, they have found that the phytoliths collected from the Caoxieshan paddy field 'have many more fixed morphotypes that are consistent with the drier signatures'. This drought-stress farming regime was likely to 'induce water stress and produce more flowers and grains' of rice plants (Weisskopf et al., 2015; also cf. Fuller, 2011).

Geophysical analysis of soil samples collected from the Chuodun paddy field has confirmed decreases in magnetic susceptibility and in the percentage of clay particles from *c.* 2 to 1.3 m below the surface. This trend is considered to be the result of long-term drainage in the field, which causes the downward leaching of clay particles (Yang et al., 2006; Yang et al., 2007). The existence of such farming practices, which emphasise the importance of drainage, is supported by a pollen study which found that the pollen assemblage was dominated

Figure 5.10 Large paddy fields at the Chenghu site (after Ding, 2004)

by *Gramineae* and *Cruciferae* species, with very few aquatic pollens (*Cyperaceae, Typhaceae*, etc.) present (Li et al., 2006). Although other factors may also be responsible for this low percentage of aquatic species in an otherwise wet environment, according to Li et al. (2006) this may be related to deliberate drainage management.

By the Songze period, the paddy fields at the aforementioned Chenghu site, for instance, had a clear division between high fields and low fields. The large pond (425 m^2) beside which the fields were located had a very skewed surface at its bottom. The difference in height between the western and eastern parts was *c.* 1.7 m. In addition, the eastern part was linked to a river (Ding, 2004; Figure 5.10). This pond had multiple functions. It provided water to irrigate the high fields, and two small pools situated right next to the high fields were used to lift water from the pond. More importantly, it served to drain water quickly from both high and low fields into the river. The use of the Chenghu paddy fields coincided with a period around 3000 BC when lakes and other natural water bodies were expanding because of increased precipitation and temperature (Ding, 2004).

Concluding remarks

In the Ningshao Plain, rice was consumed as early as the beginning of the Holocene. While cultivation or experimental farming carried on into later periods, rice did not become the dominant staple in the diet of local communities until the late Hemudu period, at the earliest. Rice fields at the Tianluoshan site were used without much labour investment in building irrigation facilities during the middle to late Hemudu period, but to what degree the rice fields were intensively cultivated, and the nature of the ecology of water-management strategies, if any, remain to be demonstrated. Liu and Chen (2012: 203–4) have rightly pointed out that the increase in the number of settlements, from around 10 in the early Hemudu period to nearly 40 in the late Hemudu period (see above), is indicative of population growth and dispersal, especially to the areas to the south of the Ningshao Plain. The role of rice farming in this late Hemudu dispersal is unclear, however, because of a lack of archaeobotanical and other environmental evidence. In the Lake Taihu region, by the early Majiabang period paddy fields had already been established at several settlements, and it is clear from excavations of related archaeological features and from interdisciplinary research on water-management practices that more labour was invested in building the larger fields and that more careful control of in-field water situations was practised. From the Majiabang period to the Songze period, we can see a clear intensification process in rice farming, characterised by both the quantity and the size of the paddy fields at the sites discussed above. We can tentatively term this intensification of rice farming 'the late Majiabang expansion'. There are several reasons for these two different developmental trajectories of rice farming in the prehistoric Lake Taihu region and the Ningshao Plain.

First, the clayey soil in the Ningshao Plain meant that it would have been more laborious to build large paddy fields with complete water-management infrastructures and to till these fields. Xie et al. (2015) compares the efficiency of stone and bone digging tools for earth working in the Ningshao Plain through an experimental study. They have found that while bone and stone tools function equally well in soft soils, stone tools 'provide significant and easily perceived advantages' when soils are hard, as the clayey soils in the prehistoric Ningshao Plain are. Bone digging tools seem to have been the main digging implements at the Hemudu culture sites (Huang, 1999). This may have constrained the inhabitants' ability to construct and maintain large-scale paddy fields as well as other earthen projects. Unlike the Hemudu culture sites, many

Majiabang culture sites were situated on higher ground, and soil conditions were much less clayey and heavy because of the loess-like parent material, which contains coarser particles and more porous and looser structures. Such soils would have been easier to cultivate and more productive, regardless of what types of tools were applied (the Majiabang culture sites seemed to contain fewer bone digging tools; see Wu, 1999).

The most important factor determining the trajectories of rice farming was the local subsistence economies, as they were restricted by the availability and seasonality of resources in the local environment. Research has shown that some of the Hemudu culture sites (e.g., Tianluoshan) were in optimal locations for hunting and gathering (L. Qin et al., 2010). The rich animal and plant resources of the sub-tropical forests and wetlands delayed the Hemudu communities' commitment to full rice agriculture; that is to say, rice did not play a predominant role in food production. Indeed, gathered wild foods remained an important component of the diet at Tianluoshan for most of the Hemudu period (L. Qin et al., 2010). In the Majiabang culture of the Lake Taihu region, natural resources were not as rich as in the coastal Ningshao Plain, and from the early Majiabang period onwards they had already developed the 'water-stress strategy' to induce greater agricultural yields. This, to a large extent, accelerated the development of rice farming throughout the Majiabang period.

By the Liangzhu period (3300–2300 BC), these divergent pathways of rice farming seem to have merged, with the dramatic increase in scale and intensity of rice farming that can be clearly illustrated by the excavation at the Maoshan site (2700–2200 BC) (Y. Zhuang et al., 2014). Here one single paddy field reached a size of 2,000 m^2 during the late Liangzhu period. This dramatic increase in the scale of farming came together with greater control of the in-field hydrology and methods of soil amendment as demonstrated by geoarchaeological and archaeobotanical investigations (Y. Zhuang et al., 2014; see also Weisskopf et al., 2015). The rice farming at Maoshan certainly is closely related to, and contributed economically to, the development of social complexity during the Liangzhu period. It is worth noting that this development was built upon the foundation laid down in the preceding Majiabang–Hemudu period.

References

Bray, Francesca. 1984. *Science and Civilisation in China. Volume 6: Biology and Biological Technology. Part II: Agriculture*. Cambridge: Cambridge University Press.

Cao, Z. H., L. Z. Yang, X. G. Lin, Z. Y. Hu, Y. H. Dong, G. Y. Zhang, Y. C. Lu, R. Yin, Y. H. Wu, J. L. Ding and Y. F. Zheng. 2007. 'Chuodun yizhi xinshiqi shiqi shuidaotian gu shuidaotu poumian zhiguiti he tanhuadao xingtai tezheng de yanjiu' (Paddy fields, stratigraphy, phytoliths and carbonised rice remains from the Neolithic site of Chuodun). *Acta Pedologica Sinica* 44, 838–47.

Chen, Qing Sho and Yu Yong Zhang. 1994. 'Taihu pingyuan diqu turang wuzhi xingcheng xintan' (New investigation of the soil parent material in the Lake Taihu region). *Chinese Journal of Soil Science* 25 (1), 7–8.

Ding, J. L. 1999. 'Majiabang wenhua shiqi de ziran huanjing yu renlei huodong' (Natural environment and human activities during the Majiabang cultural period). *Agricultural Archaeology* (3), 44–7.

Ding, J. L. 2004. 'Changjiang zhongyou xinshiqi shidai shuidaotian yu daozuo nongye de qiyuan' (Neolithic paddy fields and the origins of rice farming in the Lower River Yangtze). *Southeast Culture* (2), 19–23.

Fuller, Dorian Q. 2011. 'Pathways to Asian civilizations: Tracing the origins and spread of rice and rice cultures.' *Rice* 4 (3–4), 78–92.

Fuller, Dorian Q., Emma Harvey and Lin Qin. 2007. 'Presumed domestication? Evidence for wild rice cultivation and domestication in the fifth millennium BC of the Lower Yangtze region.' *Antiquity* 81 (312), 316–31.

Fuller, Dorian Q. and Lin Qin. 2010. 'Declining oaks, increasing artistry, and cultivating rice: the environmental and social context of the emergence of farming in the Lower Yangtze region.' *Environmental Archaeology* 15 (2), 139–59.

Fuller, Dorian Q., Ling Qin, Yunfei Zheng, Zhijun Zhao, Xuguo Chen, Lei-Aoi Hosoya and Guo-Ping Sun. 2009. 'The domestication process and domestication rate in rice: Spikelet bases from the Lower Yangtze.' *Science* 323 (5921), 1607–10.

Gu, J. X. 2003. 'Chuodun yizhi majiabang wenhua shiqi shuidaotian' (Majiabang-period paddy fields at the Chuodun site). *Southeast Culture* (1), 42–5.

Gu, J. X., H. B. Zhou, M. C. Li, L. H. Tang, J. L. Ding and Q. D. Yao. 1998. 'Dui caoxieshan yizhi majiabang wenhau shiqi daozuo nongye de chubu renshi' (Preliminary study of Majiabang-period rice farming at the Caoxieshan site). *Agricultural Archaeology* (3), 15–24.

Han, Z. Q. 1998. 'Hongzehu yanbian de lishi guocheng jiqi beijing fenxi' (Evolution of Lake Hongze and its background). *Forum of Chinese Historical Geography* (2), 61–7.

Hong, X. Q. 1991. 'Taihu de xingcheng he yanbian guocheng' (Formation of Lake Taihu). *Marine Geology and Quaternary Geology* (4), 87–99.

Huang, W. J. 1999. 'Hemudu wenhau gusi xintan' (New investigation of the Hemudu culture bone spades). *Archaeology and Cultural Relics* (6), 34–7.

Ji, Z. Q. 1983. 'Jiangsu haian qingdun yizhi' (The Qingdun site in Haian, Jiangsu). *Acta Archaeologica Sinica* (2), 147–90.

Jiang, L. P. 2007. 'Zhejiang pujiang xian shangshan yizhi fajue jianbao' (Preliminary report of the excavation at the Shangshan site, Pujiang, Zhejiang). *Archaeology* (9), 7–21.

Jiang, L. P. 2013. 'Qiantanjiang liuyu de zaoqi xinshiqi shidai ji wenhua puxi yanjiu' (Early Neolithic and cultural lineage of the Qiantang river valley). *Southeast Culture* (6), 44–53.

Jing, C. Y. 1985. 'Taihu diqu quanxinshi yilai gudili huanjing de yanbian' (Holocene palaeo-environmental changes around Lake Taihu). *Geographic Science* 5, 227–34.

Li, C. H., G. Y. Zhang, L. Z. Yang, X. G. Lin, Z. Y. Hu, Y. H. Dong, Z. H. Cao, Y. F. Zheng and J. L. Ding. 2006. 'Chuodun yizhi gu shuidaotu baofenxue tezheng chubu yanjiu' (Pollen evidence for ancient paddy fields at the Chuodun site). *Acta Pedologica Sinica* 43 (3), 452–60.

Li, X. S., D. Y. Yang and H. Y. Lu. 2001. 'Zhenjiang xiashu huangtu lidu tezheng jiqi chengyin chutan' (Particle-size analysis of the Xiashu loess in Zhenjiang and the origin of the loess). *Marine Geology and Quaternary Geology* 21, 25–32.

Lin, C. M. 1997. 'Hangzhouwan diqu 15000 yilai cengxu dicengxue chubu yanjiu' (Stratigraphical sequences in the Hangzhou Bay area since 15,000 years ago). *Geological Review* 43, 273–80.

Lin, L. G., H. Y. Gan and L. Yan. 2014. 'Jiangsu sihong shunshanji xinshiqi shidai yizhi fajue baogao' (Report of the excavation at the Neolithic site of Shunshanji, Sihong County, Jiangsu province). *Acta Archaeologica Sinica* (4), 519–61.

Lin, L. G., M. L. Tian, J. Q. Xu, H. M. Zhou, H. Zhang, Y. F. Hu, R. K. Zhou, X. T. Zhu and J. Q. Huang. 2009. 'Jiangsu yixing luotuodun yizhi fajue baogao' (Report of the excavation at the Luotuodun site, Yixing, Jiangsu). *Southeast Culture* (5), 26–46.

Liu, Li and Xingcan Chen. 2012. *The Archaeology of China: From the Late Paleolithic to the Early Bronze Age*. New York: Cambridge University Press.

Liu, Li, Gyoung-Ah Lee, Leping Jiang and Juzhong Zhang. 2007. 'Evidence for the early beginning (c. 9000 cal. BP) of rice domestication in China: A response.' *The Holocene* 17 (8), 1059–68.

Longqiuzhuang Archaeological Team. 1999. *Longqiuzhuang: Report of the Excavation at the Neolithic Site in Eastern Jianghuai Region*. Beijing: Science Press.

Lu, Y. C., X. Y. Wang, H. He, C. Gao and G. S. Zhang. 2005. 'Jiyu Landsat TM yingxiang de weitian jiegou xinxi tiqu yu fenxi yi wandongnan weitianqu weili' (Information abstraction and analysis of the poldered fields based on Landsat TM images: Evidence from the southeast Anhui poldered fields). *Remote Sensing Technology and Application* (6), 586–90.

Qin, J. G. 2006. 'Ningshao pingyuan ji linqu wangengxinshi yilai de baofenxue yanjiu ji guhuanjing yiyi' (Pollen study of the Ningshao Plain and its surrounding areas since the late Pleistocene and palaeo-environmental implications). PhD dissertation, Tongji University, Shanghai.

Qin, L. 2012. 'Zhongguo nongye qiyuan de zhiwu kaogu yanjiu yu zhanwang' (Archaeobotanical research of the origins of agriculture and perspectives in China). In Center for Archaeological Research of China (ed.), *Archaeological Research*, vol. 9, 260–315. Beijing: Cultural Relics Press.

Qin, Ling, Dorian Q. Fuller and Hai Zhang. 2010. 'Zaoqi nongye juluo de yesheng shiwu ziyuanyu yanjiu yi changjiang xiayou he zhongyuan diqu weili' (Modelling wild food resource catchments among early farmers: Case studies from the Lower Yangtze and central China). *Quaternary Research* 30 (2), 245–61.

Shen, W. S., R. Yin, X. G. Lin, H. Y. Chu, Z. Y. Hu and Z. H. Cao. 2006. 'Chuodunshan yizhi gu shuidaotu de yixie weishengwuxue tezheng yanjiu' (Microbiological properties of ancient paddy soil discovered at the Chuodun site of Kunshan City, China). *Acta Pedologica Sinica* 43, 814–20.

Shi, W., C. M. Ma, F. Jiao, C. Zhu and F. B. Wang. 2008. 'Ningshao pingyuan shiqian yizhi maicang nitan yu zhongquanxinshi haimian bianhua' (Archaeological sites, buried peat and fluctuations of sea level in the Holocene in the Ningshao Plain in Zhejiang province, China). *Acta Oceanologica Sinica* 30, 169–75.

Shu, W. J., L. P. Jiang and W. M. Wang 2013. '"Daogeng huozhong" shi junjin 8000 nian zhejiang kuahuqiao wenhua de daozuo jingji moshi ma? Yizhi poumian yu ziran zuankong baofen jilu de duibi qishi' (Was 'slash-and-burn' the rice-farming mode at the 8,000-year-old site of Kuahuqiao, Zhejiang? Comparison of archaeological stratigraphy and pollen records from natural sediments and implications). Association of Pollen Research of China (ed.), *Proceedings of the First Annual Meeting of the Ninth Committee of the Association of Pollen Research of China*. Nanjing: Association of Pollen Research of China.

Stevens, C. and D. Q. Fuller. 2017. 'The spread of agriculture in Eastern Asia: Archaeological bases for hypothetical farmer/language dispersals.' *Language Dynamics and Change* 7, 152–86.

Sun, G. P. 2013. 'Recent research on the Hemudu culture and the Tianluoshan site.' In A. P. Underhill (ed.), *A Companion to Chinese Archaeology*, 555–73. Chichester: Wiley-Blackwell.

Sun, G. P., W. J. Huang, Y. F. Zheng, Z. Y. Liu, Z. Q. Xu, K. Y. Qu, H. Z. Zhang, Y. J. Li and C. P. Xu. 2007. 'Zhejiang yuyao tianluoshan xinshiqi shidai yizhi 2004 nian fajue jianbao' (Report of the 2004 excavation season at the Neolithic site of Tianluoshan, Yuyao, Zhejiang). *Cultural Relics* (11), 4–24.

Sun, J. W., W. M. Wang, L. P. Jiang and H. Takahara. 2010. 'Early Neolithic vegetation history, fire regime and human activities at Kuahuqiao, Lower Yangtze River, East China: New and improved insight.' *Quaternary International* 227, 10–21.

Sun, Lin and Menghe Gao. 2006. 'Majiabang wenhuaqu de dili jingguan' (Geographic landscape of the Majiabang culture area). *Huaxia Archaeology* (3), 40–5.

Tan, Y. 2015. 'Taihu xibu pingyuan quanxinshi huanjing yanbian de zuankong jilu' (Pollen records of the Holocene environmental changes in the western Lake Taihu plain). MSc dissertation, Nanjing University.

Tian, M. L., M. H. Hao and H. Peng. 2009. 'Jiangsu liyang shendun yizhi fajue jianbao' (Preliminary excavation report at the Shendun site, Liyang, Jiangsu). *Southeast Culture* (5), 45–59.

Wang, K. F., Y. L. Zhang and H. Jiang. 1984. 'Jiangsu weiting caoxieshan yizhi baofen zuhe jiqi gudili' (Pollen assemblages and geographic environment at the Caoxieshan site, Weiting, Jiangsu). In Pollen Research Group of Institute of Geology & Pollen Research Laboratory of Tongji University (eds), *Pollen Study and the Quaternary Palaeo-Environment*, 78–85. Beijing: Science Press.

Wang, K. F., Y. L. Zhang, H. Jiang and Z. H. Ye. 1980. 'Songze yizhi de baofen fenxi yanjiu' (Pollen study at the Songze site). *Acta Archaeologica Sinica* (1), 59–66.

Wang, Q. and J. Y. Chen. 1999. 'Hongzehu he huaihe ru hongzehu hekou de xingcheng yu yanhua' (Formation and evolution of Lake Hongze and the River Huaihe delta). *Journal of Lacustrine Science* 11, 237–44.

Weisskopf, A., L. Qin, J. L. Ding, P. Ding, G. P. Sun and D. Q. Fuller. 2015. 'Phytoliths and rice: From wet to dry and back again in the Neolithic Lower Yangtze.' *Antiquity* 89 (347), 1051–63.

Wu, B. Y. 1985. 'Nanjing xiashu huangtu chenji tezheng yanjiu' (Characteristics of the sedimentation process of the Nanjing Xiaoshu loess). *Marine Geology and Quaternary Geology* (2), 115–25.

Wu, L., C. Zhu, C. G. Zheng, F. Li, C. M. Ma, W. Sun, S. Y. Li, T. Shui, X. H. Shao, S. X. Wang, Y. Zhou and T. T. He. 2012. 'Quanxinshi yilai Zhejiang diqu shiqian wenhua dui huanjing bianhua de xiangying' (The response of prehistoric cultures to climate environmental changes since the Holocene in Zhejiang, East China). *Acta Geographica Sinica* 67, 903–16.

Wu, R. Z. 1999. 'Majiabang wenhua de shehui shengchan wenti de tantao' (Discussion of the socio-economic production of the Majiabang culture). *Agricultural Archaeology* (3), 31–8.

Xie, L. Y., S. L. Kuhn, G. P. Sun, J. W. Olsen, Y. F. Zheng, P. Ding and Y. Zhao. 2015. 'Labor costs for prehistoric earthwork construction: Experimental and archaeological insights from the Lower Yangzi basin, China.' *American Antiquity* 80 (1), 67–88.

Yang, Y. Z., F. C. Li, Z. H. Cao, M. N. Wang, Z. D. Jin, J. Y. Dai and W. Ran. 2007. 'Kunshan chuodun gu turang lidu tezheng ji muzhi panbie' (Particle sizes of the Chuodun palaeosol and the identification of its parent material). *Chinese Journal of Soil Science* 38 (1), 1–5.

Yang, Y. Z., F. C. Li, Z. D. Jin, M. N. Wang, Z. H. Cao, J. Y. Dai and W. Ran. 2006. 'Chuodun nongye yizhi zhong cunzai zhong quanxinshi shuidaotu de xinzhengju' (New evidence of the presence of middle Holocene paddy soil at the Chuodun archaeological site). *Quaternary Research* 26, 864–71.

Zheng, Y. F., G. W. Crawford, L. P. Jiang and X. G. Chen. 2016. 'Rice domestication revealed by reduced shattering of archaeological rice from the Lower Yangtze valley.' *Scientific Reports* 6, art no. 28136.

Zheng, Y. F., L. P. Jiang and J. M. Zheng. 2004. 'Zhejiang kuahuqiao yizhi de gudao yizhi yanjiu' (Study of the ancient rice remains from the Kuahuqiao site, Zhejiang province). *Chinese Journal of Rice Science* 18, 119–24.

Zheng, Y. F., G. P. Sun and X. G. Chen. 2007. '7000 nianqian kaogu yizhi chutu daogu de xiaosuizhou tezheng' (Characteristics of rice spikelets discovered at a 7,000-year-old archaeological site). *Chinese Science Bulletin* 52, 1037–41.

Zheng, Y. F., G. P. Sun, L. Qin, C. H. Li, X. H. Wu and X. G. Chen. 2009. 'Rice fields and modes of rice cultivation between 5000 and 2500 BC in east China.' *Journal of Archaeological Science* 36 (12), 2609–16.

Zhu, N. C. 2004. 'Taihu yu hangzhouwan diqu yuanshi daozuo nongye qiyuan chutan' (Origins of rice farming in the Lake Taihu and Hangzhou Bay regions). *Southeast Culture* (2), 24–31.

Zhuang, L. N., H. N. Yu, Z. C. Qiu and L. G. Lin. 2017. 'Jiangsu sihong hanjing yizhi 2015–2016 nian fajue shouhuo' (Achievement of the 2015–216 excavation at the Hanjing site, Sihong County, Jiangsu province). State Administration of Cultural Heritage (ed.), *Major Archaeological Discoveries in China in 2016*, 12–15. Beijing: Cultural Relics Press.

Zhuang, Y., W. Bao and C. French. 2016. 'Loess and early land use: Geoarchaeological investigation at the early Neolithic site of Guobei, Southern Chinese Loess Plateau.' *Catena* 144, 151–62.

Zhuang, Y., P. Ding and C. French. 2014. 'Water management and agricultural intensification of rice farming at the late-Neolithic site of Maoshan, Lower Yangtze River, China.' *The Holocene* 24 (5), 531–45.

Zong, Y. Q. 2004. 'Mid-Holocene sea-level highstand along the southeast coast of China.' *Quaternary International* 117 (1), 55–67.

Zong, Y., Z. Chen, J. B. Innes, C. Chen, Z. Wang and H. Wang. 2007. 'Fire and flood management of coastal swamp enabled first rice paddy cultivation in east China.' *Nature* 449 (7161), 459–62.

Zong, Y. Q., Z. Wang, J. B. Innes and Z. Chen. 2012. 'Holocene environmental change and Neolithic rice agriculture in the lower Yangtze region of China: A review.' *The Holocene* 22 (6), 623–35.

Zuo, X. X., H. Y. Lu, L. P. Jiang, J. P. Zhang, X. Y. Yang, X. J. Huan, K. Y. He, C. Wang and N. Q. Wu. 2017. 'Dating rice remains through phytolith carbon-14 study reveals domestication at the beginning of the Holocene.' *Proceedings of the National Academy of Sciences* 114 (25), 6486–91.

Zuo, X. X., H. Y. Lu, J. P. Zhang, C. Wang, G. P. Sun and Y. F. Zheng. 2016. 'Radiocarbon dating of prehistoric phytoliths: A preliminary study of archaeological sites in China.' *Scientific Reports* 6, art. no. 26769.

II
Technologies across time and space

6
Recognition criteria for canals and rivers in the Mesopotamian floodplain

Jaafar Jotheri

Abstract

The ability to distinguish between the remaining traces of rivers and those of canals would greatly increase our understanding of water history and management within a given area. Such an understanding would lead in turn to a greatly enhanced understanding of the landscape, social structure, political life and economy of that area. For the Mesopotamian floodplain, intensive water-management activities, together with the frequent avulsions of the Euphrates and Tigris rivers, have rendered channel networks complex and interlocked. This complexity has long confused researchers in regard to channel origins, and whether they are natural or anthropogenic, or a combination of the two. It is a challenging task, but the present work proposes and discusses seven key differences between the two types of channels, namely topographical cross-sections, crevasse splays, marshes, meandering, cut-offs and oxbow lakes, channel patterns, and stream directions. The discussion is based on geomorphological, remote-sensing, historical and archaeological data. It is concluded that, for a given channel, these differences may be sufficient to establish its origin.

The floodplains of Mesopotamia

The floodplain of the Tigris and the Euphrates is a relatively flat area with a low topographic relief, occupying a part of the foreland basin of the Zagros fold and thrust belt (Garzanti et al., 2016). On one hand, rivers in this region are unstable as they are commonly laterally shifted, completely or partially avulsed, seasonally flooded and occasionally desiccated (Morozova, 2005; Jotheri et al., 2016). On the other hand, this region is considered one of the origins of complex societies and has been continually occupied from the Mid-Holocene; here, people have relied on natural and constructed canals for life, irrigation, transport and fortification (Wilkinson, 2003).

The origin of the ancient channels, whether natural or anthropogenic, in some cases cannot be determined with absolute confidence. This can occur when a canal has a long history of use and becomes similar to a river. However, in the present study, the focus is on channels that are now abandoned and dry, regardless of their history, in other words on the final condition of the channel when it is dried out and on whether it was modified and shaped by human activity or there are indications that it was a natural river.

There are many types of anthropological activities, but several of them have led to a re-formation of the landscape, including the creation of irrigation and trading canals, the cleaning and maintenance of the channels, the opening of channel levees, the reclamation of marshes and the strengthening of channel levees (Walstra et al., 2010; Heyvaert et al., 2012; Husain, 2016). The idea of human utilisation of surface water in southern Mesopotamia was developed over time. It started as a simple means of irrigation, such as the crevasse splay style of the fourth millennium BC, when canals of a few metres were dug to control water to supply a small farm (Wilkinson et al., 2015). This construction process did not end until large irrigation canals, such as the gigantic Nahrawan Canal, extended for hundreds of kilometres across the floodplain during the early first millennium AD.

State interventions in water and irrigation systems would lead to high levels of water management and, as a result, to the flourishing of the state and of rural communities. The challenges presented by river changes led to the development of water management over time. The three main reasons for state intervention are political, to gain more control and to tackle sudden change in the environment (Rost, 2017).

Previous work has dealt with mapping archaeological sites and ancient channels in southern Mesopotamia. It includes Jotheri and Allen (in press), Hrits and Wilkinson (2006) and Pournelle (2012). Satellite images (CORONA and DigitalGlop), historical maps and texts, maps and atlases of archaeological data have also been investigated. The present geomorphological properties of the rivers and irrigation systems have been taken into account. Consequently, in the present study several ancient channel parameters have been suggested as reliable for making distinctions between ancient rivers and ancient canals.

Key differences

In the present study, seven key differences between rivers and canals are discussed: channel patterns, topographic cross-sections, crevasse splays, marshes, meandering, cut-offs and oxbow lakes, and stream directions.

Channel patterns

Ancient rivers and canals are completely different in their patterns as a result of the way they are formed. Rivers have developed four main different types of patterns, namely meandering, braiding, anastomosing and anabranching. Ancient canals display, on landscape imagery, mostly herringbone (Figure 6.1) or dendritic (Figure 6.2) patterns. Herringbone irrigation systems are probably the oldest and most common irrigation system in the southern Mesopotamian floodplain (Wilkinson et al., 2015). This system is the most common and functional irrigation system in the present (Figure 6.3) and across floodplains. However, in some locations, people may have dug canal irrigation systems in dendritic (Figure 6.2) patterns. The dendritic pattern might be selected because of the rough topography

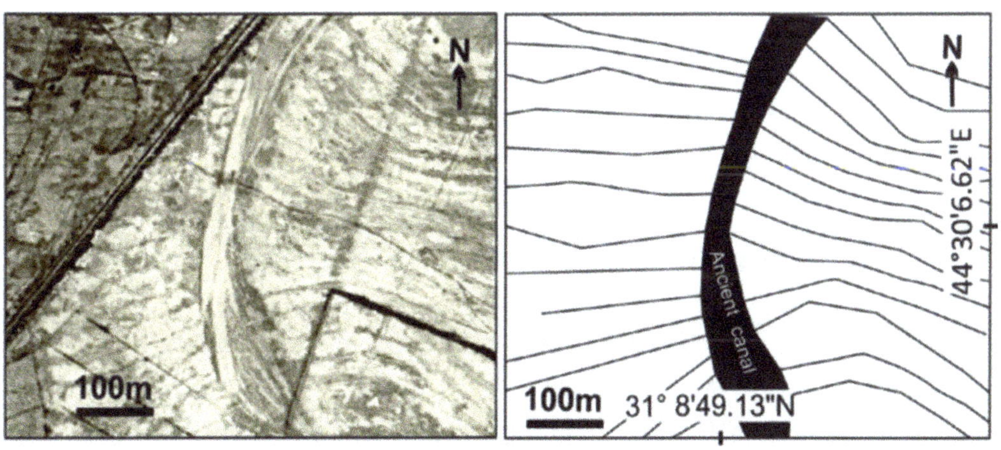

Figure 6.1 Ancient canal in herringbone irrigation system.
Source: author

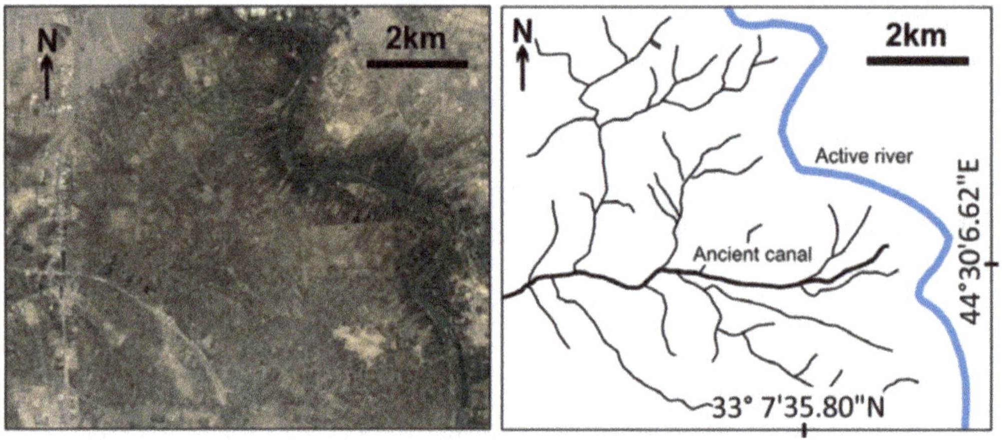

Figure 6.2 Ancient canal in dendritic irrigation system. Source: author

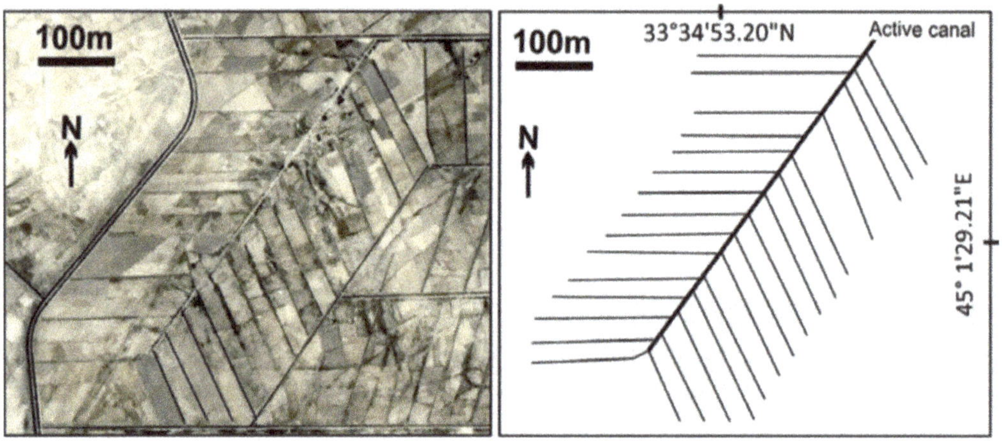

Figure 6.3 Active canal in herringbone irrigation system. Source: author

of the land or because the land is more marshy and frequently flooded; in such places people would have chosen alternative, workable patterns.

In the modern geomorphology of the Tigris and the Euphrates inside the Mesopotamian floodplain, there are no natural branches, as all the current branches are human-induced and controlled by dams or barrages (if it was uncontrolled, a single branch would overwhelm the others because of the 'avulsion' process).

Topographic cross-sections

In general, levees of rivers and canals differ in width and in their degree of slope towards the floodplain. River levees are wider and have a gentle slope, while canal levees are narrower and have a steep slope. These differences between the shapes of the levees mainly depend on how each one was formed.

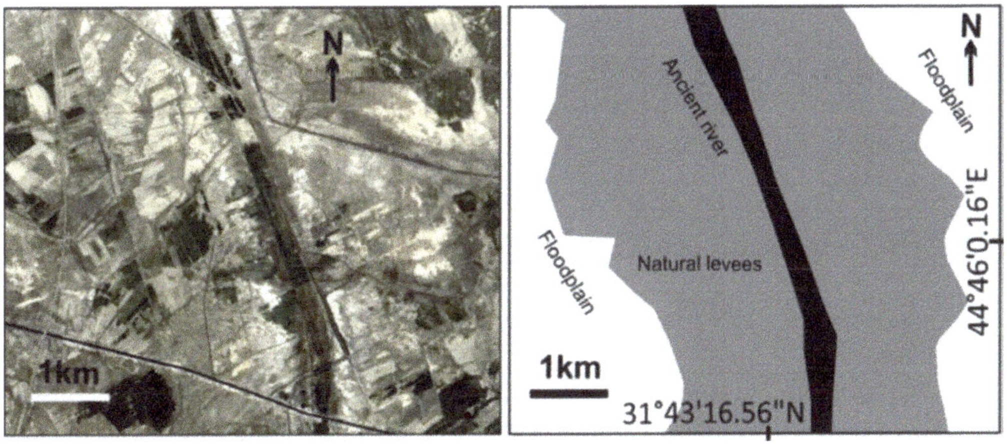

Figure 6.4 Wide natural levees with a gradual slope towards the floodplain. Source: author

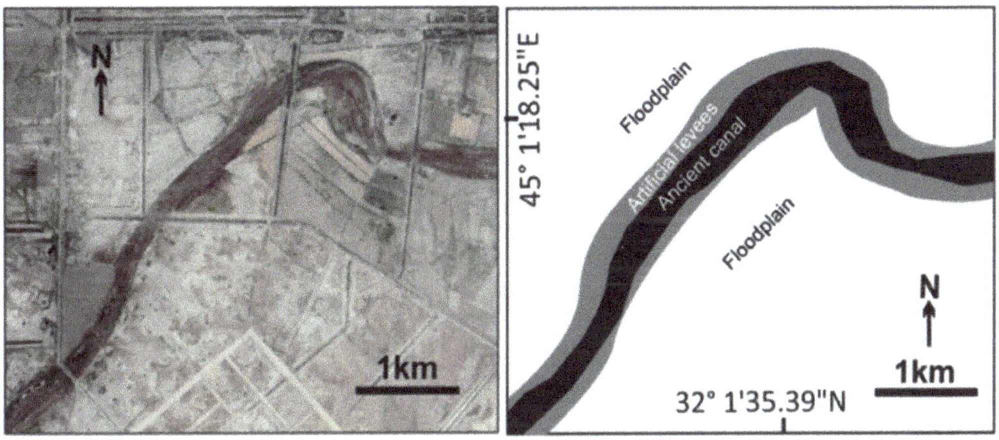

Figure 6.5 Narrow artificial levees with a sharp slope towards the floodplain. Source: author

Natural levees (Figure 6.4) are usually formed as a result of repeated river floods, where the floodwater has loaded more sediment from the river banks and the coarser particles are deposited alongside the channels, forming small elevated banks, while the lighter particles are deposited a long way from the channel that forms the floodplain (see, for example, Mohrig et al., 2000). As this process is repeated, banks (levees) are built up and confine the river.

Canals (Figure 6.5) are usually dug in a relatively flat floodplain area. Water must be confined by levees; otherwise floods will spread across the surrounding area and the purpose of digging the canal will not be served. Therefore, when people dig a canal they normally focus on strengthening the canal sides, including by putting excavated soil along both banks of the canal. They take into account the possibility of flooding or uncontrolled waves of water by increasing the elevation of the bank levees. In addition, because of the silting-up process, canals are frequently cleaned by removing sediments from the bottom and using them to firm the primary levees, and so canal levees are narrower and steeper.

Crevasse splays

One of the clues to recognising an ancient river in the southern Mesopotamian floodplain is crevasse splays, since they are easily recognisable from satellite images (Figure 6.6). Crevasse splays are fan-shaped features formed when the channel levee has been breached during stages of flooding and floodwaters have overflowed the natural levee in the floodplain through swales or breaches (Bristow et al., 1999). They occur close to the channel and in time become a feature of high elevation, but are lower than the channel levees. The flowing stream in a crevasse splay surface forms several small distributaries and braided channels. The formation of crevasse splays depends on flood frequency and the degree of levee strength (that is, they are more likely to be formed when the channel is subjected to flooding and its natural levees have not yet strengthened and become sufficiently elevated). Accordingly, in the past crevasse splays were the location in which people preferred to dig canals and control water in order to create first farms and then settlements.

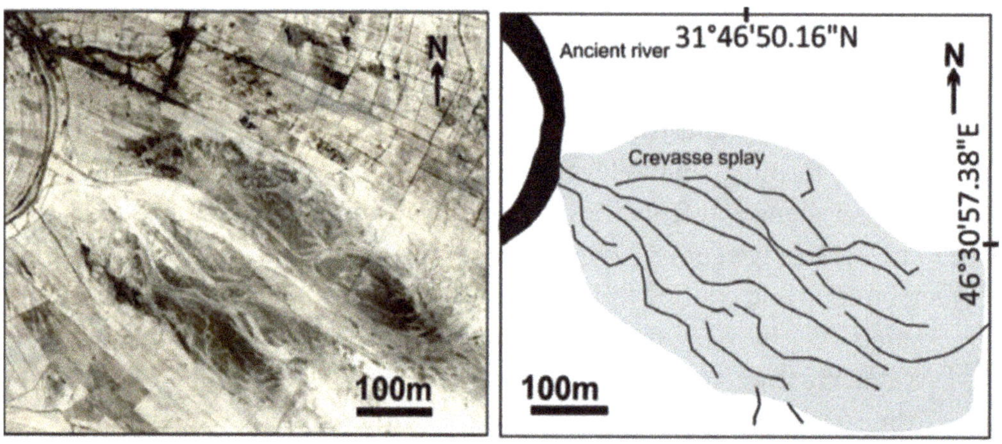

Figure 6.6 Crevasse splay associated with river. Source: author

Marshes

Marshy areas are an indication of a natural river, as they are adjacent to the stream channel which forms when water spreads out from natural levees and banks as a result of floodwaters overflowing the banks. The location of ancient marshes cannot be identified from satellite images, because such low land features are easily buried by subsequent sedimentation and there are no surface signs. However, they can easily be revealed by digging several shallow boreholes or augers in the floodplains near ancient rivers, as the marsh faces are composed of silt to sandy silt, which are rich in shells and charcoal (Jotheri et al., 2017).

Meandering

This is a more common feature identifying a river (Figure 6.7), especially in a floodplain, as channels are apt to move from side to side, forming meanders with respect to a straight course, and leaving behind scars of remains where the river channels once were (Hooke et al., 2011). Canals (Figure 6.7) are less likely to form meanders, and therefore leave no scars. However, occasionally, canals form meanders as a result of long-term use and lack of appropriate maintenance (Figure 6.8). In such cases, meanders are small (that is, their wave lengths and wave amplitude are short).

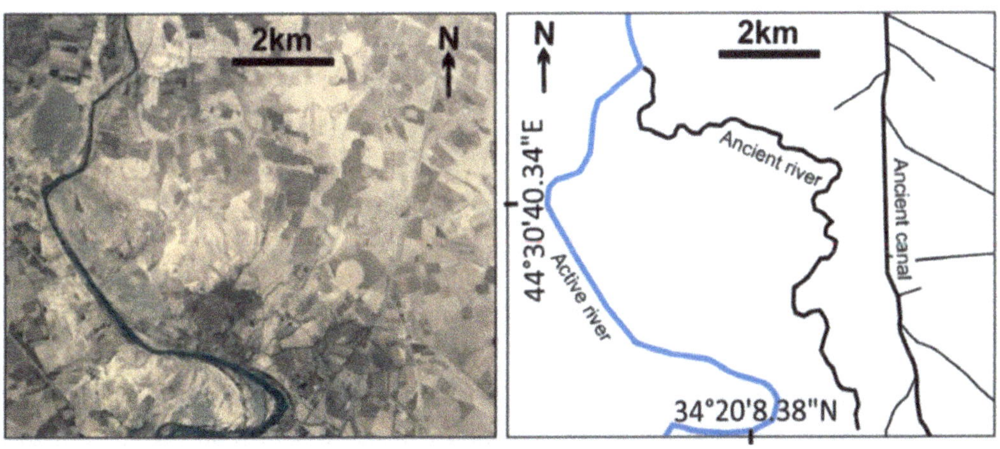

Figure 6.7 Meandering ancient river and straight ancient canal.
Source: author

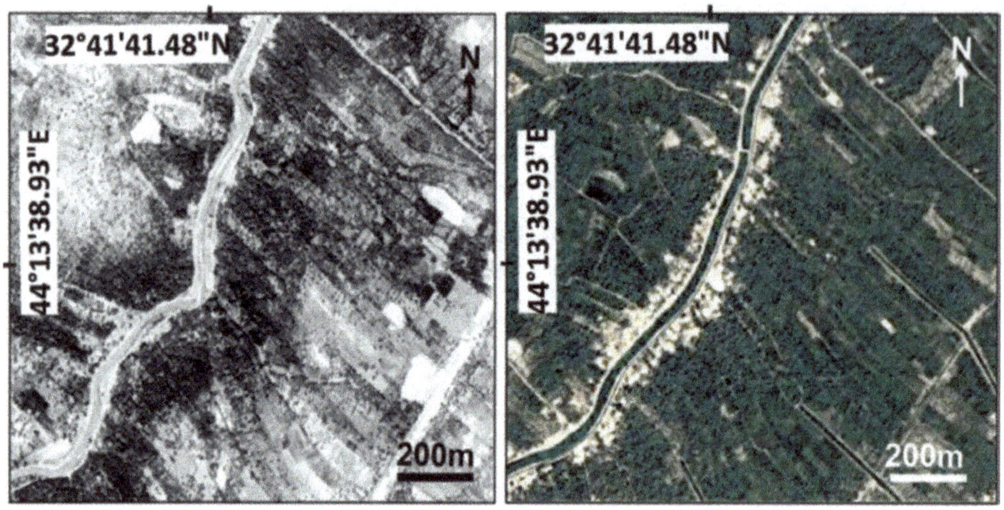

Figure 6.8 CORONA and QuickBird images of the Hindiya canal. Left, CORONA image taken in 1968 of the Hindiya canal. Right, QuickBird image of the same canal in 2006. The canal was dug straight in the nineteenth century (Jotheri et al., 2016) and then became meandering as there was no adequate cleaning operation after it was used. It became less meandering in the present time when cleaning became frequent. Source: author

Cut-offs

These features occur close to a main river (Figure 6.9) and are usually characterised by scars or meander scrolls; they are accompanied by levees of high elevation (Hooke et al., 2011). They can be clearly identified in satellite imagery. Cut-offs are old meander loops isolated from the main channel as a result of lateral shifting of the channel; that is, meanders in rivers are separated by deposition, which leaves behind oxbow lakes.

Stream direction

This characteristic can be used as an indication of a canal rather than of a river, as rivers always follow the general slope direction of the floodplain (Figure 6.10). In several cases, people have dug canal extensions in an area and in a direction that, to some extent, is not parallel to the general gradient of the area (Figure 6.10).

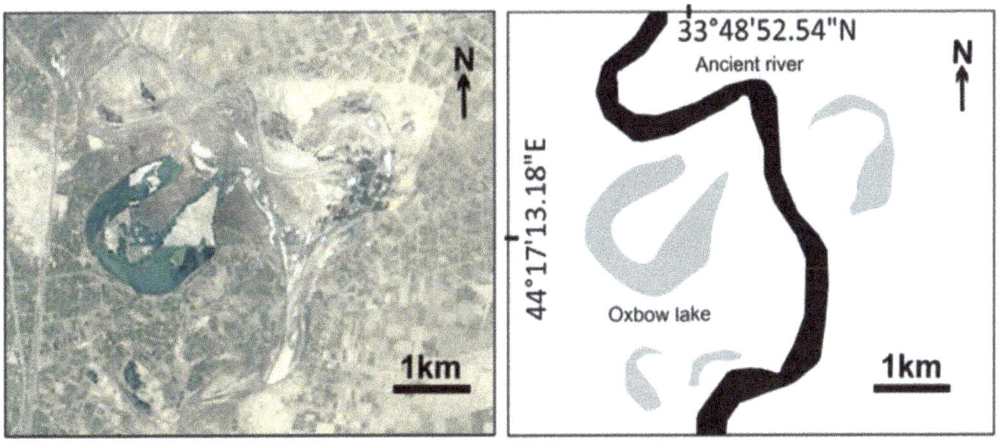

Figure 6.9 Oxbow lakes associated with ancient river. Source: author

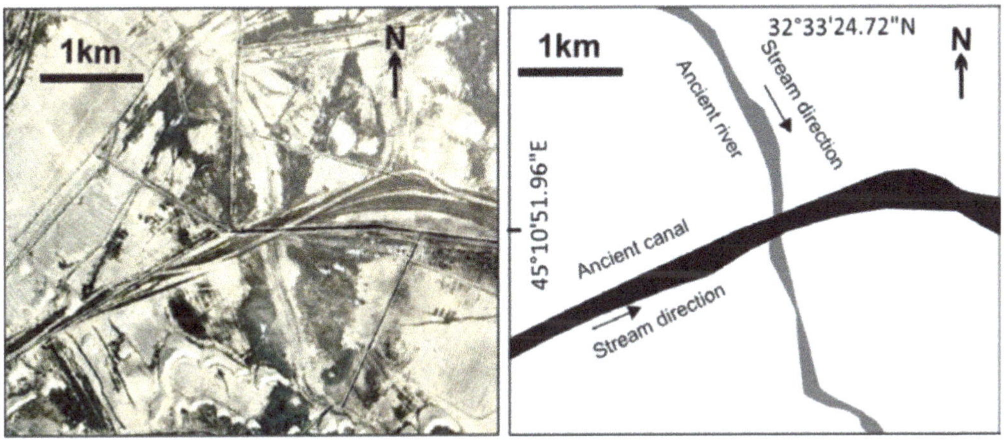

Figure 6.10 An ancient river following the general slope, and an ancient canal crossing the general slope. Source: author

A case study

A case study has been selected to evaluate the key differences between natural rivers and anthropogenic canals discussed earlier. The selected area is located between Diwaniyah and Kufa (Figure 6.11) in the centre of southern Iraq, where the branches of the Euphrates are and were the main factor controlling the hydraulic landscape (Figures 6.11 and 6.12). One of the main reasons for selecting this area as a case study is that changes in the river and irrigation systems in this area are widely described in Ottoman texts and maps. Therefore the natural rivers and canals have already been mentioned in such works. If we tried to classify the channels in this area without looking at texts and maps, it would not

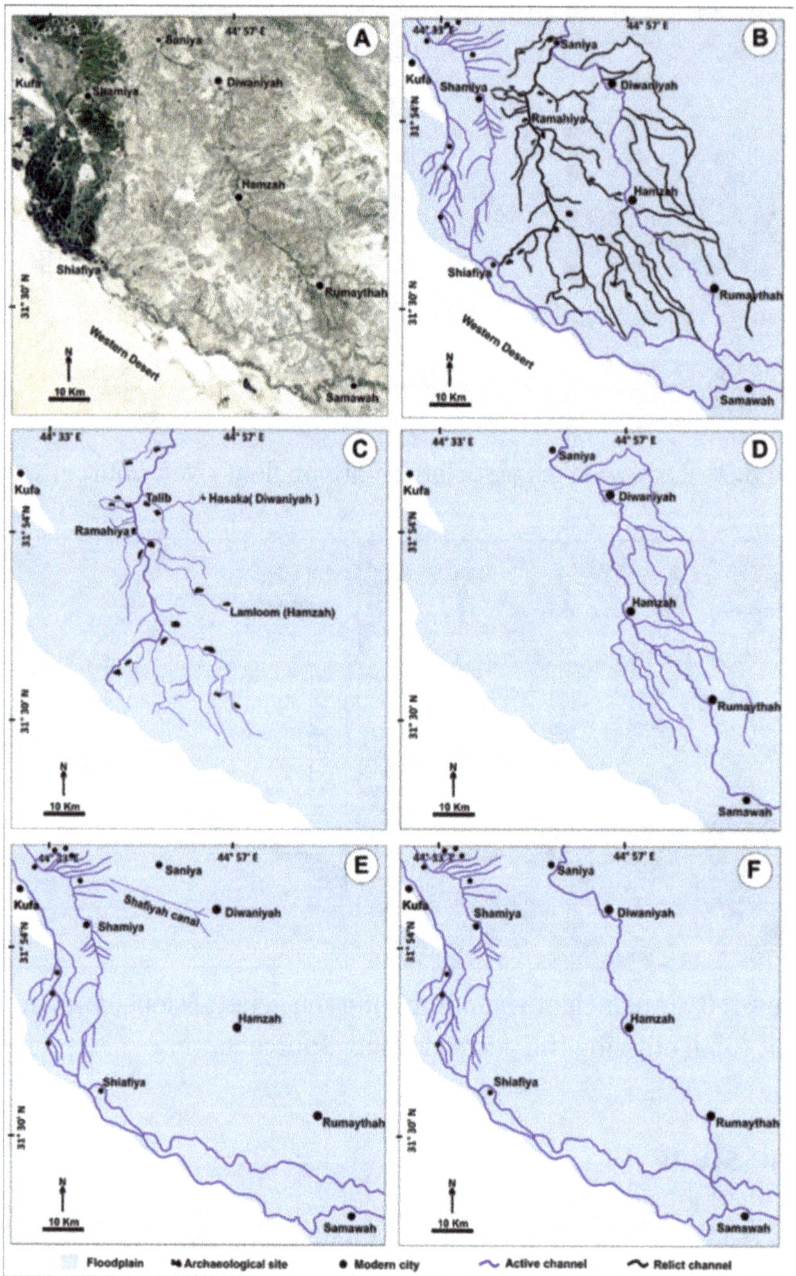

Figure 6.11 The rivers and canals in the case study. A. Satellite image of the area. B. The reconstructed rivers and canals of the area. C. The River Ramahiya and its branches (canals) that were active from the thirteenth century until 1700. D. The River Diwaniyah and its branches (canals) that were active from 1700 until 1905. E. The Kufa and Shamiya canals and their branches (canals) that were active since 1905, and the Shafiyah canal that was dug to irrigate the town of Diwaniyah. F. The current rivers and canal after the construction of the Hindiya barrier upstream of Euphrates. Source: author

be easy to distinguish between natural and anthropogenic origins. However, the key differences described help to distinguish them.

For the purposes of this chapter, only river changes in the Euphrates from the late Islamic period until the present day are taken into account, because the palaeo-channels, avulsions and hydraulic landscape of this area before the late Islamic period are discussed in detail elsewhere (Jotheri et al., 2016).

In this case study, archaeological sites and palaeo-channels have been identified and traced (Figure 6.11B) using SRTM, CORONA and QuickBirds satellite imagery and data, in addition to ground truthing carried out at several locations in the area. Historical, mainly Ottoman, texts and maps have also been reviewed.

During the Ottoman period, the agricultural and irrigation systems in the Mesopotamian floodplain were controlled by low-skilled rural community leaders such as the heads of tribes (sheikhs), peasants and local cultivators. These rural people were the most active of the players who triggered river avulsion in this part of the Euphrates. In fact, the Ottoman central bureaucratic administration had no plans or rules for irrigation systems but relied on these people's knowledge of irrigation. For example, the Ottomans gave power to tribal heads to organise labour and to suggest irrigation plans to cultivate their land, without questioning or supervising the plans.

Although it was up to rural communities to manage their water systems, the Ottoman administration encouraged peasants and sheikhs to focus on the irrigation system, which would lead to an increase in the area of cultivated land, in cases where the latter would raise tax revenues. Conversely, any work (digging or maintenance) on a waterway system that only led to villages or cities being fed was not a priority. For this reason, rivers or canals that were not used mainly for irrigation were more likely to be subject to silting up with sand and reeds. People often complained about this to the Ottoman administration, but rarely received a reply.

One of the most significant causes of conflict among ordinary farmers was the downstream flow of a canal becoming too slow and therefore incapable of supplying enough water to farms, because of excessive consumption of water upstream. Another common reason for complaint was the canal becoming silted up as a result of not having been cleaned for two or three years or more, which would lead to flooding of the upstream land and desiccation of the downstream land.

Every tribe that benefited in one way or another from a canal would be responsible for digging, cleaning and reinforcing the canal's banks along a given section. In the summer the regular water-work of a rural community was the construction of weirs across the channels to raise the water level and then the digging of canals to irrigate crops. In winter, strengthening channel levees and damming the river crevasse splays to prevent flooding were paramount.

In 1687, a farmer called Dhiyab dug a canal from the eastern bank of the River Hilla in Ciniyah village to irrigate his land, and the river began to avulse as a new channel had started to form, taking water from the main river. The new branch was called 'the Dhiyab', 'the Rashadi', 'the Hasaka', 'the Hilla' and 'the Diwaniyah', while the original channel was called 'the Ramahiyah' or 'the Ciniyah' (Husain, 2014; Jotheri et al., 2016). The avulsion was completed in 1700; the Ramahiyah dried out, while the Diwaniyah branch flooded and its floodplain became extensive marshes covering both sides of the new branch (Husain, 2016).

The main consequence of this avulsion was that the tribes that were living on the River Ramahiyah and its canal branches migrated from their farms and towns (Figures 6.11C, 6.12 and 6.13) and settled in three

Figure 6.12 Ramahiyah archaeological sites and the River Ramahiyah. See Figure 6.11C for location. Source: author

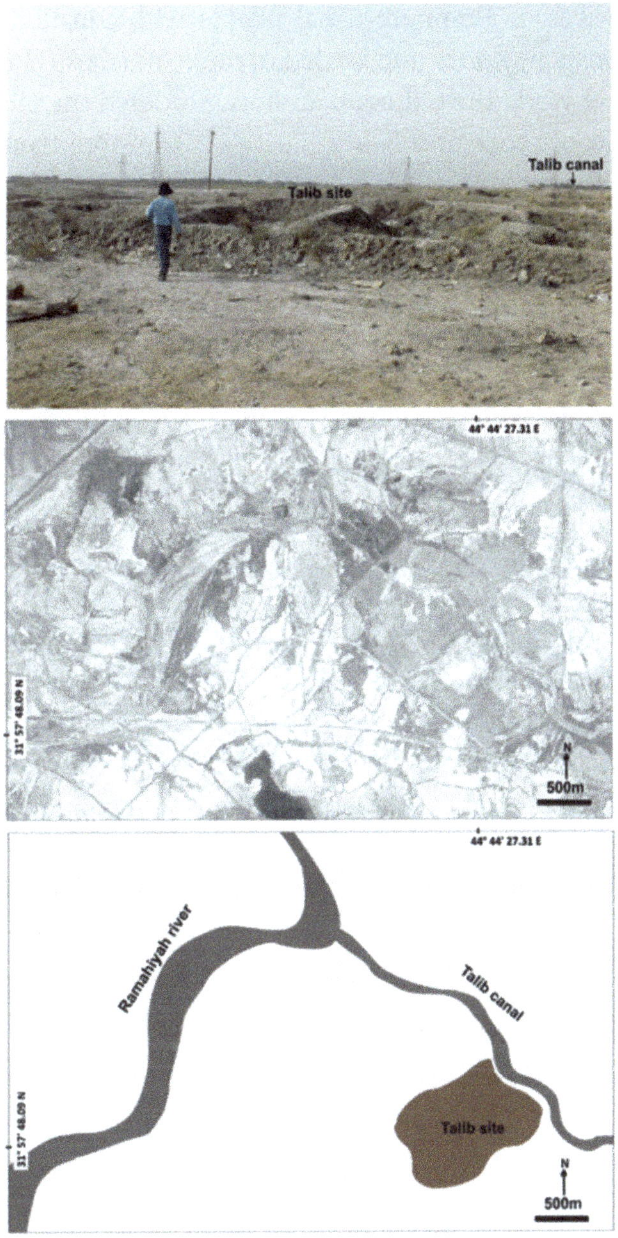

Figure 6.13 River Ramahiyah and the Talib site and canal. See Figure 6.11C for location. Source: author

areas: upstream of the avulsion node (i.e. to the north of Ciniyah village), near the new Diwaniyah branch and marshes (i.e. around Hasaka and Lamloom villages (Figures 6.11C and 6.11D), and around Kufa (Figure 6.11D).

In 1701–2, the Ottoman authority ordered the restoration of the avulsion through the cleaning of the abandoned Ramahiyah branch and the damming of the new Diwaniyah branch and its marshes, in order to keep Ramahiyah and its branches active. The other purpose of restoring the avulsion was to eliminate tribal people who constantly protested against Ottoman authority and settled in these marshes (Husain, 2014). For such water-work, the Ottoman authority ordered the local people to use large rolls of reeds and heavy pieces of palm-tree trunk, and they dumped sandbags to block the head of new channel. However, all their efforts failed to prevent the new river from continuing its ongoing avulsion and feeding the marshes. Therefore, the Ottoman authority ordered the people to block the head of the crevasse splay and all the canals taking water from the Diwaniyah branch to its floodplain. Several drain canals were dug parallel to the Diwaniyah and across to the crevasse splays and the irrigation canal to siphon water off the marshes. However, these efforts were not enough to dry out the wide marsh area rapidly and were also hampered by the frequent flooding of the River Hilla, the main branch of the Euphrates at that time. Therefore, the Ottomans continued with the same type of water-work, which led to a reduction in the discharge of water, and then the silting up the river. As a result, water diverted from upstream of Hilla started to form the Kufa and Shamiya branches. Since that time, no floods have been recorded in any season of the year, and the Hilla branch discharge has reached its lowest level. Consequently, the country-dwellers and the marsh community who had been living and thriving around Diwaniyah, Hamzah and Rumaytha for a decade downstream from the Hilla began to leave the area and spread out in other areas within southern Mesopotamia.

According to the historical texts and maps studied by Husain (2014, 2016) and Jotheri et al. (2016), including this present study, the natural channels are:

1. the main channel of the Ramahiyah (Figure 6.11C)
2. the main channel of the Diwaniyah (Figure 6.11D)

while the canals are:

1. the branches of the Ramahiyah (Figure 6.11C)
2. the branches of the Diwaniyah (Figure 6.11D)
3. the main channel of the Euphrates at Kufa and Shamiya, and its branches (Figure 6.11E, F)
4. the Shafiyah Canal (Figure 6.11E).

Having examined the key differences between naturally occurring rivers and canals, we have found that the channel patterns of the canals mentioned are herringbone and dendritic, with narrow topographical cross-sections, a lack of associated crevasse splays and a lack of adjacent marshes; they generally show a low rate of meandering and a lack of associated cut-offs or oxbow lakes, and some of the streams do not flow with the general gradient. In contrast, the natural rivers mentioned have no specific channel pattern, wide topographical cross-sections; associated with them are crevasse splays and marshes, meandering, and frequent cut-offs or oxbow lakes, and their stream directions are normally in line with the general gradient.

Conclusions

As the present study has demonstrated, the process of distinguishing between rivers and canals may require a closer consideration of geomorphological features and surface properties. From an examination of key differences in channel patterns, topographical cross-sections, crevasse splays, marshes, meandering, cut-offs and oxbow lakes, and the directions of streams, it does appear to be possible to determine whether the origin of a channel is natural or anthropogenic. An additional conclusion is that these key differences can be applied to any area within the Mesopotamian floodplain for the purpose of distinguishing between these features.

References

Bristow, C. S., R. L. Skelly and F. G. Ethridge. 1999. 'Crevasse splays from the rapidly aggrading, sand-bed, braided Niobrara River, Nebraska: Effect of base-level rise.' *Sedimentology* 46 (6), 1029–48.

Garzanti, E., A. I. Al-Juboury, Y. Zoleikhaei, P. Vermeesch, J. Jotheri, D. B. Akkoca, A. K. Obaid, M. B. Allen, S. Andó, M. Limonta, M. Padoan, A. Resentini, M. Rittner and G. Vezzoli. 2016. 'The Euphrates-Tigris-Karun river system: Provenance, recycling and dispersal of quartz-poor foreland-basin sediments in arid climate.' *Earth-Science Reviews* 162, 107–28.

Heyvaert, V. M. A., J. Walstra, P. Verkinderen, H. Weerts and B. Ooghe. 2012. 'The role of human interference on the channel shifting of the river Karkheh in the Lower Khuzestan plain (Mesopotamia, SW Iran).' *Quaternary International* 251, 52–63.

Hooke, J. M., E. Gautier and G. Zolezzi. 2011. 'River meander dynamics: Developments in modelling and empirical analyses.' *Earth Surface Processes and Landforms* 36 (11), 1550–3.

Hrits, C. and T. J. Wilkinson. 2006. 'Using Shuttle Radar Topography to map ancient water channels in Mesopotamia.' *Antiquity* 80, 415–24.

Husain, F. 2014. 'In the bellies of the marshes: Water and power in the countryside of Ottoman Baghdad.' *Environmental History* 19 (4), 638–64.

Husain, F. H. 2016. 'Changes in the Euphrates River: Ecology and politics in a rural Ottoman periphery, 1687–1702.' *Journal of Interdisciplinary History* 47 (1), 1–25.

Jotheri, J. and M. B. Allen. In press. 'Recognition of ancient channels and archaeological sites in the Mesopotamian floodplain using satellite imagery and digital topography.' In D. Lawrence, M. Altaweel and G. Philip (eds), *New Agendas in Remote Sensing and Landscape Archaeology in the Near East: Studies in Honor of T. J. Wilkinson*. Chicago: Oriental Institute of the University of Chicago.

Jotheri, J., M. B. Allen and T. J. Wilkinson. 2016. 'Holocene avulsions of the Euphrates River in the Najaf area of western Mesopotamia: Impacts on human settlement patterns.' *Geoarchaeology* 31 (3), 175–93.

Jotheri, J., M. Altaweel, A. Tuji, R. Anma, B. Pennington, S. Rost and C. E. Watanabe. 2017. 'Holocene fluvial and anthropogenic processes in the region of Uruk in southern Mesopotamia.' *Quaternary International* 483, 57–69.

Mohrig, D., P. L. Heller, C. Paola and W. J. Lyons. 2000. 'Interpreting avulsion process from ancient alluvial sequences: Guadalope-Matarranya system (northern Spain) and Wasatch Formation (western Colorado).' *Geological Society of America Bulletin* 112 (12), 1787–1803.

Morozova, G. S. 2005. 'A review of Holocene avulsions of the Tigris and Euphrates rivers and possible effects on the evolution of civilizations in lower Mesopotamia.' *Geoarchaeology* 20 (4), 401–23.

Pournelle, J. R. 2012. 'Physical geography.' In H. Crawford (ed.), *The Sumerian World*, 13–32. London: Routledge.

Rost, S. 2017. 'Water management in Mesopotamia from the sixth till the first millennium B.C.' *WIREs Water* 4 (5), e1230.

Walstra, J., V. M. A. Heyvart and P. Verkinderen. 2010. 'Assessing human impact on alluvial fan development: A multidisciplinary case-study from Lower Khuzestan (SW Iran).' *Geodinamica Acta* 23 (5–6), 267–85.

Wilkinson, T. J. 2003. *Archaeological Landscapes of the Near East*. Tucson: University of Arizona Press.

Wilkinson, T. J., L. Rayne and J. Jotheri. 2015. 'Hydraulic landscapes in Mesopotamia: The role of human niche construction.' *Water History* 7 (4), 397–418.

7

The Udhruh region: A green desert in the hinterland of ancient Petra

Mark Driessen and Fawzi Abudanah

Abstract

This chapter presents the preliminary results of an ongoing fieldwork project in the region of Udhruh (southern Jordan). It focuses on and discusses the ancient agro-hydrological activities and practices of the study area. First it gives an introduction about the history of settlements (with historical and archaeological evidence), and about the environmental and geoarchaeological settings. The second part of the chapter discusses the archaeological results pertaining to the ancient water-harvesting systems, together with the related agriculture fields, and the integrated technical and interdisciplinary approaches required to study them further.

> In the thirty-second year of the reign of ... Flavius Justinianus, ... three days before the Kalends of January, in the four hundred and [fifty]-third year of the province .... To the most respectable Flavius Valens, son of Auxolaos, tax collector of the current seventh indiction and through you to the present and future tax collectors of this city of the Petraeans ... I sold to the most God-pleasing Philoumenos, son of Geriontos, ... [one] well-watered field that belongs to me [in] the hamlet [ ]aina, near Augustopolis, called Mal-el-Amoa[?] or Mal-[al]-Etherro[ ]eïba, with every right, and I surrendered corporeally its possession. ... (I,) the above-written Theodoros, son of Obodianos, have requested that ... my person and property and account be relieved of the tax contribution assigned to me for the above-written well-watered field. ... from the total landholdings of

Augustopolis each [year, with a plot size of] one (and) one ninth *iugerum* of the (imperial) *patrimonium*.

Petra Papyrus 25, 30 January AD 559.

Introduction

Access to fresh water is one of the greatest global challenges of the twenty-first century. Scholars from different fields of research around the world are dealing with the ever-growing demand for, and the severe supply constraints upon, water. Rapid population growth and changing climatological conditions, especially in some of the most water-scarce regions of the world, result in increasing pressures on already overexploited water resources. This is especially the case in the arid and semi-arid parts of the Middle East, North and sub-Saharan Africa, where almost all the water is used for agricultural purposes (see Gleick, 2014: 227–35, Table 2). Not only is the annual precipitation in such regions low and poorly distributed over the crop-growing seasons, but a large part of it gets lost before it becomes available for agricultural use. In semi-arid regions of sub-Saharan Africa, between 30 and 50 per cent of the precipitation evaporates from the soil surface and after shallow infiltration, and can be considered non-productive water loss (Rockström and Falkenmark, 2000: 335, and references therein). A productive transpiration flow of 15 to 30 per cent of the precipitation re-enters the atmosphere via the stomata in the leaves of the vegetation coverage, while 10 to 25 per cent runs off and 10 to 30 per cent replenishes the groundwater through deep percolation. Different methods and techniques of water harvesting are employed to increase the availability of fresh water to agricultural crops in arid and semi-arid regions (Oweis et al., 2012: 3–71).

It is clear from archaeological research that ancient societies were dealing with similar problems. This chapter will shed some insights into how people practised water procurement for agricultural purposes in the hinterland of Petra (Jordan) in antiquity. After several years of archaeological fieldwork, the authors can already state that the research area around the village of Udhruh is one of the most complete and best-preserved field 'laboratories' available for studying the long-term development of innovative water-management and agricultural systems in southern Jordan, between the first century BC and the tenth century AD (Nabataean, Roman, Byzantine and early Islamic periods). Although the Udhruh Archaeological Project is still ongoing, we would already like to present some preliminary results about the ancient agro-hydrological systems.

The antique settlement of Udhruh and its environs

Background and research

The village of Udhruh, east of Petra, was an almost forgotten archaeological site until Fawzi Abudanah drew attention back to it with his large-scale surveys (Abudanh, 2006). Earlier explorations and excavations revealed that Udhruh had housed an important Nabataean settlement and a Roman legionary fortress.[1] The current village of Udhruh is dominated by and centred on the still-standing remains of this fort, which was transformed into a town in post-Roman times. Classical literary and archaeological sources point to a long-term development from Nabataean to Islamic times (Fiema, 2002; Kennedy and Falahat, 2008). The archaeological variety and perfect preservation of the area surrounding Udhruh were, in combination with the intriguing site itself, essential criteria for starting a joint archaeological project between Leiden University and the Al-Hussein Bin Talal University in 2011. The Udhruh Archaeological Project started with geographic information system (GIS)-related field surveys and small-scale excavations. This was carried out with the aim of mapping and interpreting the still-visible, standing archaeological remains and of reconstructing the geomorphology of a 48 km^2 landscape in the Udhruh region. Five years of inventory fieldwork (2011–15) revealed an actively exploited region with impressive and ingenious investments in agro-hydrological intensification, in building material procurement, communication and security networks, in military dominion and in settlement development (see Figure 7.1).

Environment and landscape

Udhruh is situated at approximately 1,200 m above mean sea level, and approximately 13.5 km east of the Nabataean city of Petra. Physiographically, the western part of the Udhruh region is located at the eastern edge of the Jabal ash-Sharah. The area is also known as the land of Edom; its height ranges from 1,500 to 1,200 m at the eastern side, extending from the Ras an-Naqb escarpment overlooking the Hisma Desert in the south to the Wadi al-Hasa canyon in the north (Cordova, 2007). The regional part of these highlands is called the Jabal ash-Sharah. The eastern part of the research area is situated in the Eastern Highland Zone.

The Udhruh environs are at the boundary of two climate zones, according to Köppen's classification: cool temperate Mediterranean climate (Csb) on the western side and cool semi-arid climate (Bsk) at

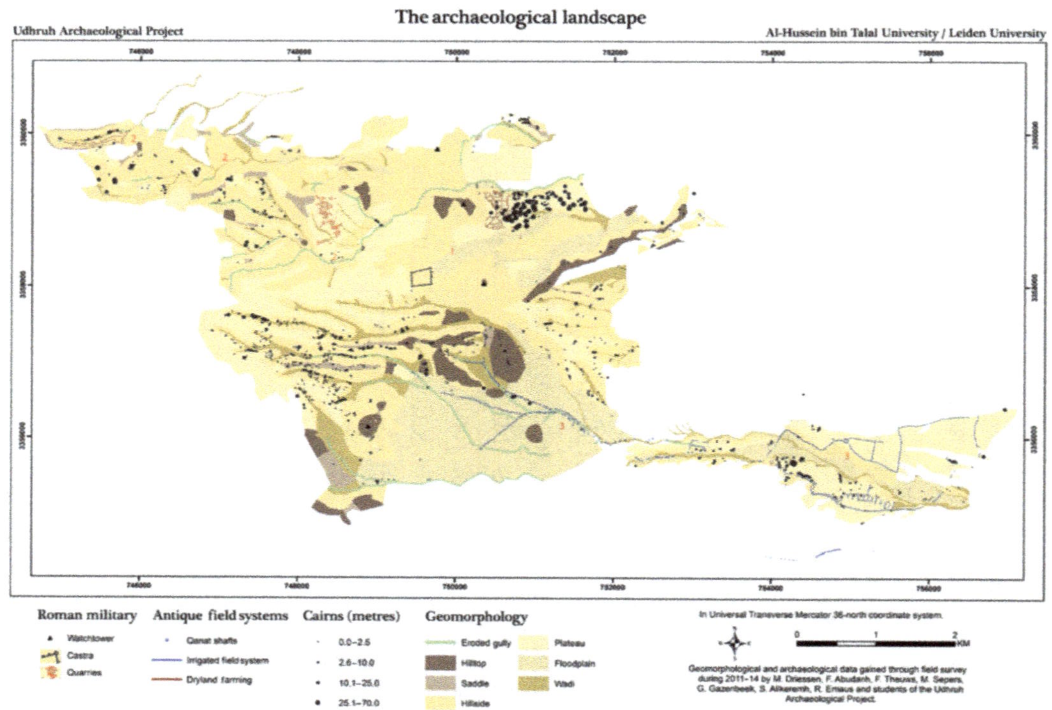

Figure 7.1 Geomorphological map of the 48 km^2 research area of the Udhruh archaeological project, showing only Roman military structures, ancient hydro-agricultural systems and cairns, based on the 2011–2015 field seasons. Three ancient agro-hydrological systems: 1. Plots irrigated by the Udhruh spring. 2. Floodwater harvesting in and towards the Jabal ash-Sharah, with the locations of the three settlements labelled '2' on the map. 3. The Udhruh *quanat* scheme. (Illustration by Roeland Emaus.)

the eastern part of the research area. The mean temperatures range from approximately 10–15 °C in winter to 30–35 °C in the summer period. In the winter it can snow, and temperatures can reach below 0 °C. Conversely, temperatures over 40 °C are not rare in the summer. Nowadays, the Udhruh region receives from 50 to 150 mm of rainfall annually, with higher rainfalls (150–200 mm) in the Jabal ash-Sharah than in the Eastern Highlands (50–100 mm) (Kouki, 2012: 104–5). This precipitation, which is sometimes received in the form of intense but short-lived downpours, falls predominantly in the months of January and February. Because nowadays the wadis lack a man-made control of these downpours, they turn into rapidly changing erosion gulleys. Afterwards the evaporation can be substantial, because of the intensity of the sun and the strong, predominantly dry, western wind coming from the Wādi Arabah (Great Rift Valley). As a result of short-lived cloudbursts, the present-day absence of man-made catchment structures, and the high infiltration capacity of some of the valley beds, only limited amounts of run-off water remain available for, for example, agricultural purposes (Bull and Kirkby, 2002).

From a climatic point of view, we can expect no dramatic changes if we compare the conditions from the Iron Age to the Byzantine period with those in present times, although it would have been more humid in those days than now. A wet phase in the classical period was followed by a drier phase in Islamic times, as was shown for instance in the research in Petra and the nearby Wadi Faynan Landscape Survey (for Petra see Besançon, 2010; for Wadi Faynan see Hunt et al., 2007 and Gilbertson et al., 2007: 406–9; for a wider geographic context see Finné et al., 2011).

The bedrock geology of the region is dominated by Cretaceous and Tertiary limestones. The western ridge towards Petra and the highest part of the water catchment area consist of outcrops of the oldest rock formations: Turonian Wādi As Sir Limestone, a dolomitic limestone deposited in a marine subtidal environment, and the Coniacian–Santonian Wādi Umm Ghudran Formation, deposited as pelagic chalk (Kherfan, 1998). The Wādi As Sir Limestone Formation is one of the highest main aquifers for the region, the topmost of which is several hundred metres below the current surface. More elevated smaller aquifers can be expected in the region, but to locate them specific geo-hydrological research is required. The western part of our research area predominantly contains Amman Silicified Limestone and Al Hisa Phosphorite formations (Santonian–Maastrichtian), consisting of phosphatic chert and limestone, concretionary chert and coquina beds. In some western areas, including Udhruh and the higher eastern parts, Muwaqqar Chalk Marl Formations

(Maastrichtian–Paleocene) dominate, while in the lower eastern areas Pleistocene fluvial, aeolian and Holocene alluvial sediments prevail. Some of the remarkable territorial markers and outcrops in the region – on top of which Nabataean-Byzantine fortlets and watchtowers were located (Driessen and Abudanah, 2013, in press) – are dominated by Umm Rijam Chert Limestone Formations (Paleocene–Eocene).

The main soil types in the Udhruh region are Inceptisols (Xerochrepts) and Aridisols (Calciorthids), while Entisols (Torriorthens and Torrifluents) are found sporadically. In relation to soil fertility and possible plant growth, such Inceptisols tend to suffer from low organic matter and nitrogen contents, while the Aridisols in the (semi-)arid regions of Jordan generally have high salt, gypsum and carbon contents, which makes them not very suitable for cultivation (Lucke et al., 2013: 76).

Literary and epigraphic sources

Udhruh is mentioned in several ancient literary references, from Roman times to the early Islamic period. The second-century AD *Geographia* (5.16) by Ptolemy is the first in which the settlement is called by its local Greek name, *Αδρου*. An early fourth-century building inscription from the legionary fortress at Udhruh does not give us the name of the settlement, but makes it clear that this Roman military camp was rebuilt by the Legio VI Ferrata in AD 303–4 (Kennedy and Falahat, 2008: 159–60). Several Byzantine literary sources tell us that Udhruh became an important town, again under its local names, *Αδροα* and *Αδαρα*.[2] The first one is from the sixth-century Beersheva Tax Edict, which states that Udhruh was assessed for a sum of 65 golden coins (*nomismata*), and thus paid the highest tax of the 18 towns of the province of Palaestina Tertia. The contemporary Stephen of Byzantium (18.18) mentions Udhruh, using its local name, as a large city in this Byzantine province. Udhruh is also known by the honorific name of Augustopolis in several bishop and town lists, relating to the fifth–early seventh centuries (Fiema, 2002: 209–10; it is generally accepted that Udhruh can be identified with the Byzantine Augustopolis; see Abel, 1938; Koenen, 1996; Nasarat et al., 2012: 111). Augustopolis is mentioned in 11 of the 49 Petra papyri.[3] The sixth-century Petra papyri not only make it clear that Petra remained an important administrative centre after the devastating earthquake of AD 551, but they also provide very interesting socio-economic, agricultural and personal information on the region, its hinterland, and the people who lived and worked here. Most of the papyri are contracts relating to the transition of property, mostly real estate, movables and even slaves, through sale, dowry, donation, inheritance and division; and some are juridical documents relating to disputes or registration of assets (Arjava, et al., 2007; Arjava et al., 2011; Frösén et al., 2002; Koenen et al., 2013). The Petra papyri provide us also with detailed information on cultivated crops, allotment sizes, water-management distribution and the different terms pertaining to structures, dwellings and agricultural use. People owned holdings, as well as related houses and structures in Augustopolis and its environs. The papyri also inform us that this Byzantine town was an administrative centre, where properties were registered, sharing the responsibility of tax collection with Petra (see Petra papyri 19, 25, 30 and 31, Arjava et al. 2007). Udhruh remained a prosperous town after the Byzantine period and is mentioned in several Islamic sources (al-Salameen et al., 2011: 239–40). The people of Udhruh agreed to pay a tribute of 100 dinars to Mohammed and his armies in exchange for a peace treaty in AD 630 (Fiema, 2002: 210 and references therein). After the Arab conquest Udhruh had a special position; an important arbitration took place there between competing Muslim parties in AD 657. This resulted in the establishment of the Umayyad state, and 'Adhruh' is mentioned as the major town of the Umayyad province of Ash-Sharah in the eighth century AD (al-Salameen et al., 2011: 233; Fiema, 2002: 210 and references therein).

Ancient water-harvesting methods in the Udhruh region

After five years of predominantly survey and exploratory fieldwork, more detailed archaeological campaigns are planned over the next few years. Three ancient agro-hydrological systems have been distinguished in the Udhruh region (Figure 7.1). These are systems that harvest water from different sources and for different purposes, although agricultural use seems to be the prevailing aim.

The central research question for the project is: What water-harvesting and agricultural adaptations can be observed in the hinterland of Petra that contributed significantly to the transformation and development of this semi-arid landscape in ancient times?

It is our aim through this project to improve the understanding of the hydro-technological innovations and major societal transformations that took place from Nabataean to Islamic times, in the wider region. It has been observed that large parts of these ancient water-harvesting and agricultural field schemes are still largely intact and partly buried by alluvial deposits. An interdisciplinary approach combining archaeological analyses, antique written sources, geophysical, geo-hydrological, socio-hydrological, bioscientific and biogeochemical soil studies will provide us with the necessary data for an integrated analysis, specifically of the ancient water-management and agricultural development of the Udhruh region and the role it played against a local, regional and supra-regional setting.

The following sections present some preliminary results concerning the three water-management systems.

The perennial spring 'Ain Udhruh and associated fields

Udhruh hosted one of the most reliable perennial springs – 'Ain Udhruh – for the entire region. Its water most probably comes from the undulating slopes of the Jabal ash-Sharah and Zebyriyā ridge towards Petra, further fed by tributaries from the Wādī Mulqan and Wādī al-Rumayi, and was transported by aquifers from the Wādi As Sir Limestone and possibly higher-lying formations. This spring was the prime location factor for early occupation, and can be linked to the continuity of human activity pre-dating the Persian period (Abudanh, 2006: 201). Udhruh housed an important Nabataean settlement and most likely developed as a second Nabataean nucleus in the hinterland of Petra, before it was redesigned as a Roman military base (for the Nabataean settlement of Udhruh; see Glueck, 1935: 76–7; M. Killick, 1990; Tholbecq, 2013: 299; Driessen and Abudanah, in press). The Nabataean people did transform the steppe region around Petra into an agricultural landscape consisting of new settlement, water-harvesting and construction works, and arable fields.[4] The beginning of Nabataean sedentary life and agricultural development can be dated to the late second–early first centuries BC (al-Salameen, 2004: 134; Kouki, 2012: 129; Schmid, 2001: 370–1; Stucky, 1996: 14–17). The Udhruh settlement and its perennial source of water were of pivotal strategic importance to the Nabataeans as they constructed an elaborate communication system for the whole region. By the time they had constructed the system, all separate watchtowers and fortlets had a direct visual connection with the higher parts of this settlement (Driessen and Abudanah, 2013, in press). This was, however, a multi-purpose system: it not only applied to military and trade objectives, but also played a role in controlling and safeguarding the newly established agro-hydrological intensifications in and around Udhruh (Driessen and Abudanah, in press). This communication system was kept in use for the protection of field systems established later, during the Roman and Byzantine periods.

The spring of Udhruh was an important factor in the choice of location for the Roman camp (for locational analyses of Roman military sites based on classical sources and archaeology, see Driessen, 2007: 28–35 and tables 2.1–3). In Roman times access to this water resource was at the north-east side of the fortress, where a natural depression leads to the present-day spring. This connection to the spring and to the control of this important water source is most probably the reason why this side of the *castra* has an atypical trapezoidal shape. Another unusual feature that struck us immediately was the slope on which the fort was built. These somewhat odd characteristics were necessary both to incorporate the source of water, and to provide a territorial marker connecting to all the watchtowers in the surrounding region. The layout of the fortress, therefore, was thought out very well. The towers were constructed with the intention that they would withstand potential earthquakes, and with measures to enable good air circulation. There were even large cisterns in some of them. The south-east corner tower used to house a large antique cistern, which was in use until recently, when it was closed because of the risk to playing children. A cistern excavated by us in part of the eastern gate towers was covered up again for the same reason a few years ago.

In an aerial picture taken in 1939 by Sir M. Aurel Stein, a system of well-watered and dark-coloured allotments, probably compound gardens, is observable at the north-eastern side of Udhruh's Roman fortress (see Figure 7.2). The visible network of water-distribution channels, diversion structures, barriers and even spillways gives a good impression of how the system must have operated in those days. Three other aerial pictures, taken on 11 April 1939, do not show any more houses or dwellings than those visible in Figure 7.2. This corresponds with the results of our oral history project, through which elderly members of three extended Udhruh families told us that they all held several agricultural allotments here, while still living a predominantly semi-nomadic Bedouin lifestyle. The 'Ain Udhruh spring dried out in the 2010s, but was used as a source of water distribution to feed (via still visible channels) the field systems east of Udhruh. In 2014 we saw this water-distribution system in operation for the last time, no longer fed by the ancient spring but through a modern water supply system. The distribution was the responsibility of the men, who worked out a water-management schedule at the community level. 'Ain Udhruh is considered of high importance because of its creation and maintenance of a shared sense of identity and spiritual meaning for the people of Udhruh (Hageraats, 2014). The spring fell dry as the result of a combination of changing climatic conditions, sinking groundwater levels and, according to local people, construction failures

Figure 7.2 Aerial picture of Udhruh (taken on 11 April 1939). On the left the Roman fortress of Udhruh and on the right irrigated plots (picture by Sir M. Aurel Stein: APAAME_19390411_Stein-BA-ASA-3–0510, Aerial Photographic Archive for Archaeology in the Middle East)

by the government when they built the nearby road. The groundwater levels will be further lowered by recent attempts at 'hit-and-run agriculture', a problem that is noticeable over the whole region. This consists of pumping up excessive quantities of very deep underground water for watering annual food crops. We call it 'hit-and-run agriculture' because it is definitely not permanent, nor sustainable, and the growers seem to be interested only in a quick profit. Some of the more fertile soils were selected for this activity for large-scale one-year cultivation sites; an enormous waste of precious water appeared to take place, probably resulting in serious salinisation of the soils.

Till 2014, when the irrigation scheme fed by 'Ain Udhruh was in operation for the last time, we thought it inappropriate to venture into these fields for our archaeological field surveys. We asked local permission in the following years, however, to do some pilot surveys and have a closer look at the scheme. From these it became clear that older stone walls and bunds were laid out under and next to the more recent structures. We have distinguished 127 of these older allotments so far, varying in size from 275 to 5,500 m^2 and covering a total area of 0.735 km^2. In these fertile plots we encountered several concentrations of ceramics, dating predominantly to the Nabataean, Roman and Byzantine periods, and mainly located at the northern and eastern elevated boundaries of the small valley. We also picked up several pottery fragments of the same periods when surveying across the field system. The dating of this material culture corresponds to the major settlements of Udhruh in antiquity, and their spatial correlation with the perennial spring. These observations make it very plausible that this fertile land was used in antiquity for orchards or compound gardens. Such field, irrigation and water-distribution techniques were widely practised in the Petra region in the Nabataean period. Several archaeologists have noticed that the Nabataean irrigation structures were often reused in later periods, and some are still in use by local communities (al-Muheisen, 2009: 142–3; Beckers et al., 2013: 346–7; Beckers and Schütt, 2013: 321; ‘Amr et al., 2000: 234, 239; Kouki, 2012: 108, 123–5; Nasarat et al., 2012: 107–9). It is not only archaeology that shows that Udhruh and its environs were turned into a prosperous agricultural region. The literary sources provide us with intriguing examples as well, especially for the Byzantine and early Islamic periods. In the light of this the Petra papyrus (39) is of special interest. It describes a dispute between two neighbours in a nearby settlement over water-draining rights from a spring and the theft of building materials and the construction of water channels (Arjava et al., 2011: 48–56). These include structures containing intact mortar samples that include charred twigs, which we intend to use for 14C dating.

Flood-water harvesting in the Jabal ash-Sharah

The best conditions for cultivating crops in the hinterland of Petra are to be found in the Jabal ash-Sharah area (Besançon, 2010: 42). This is the area west of Udhruh and is characterised by a landscape with gentle slopes interspersed with wadis.

Settlements

During our field surveys we encountered three ancient settlements in the Jabal ash-Sharah area. The first is located 2.3 km west of Udhruh, in the Wādī al-Harab, north of the Udhruh–Wadi Musa road (this is site no. 27, called Wādī al-Harab in Abudanh, 2006: 411). The settlement is built on the slopes of a curved tributary of two wadis, where a diversion dyke with connected conduit walls was built, probably to avoid water damage in the rainy season. The large settlement, of about two hectares, is surrounded by low enclosure walls, and is occupied by some traditional houses which were reportedly built in the mid-twentieth century by Sheikh 'Abd Allah Dhyab Harb Al Jazi, who was about 90 years old when the authors interviewed him in 2014. These houses were built on older foundations that lie underneath a dense layout of dwellings and other structures that fill the northern side of the settlement. Two cisterns and four threshing floors were uncovered on the northern slopes of the settlement. One cistern was inaccessible, but the other one (7.2 m long × 3.5–4.8 m wide × > 3 m high) had a cover stone with a neatly cut square opening, and was found to be plastered with a remarkable mortar lining. The mortar contains fragments of Nabataean pottery, and resembles the mortar linings found in several structures in Petra, like those of the water basins of the 'Garden Temple' complex. The threshing floors are mostly oblong-oval in shape (roughly 6 ×16 m, 8 × 27 m, 12 × 29 m and 14 × 28 m), and have low dry-stack sidewalls on the southern wadi sides. They also seem to be determined by the existence of natural flat outcrops. Such threshing floors are impossible to date, but are found at other archaeological sites around Petra and are mentioned in the Petra papyrus 17 (Abudanh, 2006: 203; Fiema, 2002: 205–6; Frösén et al., 2002: 312–13; Glueck, 1935: 74–5; Koenen, 2004: 355; Koenen et al., 2013: 1–2, 107, 126–7, 142). Intensive surveys of this settlement – an area of over 61,200 m^2 and with a grid of 20 × 20 m blocks – produced almost 1,200 ceramic surface finds (Table 7.1), which can be assigned to the Nabataean, Roman and Byzantine periods.[5]

Two other settlements were uncovered in the more westerly part of the research area (Figure 7.1) that contained house structures comparable to the ones found in the first settlement.

Pilot surveys at these settlements and in surrounding fields revealed ceramic sherds, predominantly Nabataean. The assemblages also revealed a decline in ceramics in the late Roman period, but with an upturn in the Byzantine period (Wenner, 2015: 137–65).

Table 7.1 Ceramic evidence (number of sherds per period) of intensive surveys at Wādī al-Harab settlement

Nabataean	Nabataean-Roman	Roman	Roman-Byzantine	Byzantine
435	302	141	37	145

Run-off water harvesting

The slopes and valleys of the wadis in the western part of the research area (the Wādī al-Harab, the Wādī Zubayra, the Wādī al-Rumayi, the Wādī Mulqan and its tributaries) are covered with structures relating to a combination of, or adjacently operated, hillside conduit and flood-water farming systems. Terrace walls and conduit channels were constructed on hillsides in order to collect and direct the run-off rainwater to the connected levelled fields on the lower slopes and the bottoms of the wadis. At the bottom of the wadi beds west of Udhruh many dams and dykes have been uncovered. Several dams, whose lengths varied between 15 and 130 m to match the width and depth of the valley bottoms, were neatly built at irregular distances in almost every wadi that we surveyed. The surviving height of these dams was never more than 1.5 m, but on some the original spillways could be distinguished. Such dams were probably constructed both to reduce the water velocity after the intense downpours of the rainy season, and to collect this water. But they probably also encouraged eroded sediments to settle and thus improve the arable land in the wadi-bed (see Oweis et al., 2012: 58–62, 66–9 for such systems, which are still employed today). These wadi-bed water-harvesting and hillside conduit systems (with both macro- and micro-catchments) have also been found at other archaeological projects in the Petra region, including in the Wadi Faynan and in the Negev (for the Petra region, see Alcock and Knodell, 2012: 7; Beckers et al., 2013; ʻAmr et al., 1998; Lavento et al., 2004: 166–7; Urban et al., 2013; for Wadi Faynan, Newson et al., 2007; for Negev see Bruins, 1986). The dating of these schemes is complicated by a lack of directly associated dating material, possibly by use over a very long period, and by the fact that similar structures are used for current agriculture. From the dating of ceramic finds from the fields integrated in or adjacent to these wadi-bed water-harvesting and hillside conduit systems, and the dating of nearby settlements, a Nabataean origin can be suspected. Our field surveys also lead us to expect that these flood-water-harvesting systems in the Jabal ash-Sharah region experienced an increase in the Byzantine period. OSL and radiocarbon dating were employed, by another research team in the vicinity of Petra, on samples from several similar agricultural terraces. This took place in the Wadi al Ghurab catchment area, and the results suggested a Nabataean origin with a possible continuity of use until the eighth century AD (Beckers and Schütt, 2013: 321; Beckers et al., 2013: 346–7). These results are similar to our analyses of the ceramic assemblages for the Udhruh region. We were initially reluctant to apply these scientific dating techniques to dams and walls, as we noticed many cases of natural undercutting and backfilling. We will, however, continue our search for intact interiors inside standing dams and terrace walls. This is because we think that using samples from such contexts diminishes the risk of post-dating errors. The Udhruh surveys were initially started to reconstruct the geomorphology of the landscape and to provide GIS-related locations of all (old) man-made structures. More detailed information has been gathered over the last few years by measuring 3D-points, angles and distances between the structures. This includes the measurement of hillside conduit and flood-water-farming systems, nearby settlements and the related landscape, using Total Stations and a 3D-scanner.[6] With these results it is planned to make modelling calculations and then reconstructions of the hydrological capacity, abilities and effectiveness of the systems in several locations (the side-arms and catchment areas of the Wādī al-Harab, Wādī Zubayra, Wādī al-Rumayi and Wādī Mulqan).[7]

These flood-water-harvesting schemes were probably used for cereal cultivation. This main staple fits well with the large-scale set-up on all hillsides for a whole region, the seasonal water availability, and the evidence from the excavated threshing floors. Sampling for archaeo-botanical and pollen analyses is not useful here, however, as these fields are still regularly used for cereal growing by local Bedouin families. This is also true of similar ancient run-off water systems for the wider regions (Alcock and Knodell, 2012; Beckers et al., 2013; Bruins, 1986; ʻAmr et al., 1998; Lavento et al., 2004: 166–7; Newson et al., 2007; Urban et al., 2013). Dozens of quernstone fragments used for milling grain have been found during surveys in the research area. Such grinding stones were retrieved from Roman-Byzantine archaeological contexts, but also as surface finds. The problem is that these stones were also used by Bedouin throughout the nineteenth and twentieth centuries, and it is difficult to distinguish the later ones from the ancient ones.

These water-harvesting systems in the Jabal ash-Sharah part of the research area were employed to hold and redirect run-off water that would otherwise have been lost. However, these systems do not allow for a more continuous agricultural crop rotation, as the captured and diverted water would only be available for a limited period of time after the seasonal precipitation. In antiquity solutions were also employed to make use of deep percolation water that would otherwise be lost and to prevent evaporation, as will be illustrated in the following section.

The Udhruh *qanat* and connecting field systems

To the south-east of Udhruh lies an impressive network of well-preserved ancient subsurface and surface-water conservation measures and connected irrigated fields – a *qanat* system – was recorded in a large floodplain largely covered by alluvial deposits (Figure 7.1) (for earlier observations of structures related to this *qanat* system see Abudanh, 2004: 488–489, 492–493; 2006: 71–81; Abudanh and Twaissi, 2010: 69–70; A. Killick, 1987: 28; Stein, 1940: 435). *Qanats* were most probably first used in the Armenian-Persian region, date back to around 700–600 BC (Lightfoot, 1996: 324), and are the subject of many other studies. The Udhruh *qanat* consists roughly of three components: 1) the subterranean water system consisting of vertical *qanat* shafts and horizontal underground water conduits; 2) the surface part of the water system comprising outlet(s), channels, distribution structures and large reservoirs; 3) agricultural field systems with irrigation channels.

Subterranean structures

For the subsurface part, three lines of vertical *qanat* shafts (totalling more than 200 individual shafts) were dug and hacked out of the limestone bedrock, 1.1–3.9 km south-east of the fortress of Udhruh. These shafts are currently filled in, and recognisable by circular mounds or hollows, probably created during the construction and maintenance of the shafts, and later filled in with blown- or filled-in debris. We have already counted 243 of these circular mounds, positioned at regular intervals of about 25–30 metres apart. There must have been more though, as some parts of the lines lack such mounds and some shafts are losing their surface visibility as a result of erosion and modern cultivation. At the western side – near Tell Abara – the *qanat* tapped into three or four 'mother wells' which are probably fed by groundwater from elevated aquifers. A modern water station is situated to the west of Tell Abara and close to the mother well in the southernmost line of shafts. We noticed over several campaigns that a small but continuous flow of pulsating water was coming out of a surface pipe near this station, which was probably drilled into the subterranean water level. Therefore it is assumed that the aquifers must still contain water. The three lines seem to merge near the modern Udhruh-Ma'an road approximately 2.8 km south-east of Udhruh. This intersection is uncertain, however, because of the observable distortions in this area, and further east two parallel lines of shafts are seen heading towards the Wādī el-Fiqay. Along with construction and maintenance purposes, the shafts play an important role in ventilation of the *qanat* (for such use see Lambton, 1989: 7). The horizontal underground tunnels of the Udhruh *qanat* have not been investigated yet, as the vertical shafts are filled in. One of the men we interviewed from Udhruh told us that some of the shafts near the Udhruh-Ma'an road were still (partly) open a few decades ago. But because someone fell in and was killed they were backfilled. A start has been made on the archaeological excavation of one vertical shaft, which was stopped unfortunately after four metres because we reached the maximum depth with the ladders we had available. It was already clear that the more or less square shaft (about 2 × 2 m) was cut through the Muwaqqar Chalk Marl rock formation, with evenly distributed man-made holes in the opposite walls, which were probably used to hold bars for a ladder construction. A large tripod and military rope hoists will be used to further excavate this *qanat* shaft. Soil samples will be taken to date the period during which the *qanats* went out of operation, and also for micro-morphological research in order to find out if the backfilling started through natural processes or anthropogenic activities. Reaching the bottom of this shaft will also allow us to access and gain knowledge about the horizontal tunnel and channel construction, hopefully with datable material culture and mortar samples. The aim is eventually to excavate at least one 'mother well' shaft and two other additional shafts to obtain data, especially on the hydrological capacities of the channels. Research at other *qanats* makes it clear that the gradient of the shorter conduits varies from 1 to 5 per mille, while longer ones are almost horizontal (Lambton, 1989: 7). Our observations of the surface parts of the Udhruh *qanat* show that a gradient of less than 2 per mille was utilised for this. The *qanats* with such a gradient not only provide a steady year-round flow of non-turbulent water, but also furnish another advantage: evaporation from these subsurface conduits is limited.

Surface constructions

The subsurface parts of *qanats* transport water from a mother well to a surface outlet. After travelling 3.1–4.5 km through underground conduits, the water probably reaches a surface outlet. An erosion gully in the alluvial deposits of the Wādī el-Fiqay unearthed two parallel surface channels of different construction only 200 metres from the most north-eastern *qanat* shaft. An outlet point can therefore be expected somewhere in these environs. The surface parts of the water scheme in this wadi are covered with thick alluvial deposits in some places, while at other spots they stand above the current ground level. The surface water conduits – with total lengths of 1.9 and 2.6 km – end at two large reservoirs connected to agricultural field systems. Figure 7.3 uses actual field observations to show the current state of research. The two parallel surface channels seem to be pivotal in this agro-hydrological scheme. One consists of two straight side walls (0.2–0.3 m wide) made of roughly cut limestone blocks with a floor of the same material, creating a 0.35 m-wide and 0.9 m-high conduit. A channel with similar dimensions and layout was found in a 3 m-deep erosion gulley (this is the point with water-level height of 1,001 m in Figure 7.3). At this location it became clear that this channel was originally covered with slabs of brecciated chert. This was done to prevent the channels from silting up with alluvial, colluvial and aeolian deposits, to reduce the evaporation of the transported water, or a combination of the two. This covering made the interior dimensions of the water channel clear. The second channel is made up of two parallel 0.5–0.6 m-wide and 0.65 m-high side walls of concrete and large limestone blocks, on top of a 2.3 m-wide and 1.5 m-high foundation of the same construction. The water conduit itself, 0.4 m wide and 0.55 m high, is made up of a perfectly smooth 3–4 cm-thick mortar lining. A channel of similar layout was partially excavated in 2016 (Figure 7.4).

Two large reservoirs were constructed at the eastern end of the water-harvesting and transportation scheme. The most northern 50 × 50 m reservoir – with standing walls still up to 1.5 metres – was partly dug with 2.5–3 m-wide earthen ramparts, probably to withstand the internal water pressure. The side walls are made up of large cut coquina blocks, resembling in size and provenance the ones applied to the curtain wall of the Roman fort, upon which a 0.3 m-thick layer of larger stones filled up with mortar and charred particles was placed. A 7–8 cm-thick concrete lining which included small rounded pebbles and pottery fragments was applied to make it waterproof. Measuring the lower water levels of the different structures (see Figure 7.3) also enabled the capacity of the reservoir (which has an inside surface area of 2,238 m^2) to be calculated: it comes to a maximum of about 3.3 million litres of water. The southern 34 × 33 m reservoir was dug into the slope of a small hill (on the northern, western and southern sides). The dry-stack side walls of this reservoir are still standing up to 2.7 metres high; they were made of large limestone blocks plastered with a 3 cm-thick concrete lining. The same plaster lining, to which red pottery fragments and small pebbles were added, was also applied to the bottom of the reservoir. On the eastern side an earthen rampart, with special reinforcements, which proved to be part of later renovations, was added to prevent this side from collapsing as a result of the water pressure. The maximum capacity of this reservoir (with an inside surface of 988 m^2) was about 2.7 million litres of water.

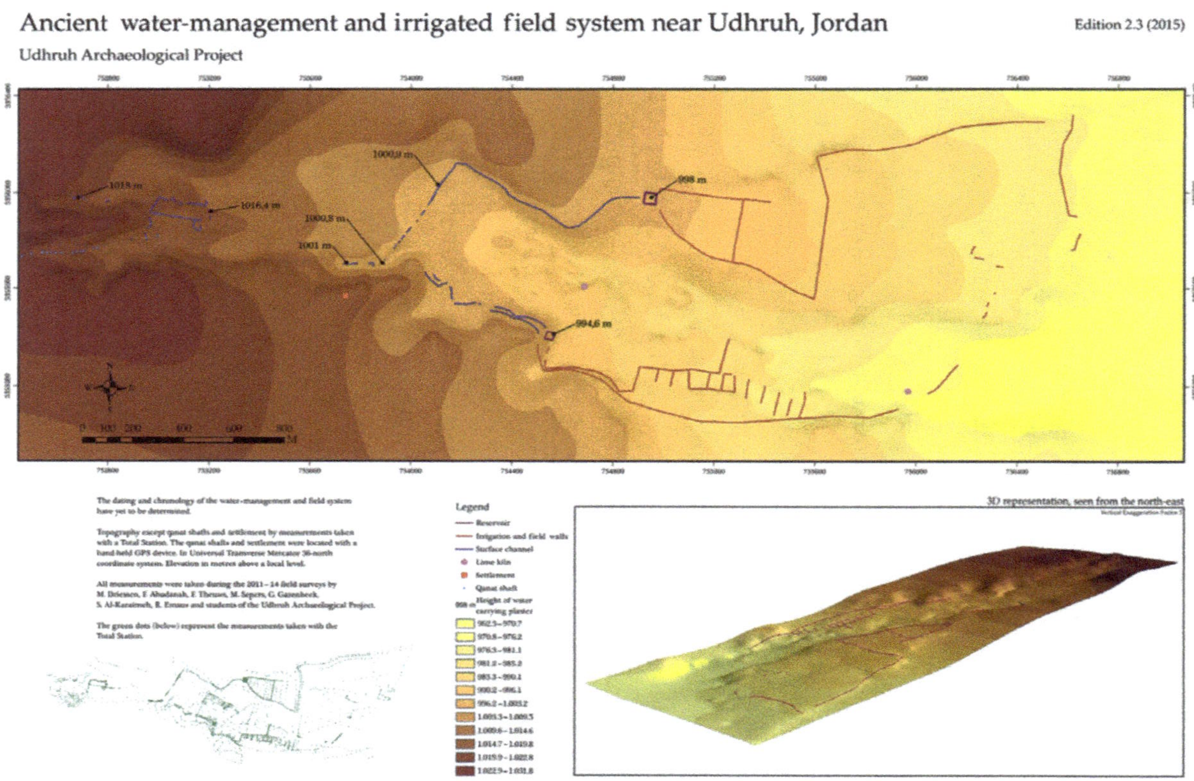

Figure 7.3 The ancient water-management and irrigated field systems in the Wādī el-Fiqay, which is fed by the Udhruh *qanat* system (drawing by Roeland Emaus)

Figure 7.4 3D reconstruction of a surface channel of the Udhruh *qanat* scheme (drawing by Maarten Sepers)

No features were found that may relate to the reduction of evaporation from these large open water surfaces. The basic concept of the water cycle was already understood in classical times, and the evaporation of rainwater was a known concept mentioned by Hippocrates in the fifth century BC.[8] It seems logical that measures against evaporation were installed, especially if one considers all the effort and investment that were already made to get the water here. These, however, do not have to consist of a complete roofing; effective precautions, archaeologically almost or completely untraceable, may have been applied.[9]

By means of small-scale excavations, together with aerial pictures, 3D-scanning, and non-destructive, geophysical, ground-based and airborne exploration methods, we plan to complete the survey of the layout of the surface part of the water transportation system.[10] Regarding the non-destructive, ground-based techniques, some initial work with a ground-penetrating radar (GPR) and a magnetometer have been carried out with promising results. These results have already revealed some new information on the layout of the scheme, revealing, for instance, the dimensions of a third reservoir (14 × 14 m). The construction of the reservoir walls and the typical form of the lower outlet on the eastern side make it clear that this reservoir had a dual function: water distribution and water treatment through sedimentation.

Many different types of mortar came to light during the surveying and excavation of the different structures of the hydrological scheme in the Wādī el-Fiqay. The mortar linings of the reservoirs and channels are solid, smoothly finished and loaded with small rounded pebbles and small ceramic fragments, which make the linings strong and waterproof. The foundations of the large reservoirs were made of larger stone blocks held together by loamy mortars with included charred wood and twig particles. These were applied to create a solidity or sturdiness able to withstand the pressure of large quantities of water. The mortars of the channel foundations, on the other hand, are lightweight, with elastic properties, and seem to be made to withstand tensile forces. We encountered at least nine different types of mortar. It seems that people were making these mortars with different physical capacities for different purposes; of these, the very elastic mortars were the most remarkable, and were probably used in a trial to withstand the earthquakes that ravaged these regions in antiquity.[11]

Two large lime kilns were located (Figure 7.3) that shed light on the mortar production for the construction, maintenance and renovation of the conduits and reservoirs. Vast amounts of combustibles were needed for the lime-burning process to make these mortars. Heaps of ashes and charcoal particles near the kilns – with a diameter of around 10 m – may suggest that firewood was used. Intensive surveys on and around the fields provided us with a wide variety of material culture. This includes several Upper and Middle Palaeolithic hand-axes, blades and arrowheads, together with substantial amounts of Nabataean, Roman and Byzantine pottery, smaller quantities of early Islamic and Ottoman pottery, fragments of Mamluk glass bracelets and a wide variety of remains from the Great Arab War. This broad spectrum does not help us to get a balanced, well-founded grip on the chronological framework, but it makes it clear that people have been using this area for vast periods of time. The ceramic evidence from pilot surveys at a nearby settlement – Khirbet el-Fiqay (Figure 7.3) – shows high usage during the Nabataean, Roman and Byzantine periods, with possible reuse during Ayyubid-Mamluk times.[12] This settlement is located 1.0 and 0.6 kilometres – as the crow flies – from the northern and the southern reservoirs respectively. Its position is around 10 metres higher than the nearest channel, which was found 50 metres away in the erosion gully. No other settlements have been found near or next to the reservoirs and fields, so uses for the *qanat* system other than for agricultural purposes can be excluded at this stage. The OSL dating of the mortars used for the two large reservoirs shows that the northern reservoir was built during the Nabataean period, with adaptations or renovations in Roman times (Table 7.2). The southern

Table 7.2 OSL and 14C dating of Udhruh *qanat* reservoirs in Wādī el-Fiqay (OSL datings by Alice Versendaal and Jakob Wallinga (Centre for Luminescence Dating – Wageningen University; 14C datings by Hans van der Plicht and Sanne Palstra (Centre for Isotope Research – University of Groningen))

Feature description	Feature number	Sample number	Palaeo-dose (Gy)	Dose rate (Gy/ka)	Age (ka)	Model	Comments	Date (from OSL)
North reservoir – mortar lining bottom	9001	875	3.00 ± 0.12	1.69 ± 0.06	**1.78 ± 0.10**	Iterative	Fraction 180–212	237 AD ± 100
North reservoir – mortar lining S wall	9001	876	1.37 ± 0.03	0.78 ± 0.04	**1.77 ± 0.09**	Iterative (1.5 SD)	Fraction 63–90	247 AD ± 90
North reservoir – mortar lining wall SE corner	9001	1450 – 1451	3.05 ± 0.14	1.46 ± 0.06	**2.09 ± 0.13**	Iterative	Fraction 90–250	73 BC ± 130
North reservoir – basic layer bottom	9001	884	3.01 ± 0.28	1.53 ± 0.06	**1.97 ± 0.20**	MAM	Fraction 90–250	47 AD ± 200
South reservoir – mortar lining wall SW corner	9479	1440	3.37 ± 0.10	1.84 ± 0.07	**1.83 ± 0.09**	Iterative	Fraction 90–250	187 AD ± 90
South reservoir – wall NE corner – upper part	9479	1441 – 1444 – 1445	3.12 ± 0.21	1.87 ± 0.14	**1.66 ± 0.17**	MAM	Fraction 90–250	357 AD ± 170
South reservoir – renovation/enforcement NE corner abutment	9479	1453			**1.325 ± 0.03**			655–90 AD / 750–76 AD (14C)
South reservoir – renovation/enforcement NE corner abutment	9479	1447			**1.4 ± 0.02**			630–60 AD (14C)

reservoir was of later date; it was constructed during Roman times with adjustments in the late Roman or Byzantine period. The 14C analyses of charred twigs found in the mortar of a later outer reinforcement from the southern reservoir show that this was accomplished in the Umayyad period. An analysis of the scientific dating in relation to the material culture makes it clear that a Nabataean-Roman origin, with later reuse in late Roman, Byzantine and Islamic times, is quite plausible and fits with the broader picture.

Agricultural fields

This subtle system for harvesting deep percolation and groundwater is designed to irrigate an extended agricultural field system with at least 35 hectares of tilled land, east of the reservoirs. Two large fields (approximately 20 hectares in total) connected to the northern reservoir are still completely level with surrounding walls. These walls were used as water conduits. Older and more recent aerial pictures reveal a delicate chequerboard pattern within these fields, a possible indication of a regular block field system. In more recent times, such field systems in the Iranian Plateau are seen as the result of more efficient irrigation methods and agricultural intensification, arising from increased population pressure (Bonine, 1989: 35–8).

Next to the southern reservoir we found an old field of about 4 hectares, with a connected field system of 10 smaller terraced fields ranging in size from 2,800 to 5,400 m². Some of the walls or dams between these fields still have visible spillways. The smallest of these plots is of particular interest because the size is about the same as the 1⅑ *iugerum* well-watered field sold by Theodoros to Philoumenos somewhere near Udhruh in January AD 559, as mentioned in the Petra papyrus 25.

The tilled fields of the northern reservoir were connected to those of the southern reservoir by a surrounding enclosure (Figure 7.3) – east of both reservoirs – comprising an area of more than 1.5 km². Although the northern reservoir was of Nabataean construction, at some time in the Roman and Byzantine period the *qanats* and connected fields became part of one large agro-hydrological scheme, covering a total area of more than 6.5 km², that made use of water that would otherwise have been lost because of deep percolation. Water was protected from evaporation so that it could be accessed not only for a few months of the year (just after the rainy season), but also more continuously over longer periods. This system allowed for differing kinds of agricultural crop rotations (resulting, for example, in more than one yield of cereals a year), or irrigated the fields for newly introduced perennial plants. This must have led to a serious agricultural transformation, with new farming strategies and equivalent technologies in processing and handling. It must also have affected the human agents, thinking only of changes in the seasonal life cycles of the communities involved. Such a large scheme as this can only be part of a programme of agricultural intensification: its scale and technical innovations point to an authority with means, vision, labour capacity and level of organisation, the Roman state for instance. It is tempting to look, therefore, at a nearby Roman legionary fortress, also constructed with great ingenuity, that housed more than 1,000 well-trained soldiers – young men whom one does not want to sit idle, as ancient literary sources provide plenty of examples of Roman military mutinies, the result of boredom and idleness among the troops.

We were eagerly anticipating finding out what crops were grown in the ancient fields watered by the *qanat*, and what the background vegetation looked like. Samples were taken for archaeobotanical and pollen analyses. Several samples from the northern fields were processed. Unfortunately, they contained neither macrobotanical remains, nor pollen, nor non-pollen palynomorphs.[13] More samples will be collected, however, and processed, as some mortar samples have provided us with exciting macrobotanical remains like grape and almond seeds. In the meantime, clues about the cultivated crops might be retrieved from artefacts such as the already described quern and milling stones, which were also found in this part of the research area. A cylindrical crushing roller (diameter 1.02 m and width 0.41 m) made of very hard limestone, showing marks of continuous rolling, was found in the centre of the Roman fortress of Udhruh. Such rollers, which formed part of olive oil presses, have been found in predominantly Roman and Byzantine contexts, although some Nabataean and Ottoman contexts have also been identified (Frankel, 1999; Kouki, 2012: 109).

Interdisciplinary approaches

The two main soil types of the Udhruh region (as seen in the subsection ‘Environment and landscape’, above) bear some negative characteristics that make them not very suitable for agriculture. The sections presented above make it clear that investments of great effort, ingenuity, knowledge and experience were made to plan and construct the Udhruh *qanat* system. The many kilometres of long subterranean and surface water channels, together with connecting reservoirs, would not have been built here if the soils of the related fields had not been suitable for cultivation. Further tests are planned to find out more about the chemical soil quality and characteristics. Two soil samples taken from the fields east of the northern reservoir have already been analysed as a preliminary test for such characteristics.[14] The result for electrical conductivity shows that these samples have low salinity levels, and therefore no restrictions on plant growth are to be expected. The soil pH (KCl) is about 8, only slightly above the optimum for many plant species. Soil N (primarily as organic N) and Soil P are high when compared with the values expected for this area, although the latter tends to decrease with the depth. Most of the carbon in the samples is present as inorganic C, but the organic C is (still) high compared with soils typical of this climate and landscape. The organic component is quite old and probably consisted of recalcitrant organic matter, as was shown by the soil C:N ratios. These results correspond with archaeological data that suggests that these fields were not used for agriculture throughout the last centuries. The oral history project underlines this as well. During a 2014 site visit, the 90-year-old Sheikh 'Abd Allah Dhyab Harb Al Jazi told us that these fields were not cultivated by his ancestors, and, according to his oral tradition, nor were they by the Ottomans. These soils – according to him – are too hard to work, and could have only been cultivated by the 'Romani'. Archaeological surveys show that the Ottoman water piping from Udhruh did reach to other fields quite remote from the field systems we investigated. Concluding the results for these still very preliminary soil tests, it was stated that, '[e]ven though the old organic matter is expected to be resistant to further decomposition, recent research has shown that an external source of labile organic matter, e.g. fresh plant litter, may "prime" old organic matter and release nutrients like N and P through mineralisation'. The preliminary results of the soil analyses are therefore promising, and justify further research into the extent to which they apply to all the ancient agricultural fields in the region. This research raises the question of how to revive these soils for possible future use.

In order to obtain a well-founded insight into the hydrological capacities of the *qanat* system, we must measure all the ancient water levels and dimensions of the structures, together with the fields. A pilot study had already been undertaken, which calculated the abilities of the system to provide enough water to the related fields. This made a set of assumptions about evaporation, infiltration, seepage and roughness coefficient. Several scenarios were prepared that made use of the AquaCrop, the crop growth model developed by the Food and Agriculture

Organization (FAO) of the United Nations.[15] This preliminary research already shows that the system worked properly; that is, the *qanats* must have been able to provide a water flow rate which was more than adequate to irrigate the connected field systems.

Conclusions and future research

A large part of the unevenly distributed annual rainfall in arid and semi-arid regions gets lost without becoming available for agricultural use. Most of this loss is the result of evaporation (30–50 per cent), run-off (10–25 per cent) and deep percolation (10–30 per cent). In antiquity different techniques of water collection were already being applied in order to decrease such water loss.

The well-preserved ancient landscape around Udhruh shows that such techniques were being practised in this region; water-harvesting measures relating to run-off loss were already employed in the hilly Jabal ash-Sharah. These harvested waters were probably used for cereal production, and had an origin in Nabataean days but experienced an increase in the Byzantine era.

The Udhruh *qanat*, with its related field scheme, is probably one of the best-preserved field 'laboratories' for the study of the long-term development of innovative water-management and agricultural systems in southern Jordan. It is also one of the most complete *qanats* in the wider region, and one that has not been modified for some centuries, as can be observed at other examples in Syria, Iran and Saudi Arabia. It has its origins in the first century AD, then develops into a programme of agricultural intensification in the following centuries, making use of water that would otherwise have been lost as a result of deep percolation. Water thus became available throughout the year, which must have led to the development of other farming strategies, together with newly introduced perennial plants, changes in processing technologies, and transformations in the seasonal life cycles of the communities involved. This state of affairs could only have been established under the supervision of a central authority with adequate vision and technical background, an authority that was able to control and organise the required means and labour. The system was probably very successful, as it was renovated and adjusted in the Byzantine and Umayyad periods and eventually covered a time span of at least six centuries. The long-term use of such a system was only possible if people were practising what is nowadays labelled sustainable agricultural and hydrological management. Overwatering, as observed in so many current irrigation schemes, would have led to detrimental salinisation of the soils and short-term use. Trying to unravel the exact *modus operandi* of this *qanat* scheme and its long continuity of use will be one of the greatest challenges for our research project in the coming years. This can only be accomplished by practising an interdisciplinary approach in which archaeological research is integrated with historical, geophysical, geo-hydrological, socio-hydrological, bioscientific and biogeochemical soil studies. On the one hand we think that this approach will help us to reconstruct the development of the agro-hydrological landscape in the Udhruh region over the *longue durée*. On the other hand we hope that the interdisciplinary approach – the first author is both an archaeologist and an agronomist – will not only result in a better comprehension of the ancient, semi-arid landscape management strategy from a diachronic perspective, but also lead to translational and innovative thinking, which may contribute to sustainable agricultural and water-management solutions for future use in these regions.

Acknowledgements

The Udhruh Archaeological Project and the preliminary results in this paper would not have been possible without the hard work and assistance of our team. The Project's core team consists of Roeland Emaus, Guus Gazenbeek, Sufyan al Karaimeh, Maarten Sepers, Frans Theuws, Sarah Wenner, Fawzi Abudanah and Mark Driessen. We would also like to thank all the students who participated in the project, and all the researchers who provided us with new and additional results. We sincerely hope for a continuing academic relationship with them and all our other partners, and would like to additionally thank the following: Alice Versendaal and Jakob Wallinga (OSL dating – Wageningen University), Hans van der Plicht and Sanne Palstra (14C dating – Groningen University), Maurits Ertsen (hydrology and water resources management – Delft University), Dominique Ngan-Tillard and Martijn Warnaar (GPR – Delft University), Khaled al-Bashaireh (stone provenance determination – Yarmouk University), Marcel Hoosbeek and André van Leeuwen (soil quality analyses – Wageningen University), Joanita Vroom (Byzantine–early Islamic ceramics – Leiden University), Willem Willems† (CS – Leiden University), Mohammad al-Marahleh and Aktham Oweidi (Department of Antiquities, Jordan), and of course the people of Udhruh. Our gratitude also goes to the APAAME team of David Kennedy, to Fenno Noij and Marcus Roxburgh for their assistance in editing, and to Mark Altaweel and Yijie Zhuang for the invitation to participate in the conference 'Comparative Water Technologies and Management: Pathways to Social Complexity and Environmental Change'.

Notes

1. The Nabataean period in this region traditionally dates from the third to the second centuries BC till AD 106, the date of the establishment of the Roman province of Arabia, although the material culture and layout of structures remain dominantly Nabataean through large parts of the second and third centuries AD. For earlier expeditions and research in Udhruh, see, for example, Brünnow and Domaszewski, 1904: 429–62; Glueck, 1935: 76; M. Killick, 1990: 249–50.
2. Respectively from the Beersheva Edict (Segni, 2004: 151–2) and Stephen of Byzantium (18.18).
3. Augustopolis is mentioned in papyri 3, 7, 8, 9 and 10 (Frösén et al., 2002: 19, 25, 30, 31, 32, 36; Arjava et al., 2007, but it must be noted that not all of them are well enough preserved to provide such geographical information).
4. Diodorus Siculus (1979: 2.48.3–4) says of the Nabataeans that they have strategically located and hidden wells, and Strabo (1983: 16.4.21) mentions that they have springs which they use both for domestic purposes and for watering their agricultural plots.
5. The great majority of the pottery was picked up near the foundations of the older dwellings. The ceramic analyses were carried out predominantly by Sarah Wenner during the 2013 and 2014 Udhruh campaigns. For further reading see Wenner, 2015.
6. For this the TopCon Total Stations of the Al-Hussein Bin Talal University were used and a Leica P30 Scan-station was brought from the Netherlands.
7. Parts of these will be carried out at the Technical University of Delft (Water Resources Management, Civil Engineering and Civil Sciences) under the supervision of Maurits Ertsen.
8. Hippocrates (1923, VIII). Hippocrates warns here of the resulting degradation in taste, which might be made even worse when this water was blended with water from other sources. This view of the blending of water from different origins remained current in Roman times, as can be read in Vitruvius (1934, VIII.1–3).
9. When working in agriculture in Zambia and Zimbabwe in the 1990s I noticed people who took simple but effective measures to prevent evaporation from dams and reservoirs by making coverings of wicker, banana leaves and sturdy grasses (e.g. *Pennisetum purpureum*).
10. For the ground-penetrating radar (GPR) we have a collaboration with the Technical University of Delft, and an FM256 dual-Fluxgate Gradiometer is owned by the Al-Hussein Bin Talal University.
11. Further technical analyses of these mortars would make an interesting addition, for which we are thinking of chemical analysis, X-ray fluorescence and diffraction, compression testing and micro-morphological research. Most of this can be executed at our Laboratory for Material Culture Studies in the Faculty of Archaeology at Leiden University.
12. This was a result from both recent surveys and older ones carried out in the early 2000s (Abudanh, 2006: 73).
13. As was carried out by Erica van Hees of the archaeo/palaeobotany laboratory of the Faculty of Archaeology, Leiden University.
14. These tests were carried out and analysed by André van Leeuwen and Marcel Hoosbeek at the Department of Soil Quality, Wageningen University.
15. This was executed by a student of Civil Engineering and Geosciences at the Technical University Delft, under the supervision of Maurits Ertsen.

References

Ancient sources

Diodorus Siculus. 1979. *Diodorus of Sicily in Twelve Volumes*, books II.35–IV.38. English translation by C. H. Oldfather. Loeb Classical Library, 303. Cambridge, MA: Harvard University Press.

Hippocrates. 1923. *Ancient Medicine. Airs, Waters, Places. Epidemics 1 and 3. The Oath. Precepts. Nutriment*. English translation by W. H. S. Jones. Loeb Classical Library, 147. Cambridge, MA: Harvard University Press.

Strabo. 1983. *The Geography of Strabo in Eight Volumes*, VII, books XV–XVI. English translation by H. L. Jones. Loeb Classical Library, 241. Cambridge, MA: Harvard University Press.

Vitruvius. 1934. *On Architecture*, vol. 2: Books 6–10. English translation by Frank Granger. Loeb Classical Library, 280. Cambridge, MA: Harvard University Press.

Modern sources

Abel, Félix-Marie. 1938. *Géographie de la Palestine*, 2 vols. Paris: Lecoffre.

Abudanh, Fawzi. 2004. 'The archaeological survey for the region of Udhruh 2003, preliminary report.' *Annual of the Department of Antiquities of Jordan* 48, 485–96.

Abudanh, Fawzi. 2006. 'Settlement patterns and military organisation in the region of Udhruh (southern Jordan) in the Roman and Byzantine periods.' Unpublished PhD thesis, Newcastle upon Tyne University.

Abudanh, Fawzi and Saad Twaissi. 2010. 'Innovation or technology immigration? The Qanat systems in the regions of Udhruh and Ma'an in southern Jordan.' *Bulletin of the American Schools of Oriental Research* 360, 67–87.

Alcock, Susan E. and Alex R. Knodell. 2012. 'Landscapes north of Petra: The Petra area and Wadi Silaysil Survey (Brown University Petra Archaeological Project, 2010–2011).' In Laila Nehmé and Lucy Wadeson (eds), *The Nabataeans in Focus: Current Archaeological Research at Petra: Papers from the special session of the seminar for Arabian studies held on 29 July 2011*, 5–15. Oxford: Archeopress.

al-Muheisen, Zeidoun. 2009. *The water engineering and irrigation system of the Nabataeans*. Irbid: Publications Office, Faculty of Archaeology and Anthropology, Yarmouk University.

al-Salameen, Zeyad. 2004. 'The Nabataean economy in the light of archaeological evidence.' Unpublished PhD thesis, University of Manchester.

al-Salameen, Zeyad, Hani Falahat, Salameh Naimat and Fawzi Abudanh. 2011. 'New Arabic-Christian inscriptions from Udhruh, southern Jordan.' *Arabian Archaeology and Epigraphy* 22 (2), 232–42.

'Amr, Khairieh, Ahmed al-Momani, Suleiman Farajat and Hani Falahat. 1998. 'Archaeological survey of the Wadi Musa water supply and wastewater project area.' *Annual of the Department of Antiquities of Jordan* 42, 503–48.

'Amr, Khairieh, Ahmed al-Momani, Naif al-Nawafleh and Sami al-Nawafleh. 2000. 'Summary results of the archaeological project at Khirbat an-Nawafla/Wadi Musa.' *Annual of the Department of Antiquities of Jordan* 44, 231–55.

Arjava, Antti, Matias Buchholz and Traianos Gagos (eds). 2007. *The Petra Papyri*, vol. 3. Amman: American Center of Oriental Research.

Arjava, Antti, Matias Buchholz, Traianos Gagos and Maarit Kaimio (eds). 2011. *The Petra Papyri*, vol. 4. Amman: American Center of Oriental Research.

Beckers, Brian and Brigitta Schütt. 2013. 'The chronology of ancient agricultural terraces in the environs of Petra.' In Michel Mouton and Stephan G. Schmid (eds), *Men on the Rocks: The Formation of Nabataean Petra*, 313–22. Berlin: Logos Verlag.

Beckers, Brian, Brigitta Schütt, Sumiko Tsukamoto and Manfred Frechen. 2013. 'Age determination of Petra's engineered landscape – optically stimulated luminescence (OSL) and radiocarbon ages of runoff terrace systems in the Eastern Highlands of Jordan.' *Journal of Archaeological Science* 40 (1), 333–48.

Besançon, Jacques. 2010. 'Géographie, environnements et potentiels productifs de la région de Pétra (Jordanie).' In Pierre-Louis Gatier, Bernard Geyer and Marie-Odile Rousset (eds), *Entre nomades et sédentaires: Prospections en Syrie du Nord et en Jordanie du Sud*, 19–71. Lyon: Maison de l'Orient et de la Méditerranée.

Bonine, Michael E. 1989. 'Qanats, field systems, and morphology: Rectangularity on the Iranian Plateau.' In Peter Beaumont, Michael Bonine and Keith McLachlan (eds), *Qanat, Kariz and Khattara: Traditional Water Systems in the Middle East and North Africa*, 35–58. London: Middle East Centre, School of Oriental and African Studies in association with Middle East and North African Studies Press.

Bruins, Hendrik J. 1986. *Desert Environment and Agriculture in the Central Negeb and Kadesh-Barnea during Historical Times*. Nijkerk: Midbar Foundation.

Brünnow, Rudolf E. and Alfred von Domaszewski. 1904. *Die Provincia Arabia: Auf Grund zweier in den Jahren 1897 und 1898 unternommenen Reisen und der berichte früherer Reisender. Volume 1: Die Römerstrasse von Mâdebâ über Petra und Odruh bis El-'Akaba*. Strasbourg: Verlag Trübner.

Bull, Louise J. and Michael J. Kirkby. 2002. 'Dryland river characteristics and concepts.' In L. J. Bull and M. J. Kirkby (eds), *Dryland Rivers: Hydrology and Geomorphology of Semi-Arid Channels*, 3–15. Chichester: John Wiley and Sons.

Cordova, Carlos E. 2007. *Millennial Landscape Change in Jordan: Geoarchaeology and Cultural Ecology*. Tucson: University of Arizona Press.

Driessen, Mark J. 2007. 'Bouwen om te Blijven: De topografie, bewoningscontinuïteit en monumentaliteit van Romeins Nijmegen.' PhD thesis, University of Amsterdam. Amersfoort: Rapportage Archeologische Monumentenzorg 151.

Driessen, Mark J. and Fawzi Abudanah. 2013. 'The Udhruh lines of sight: Connectivity in the hinterland of Petra.' *Tijdschrift voor Mediterrane Archeologie* 50, 45–52.

Driessen, Mark J. and Fawzi Abudanah. In press. 'The Udhruh intervisibility: Antique communication networks in the hinterland of Petra.' *Studies in the History and Archaeology of Jordan* 13.

Fiema, Zbigniew T. 2002. 'Petra and its hinterland during the Byzantine period: New research and interpretations.' In J. H. Humphrey (ed.), *Roman and Byzantine Near East: Some Recent Archaeological Research*, vol. 3 (*Journal of Roman Archaeology* Supplementary Series 49), 191–252. Ann Arbor, MI: Journal of Roman Archaeology.

Finné, Martin, Karin Holmgren, Hanna S. Sundqvist, Erika Weiberg and Michael Lindblom. 2011. 'Climate in the eastern Mediterranean, and adjacent regions, during the past 6000 years: A review.' *Journal of Archaeological Science* 38 (12), 3153–73.

Frankel, Rafael. 1999. *Wine and Oil Production in Antiquity in Israel and Other Mediterranean Countries*. Sheffield: Sheffield Academic Press.

Frösén, Jaakko, Antti Arjava and Marjo Lehtinen (eds). 2002. *The Petra Papyri I*. Amman: American Center of Oriental Research.

Gilbertson, David, Graeme Barker, David Mattingly, Carol Palmer, John Grattan and Brian Pyatt. 2007. 'Archaeology and desertification: The landscapes of the Wadi Faynan.' In Graeme Barker, David Gilbertson and David Mattingly (eds), *Archaeology and Desertification: The Wadi Faynan Landscape Survey, Southern Jordan*, 397–421. Oxford: Oxbow Books/London: Council for British Research in the Levant.

Gleick, Peter H. (ed.). 2014. *The World's Water. Volume 8: The Biennial Report on Freshwater Resources*. Washington, DC: Island Press.

Glueck, Nelson. 1935. 'Explorations in Eastern Palestine, II.' *Annual of the American Schools of Oriental Research* 15, ix, 1–149, 151–61, 163–202.

Hageraats, Carlijn. 2014. 'Who says myths are not real? Looking at archaeology and oral history as two complementary sources of data.' Unpublished MA dissertation, Leiden University.

Hunt, Chris O., David D. Gilbertson and Hwedi A. El-Rishi. 2007. 'An 8000-year history of landscape, climate, and copper exploitation in the Middle East: The Wadi Faynan and the Wadi Dana National Reserve in southern Jordan.' *Journal of Archaeological Science* 34 (8), 1306–38.

Kennedy, David L. and Hani Falahat. 2008. 'Castra Legionis VI Ferratae: A building inscription for the legionary fortress at Udruh near Petra.' *Journal of Roman Archaeology* 21 (1), 150–69.

Kherfan, Ali M. 1998. *Geological Map of Bir Khidad, Map Sheet 3150 IV*. Amman: Geological Mapping Division, National Resources Authority, Geology Directorate, Hashemite Kingdom of Jordan.

Killick, Alistair. 1987. *Udhruh: Caravan City and Desert Oasis: A Guide to Udhruh and its Surroundings*. Romsey, Hants: A. C. Killick.

Killick, Marie. 1990. 'Les Nabatéens à Udhruh.' *ARAM Periodical* 2(1–2), 249–52.

Koenen, Ludwig. 1996. 'The carbonized archive from Petra.' *Journal of Roman Archaeology* 9, 177–88.

Koenen, Ludwig. 2004. 'Sprachliche Bemerkungen zu *P. Petra* 17 (inv. 10).' In Anton Bierl, Arbogast Schmitt and Andreas Willi (eds), *Antike Literatur in neuer Deutung: Festschrift für Joachim Latacz anlässlich seines 70. Geburtstages*, pp. 353–72. Munich: K. G. Saur.

Koenen, Ludwig, Jorma Kaimio, Maarit Kaimio and Robert W. Daniel (eds). 2013. *The Petra Papyri II*. Amman: American Center of Oriental Research.

Kouki, Paula. 2012. 'The hinterland of a city: Rural settlement and land use in the Petra region from the Nabataean-Roman to the Early Islamic period.' Unpublished PhD thesis, University of Helsinki.

Lambton, Ann K. S. 1989. 'The origin, diffusion and functioning of the qanat.' In Peter Beaumont, Michael Bonine, Keith McLachlan (eds), *Qanat, Kariz and Khattara: Traditional Water Systems in the Middle East and North Africa*, 5–12. London: Middle East Centre, School of Oriental and African Studies in association with Middle East and North African Studies Press.

Lavento, Mika, Mika Huotari, Henrik Jansson, Sarianna Silvonen and Zbigniew T. Fiema. 2004. 'Ancient water management system in the area of Jabal Haroun, Petra.' In Hans-Dieter Bienert and Jutta Häser (eds), *Men of Dikes and Canals: The Archaeology of Water in the Middle East*, pp. 163–72. Berlin: Marie Leidorf.

Lightfoot, Dale R. 1996. 'Syrian qanat Romani: History, ecology, abandonment.' *Journal of Arid Environments* 33, 321–36.

Lucke, Bernhard, Feras Ziadat and Awni Taimeh. 2013. 'The soils of Jordan.' In Myriam Ababsa (ed.), *Atlas of Jordan: History, Territories and Society*, 72–6. Beirut: Presses de l'Ifpo, Institut français du Proche-Orient.

Nasarat, Mohammed, Fawzi Abudanh and Slameh Naimat. 2012. 'Agriculture in sixth-century Petra and its hinterland: The evidence from the Petra papyri.' *Arabian Archaeology and Epigraphy* 23 (1), 105–15.

Newson, Paul, Graeme Barker, Patrick Daly, David Mattingly and David Gilbertson. 2007. 'The Wadi Faynan field systems.' In Graeme Barker, David Gilbertson and David Mattingly (eds), *Archaeology and Desertification: The Wadi Faynan Landscape Survey, Southern Jordan*, 141–74. Oxford: Oxbow Books/London: Council for British Research in the Levant.

Oweis, Theib Y., Dieter Prinz and Ahmed Y. Hachum. 2012. *Rainwater Harvesting for Agriculture in the Dry Areas*. Leiden: CRC Press/Balkema.

Rockström, Johan and Malin Falkenmark. 2000. 'Semiarid crop production from a hydrological perspective: Gap between potential and actual yields.' *Critical Reviews in Plant Sciences* 19 (4), 319–46.

Schmid, Stephan G. 2001. 'The Nabataeans: Travellers between lifestyles.' In Burton MacDonald, Russell Adams and Piotr Bienkowski (eds), *The Archaeology of Jordan*, 367–426. Sheffield: Sheffield Academic Press.

Segni, Leah di. 2004. 'The Beersheba tax edict reconsidered in the light of a newly discovered fragment.' *Scripta Classica Israelica* 23, 131–58.

Stein, M. Aurel. 1940. 'Surveys on the Roman frontier in 'Iraq and Trans-Jordan.' *Geographical Journal* 95 (6), 428–38.

Stucky, Rolf A. 1996. 'Die nabatäischen Bauten.' In Andrea Bignasca et al. (eds), *Petra, Ez Zantur. Volume I: Die Ergebnisse der Schweizerisch-Liechtensteinischen Ausgrabungen 1988–1992*, 13–50. Mainz am Rhein: Philipp von Zabern.

Tholbecq, Laurent. 2013. 'The hinterland of Petra (Jordan) and the Jabal Shara during the Nabataean, Roman and Byzantine periods.' In Michel Mouton and Stephan G. Schmid (eds), *Men on the Rocks: The Formation of Nabataean Petra*, 295–311. Berlin: Logos Verlag.

Urban, Thomas M., Emanuela Bocancea, Clive Vella, Susan N. Herringer, Susan E. Alcock and Christopher A. Tuttle. 2013. 'Investigating ancient dams in Petra's northern hinterland with ground-penetrating radar.' *The Leading Edge* 32(2), 190–2.

Wenner, Sarah E. 2015. 'Petra's hinterland from the Nabataean through Early Byzantine Periods (ca. 63 BC–AD 500).' Unpublished MA dissertation, North Carolina State University.

8

Flowing into the city: Approaches to water management in the early Islamic city of Sultan Kala, Turkmenistan

Tim Williams

Abstract

This chapter explores the demand, organisation and delivery of water into the early Islamic city of Sultan Kala (better known as Merv), in modern-day Turkmenistan. It discusses the way pre-existing agricultural irrigation systems were adapted, and how these water systems shaped the urban space. It considers consumption in the context of spiritual (mosques and madrassahs) and practical (domestic, industrial and sanitary) needs, and how the provisioning of these impacted on the development of urban fabric and communities. This provides a context for considering the broader impacts of water management on society and the role of the state and communities in enabling this.

Introduction

This chapter explores the specifics of the demand for water in the early Islamic city of Sultan Kala (ancient Merv), in modern-day Turkmenistan, and how it was organised and delivered. This discussion sits within a broader discourse on water, social complexity and religion (e.g. Dukhovny and de Schutter, 2011; Manuel et al., 2017; Smith, 2016; Willems and van Schaik 2015), as well as on the much-debated issue of water and the Asian state, which rose to prominence through Wittfogel's hydraulic explanation of 'Oriental despotism' (Wittfogel, 1957, but see Davies, 2009; Harrower, 2009; Stride et al., 2009; Wilkinson and Rayne, 2010).

The Ancient Merv Project from UCL has been undertaking research, in partnership with the Ministry of Culture of Turkmenistan and the Ancient Merv Archaeological Park, since 2001. During this time, the project undertook wide-scale surveys, including aerial photography and topographic modelling, which have been merged with targeted surface collections and excavations. The latter included substantial excavations across the main canal system from 2009 to 2013.

The Murghāb Delta

Ancient Merv lies at the heart of a large fertile dendritic delta, fed by the river Murghāb which drains into the Karakum Desert in modern-day Turkmenistan (Badaev, 1994).[1] The Murghāb rises in the Paropamisus Mountains, one of the northernmost corrugations of the Hindu Kush, some 350 km to the south of Merv, in modern-day Afghanistan, and is joined by two smaller rivers, the Kaysar and the Kachan, before arriving in the delta (Maksimov, 1914). Today the fertile delta covers an area of *c.* 6,000 km^2, although two major changes took place during the Holocene. In antiquity, the delta extended *c.* 25–40 km further north, an area extensively settled in the Bronze Age (Sarianidi, 2002), before the intensification of agriculture and water extraction in the south led to desertification, perhaps from *c.* 1000 BC (Cattani et al., 2008: 40). Much later, the western side of the delta was significantly reshaped by the destruction of the headwater dam in AD 1785 (Skrine and Ross, 1899: 206), which led to the main river channel changing course and caused it to flow some 30 km to the west of its ancient course near Merv. This later event created new fertile areas, particularly in the north-west of the delta.

Precipitation mainly falls as rain or snow, in winter or early spring, with virtually none in the summer months (Orlovsky, 1994). The current average rainfall in the delta is 157 mm, while the daytime temperature ranges from *c.* 2 °C in the winter to *c.* 40 °C in the summer.[2] Potential evaporation rates vary from 1–2 mm per day in winter to a very significant 10–15 mm in summer, giving an annual evaporation rate of *c.* 2,500–3,000 mm, much higher than precipitation (Nesbitt and O'Hara, 2000: 106). Köppen and Geiger classify the climate as BWk: arid, desert, cold arid (Kottek et al., 2006). The fertility and agricultural productivity of the delta, therefore, clearly depended upon the river system.

The Murghāb, fed by winter rain and snowmelt, today has a maximum discharge in the spring, reaching *c.* 180 m^3 per second in May and dropping to 26.5 m^3 per second in August.[3] Water management, therefore, through the long hot summer, would have been vital for both the city and the agriculture of the region.

Urbanisation

After the collapse of the major Bactria-Margiana Bronze Age polities in the northern delta (in the early second millennium BC) (Sarianidi, 2002), significant urbanisation in the delta did not occur until the expansion of the Achaemenid into the region in the mid-sixth century BC (Genito, 1998). The first oval city, Erk Kala (Herrmann, 1997: 8–10), encompassed some 12 ha, and was constructed on the right (east) bank of one of the main river channels at Merv, which was crucial in the location of Achaemenid and later cities (Figure 8.1). It has been suggested that the Achaemenid period also saw the first manipulation of the river system of the delta through the use of dams (Cerasetti, 2002: 22). The scale of urbanism radically changed with the construction of the Hellenistic city of Antiochus Margiana,[4] probably a Seleucid foundation of

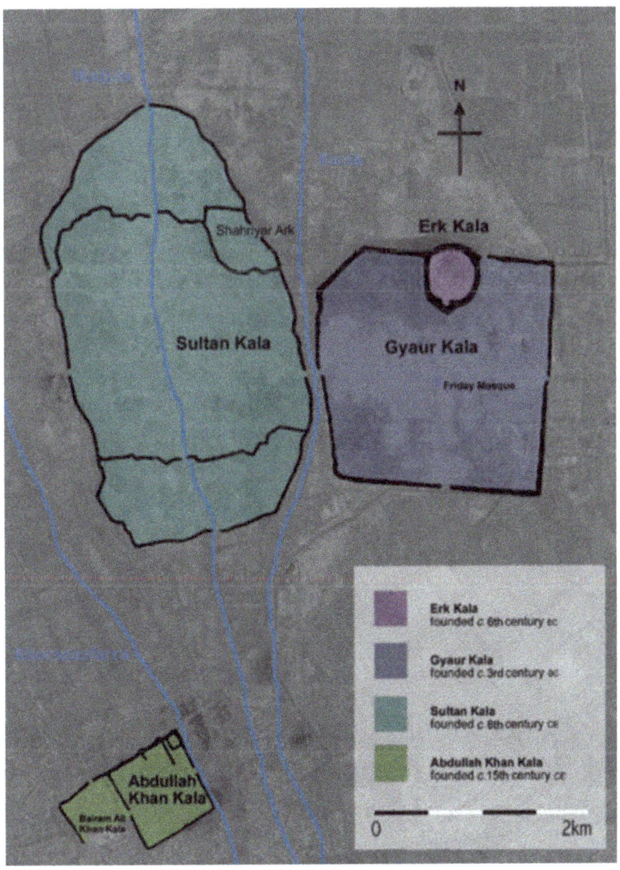

Figure 8.1 The cities of Merv, showing three of the main canals.
Source: author

Antiochus I Soter (280–261 BC) (Herrmann, 1997: 10–17) (Figure 8.1). The new city covered some 337 ha, reflecting an explosion of urban population and concomitant pressures upon the agricultural productivity of the delta. This city, occupied and developed over the next nine centuries, through the Greco-Bactrian, Parthian and Sasanian periods, would have sustained the demand for water, for both agricultural production and the fulfilment of the city's requirements.

The expanding Islamic polity occupied the Sasanian city in AD 651. It appears that Merv peacefully surrendered, rather than being conquered (Kennedy, 1999: 28): as a result, there would have been no evictions of residents, and space for the incoming soldiers would have had to be provided by billeting and perhaps temporary camps. Some modifications of the urban landscape occurred, including the construction of a Friday mosque and administrative complex at the centre of the ancient town. However, in AD 671 a large number of settlers were sent from Basra and Kufa to Merv (Al-Ṭabarī, 1879–1901, ii, 81), which may have put a strain upon the available urban housing. A walled camp, Shaim Kala, constructed to the east of the city, may have temporarily accommodated some of these settlers.

By this time the streets of the old city sat on *c.* 11–13 m of stratigraphy, a palimpsest of the earlier cities. This city lay around 13–15 m above the water table, which would have had significant logistical consequences for moving water into the town (see Figure 8.5). Water collection and transportation cycles can be very time-consuming, a daily activity taking considerable energy.[5] Sardobas (enclosed reservoirs) could be used to capture rainwater, primarily in the winter/spring, but for most of the year they would still have needed to be filled manually.

In the mid-eighth century AD a new city, Marv al-Shāhijān (Merv the Great, today Sultan Kala), developed on the left bank (west) of the river channel, opposite the old city of Gyaur Kala (Figure 8.1) (Williams, 2007). It seems probable that the provision of an effective urban water supply was at least part of the imperative for creating this new city.

Demand

The demand for water within urban areas is in part shaped by individual domestic needs, but also by the need to supply water for communal facilities, bathhouses, religious uses, industrial or craft processes and even garden spaces. The Islamic city would have increased the demand for fresh water for spiritual purposes (clean water in mosques and madrassahs), and the rejuvenated significance of bathhouses/*hammams* for the urban community, on top of the urgent needs of daily life (Blair and Bloom, 2009; Deguilhem, 2008; Kamash, 2006; Naff, 2009).

Modern data on per capita urban water consumption by the United Nations Development Programme (UNDP) stated that '20 litres of water a day per person [is] required to meet the most basic human needs' (Watkins, 2006: 2), and '[f]or people with access to a water source within 1 kilometre, but not in their house or yard, consumption typically averages around 20 litres per day' (Watkins, 2006: 35). Modern per capita consumption varies enormously in Asian cities (McIntosh, 2003): for example, litres per capita per day (lpcd) consumption for communities without piped water in Kathmandu is *c.* 65 lpcd and in Jakarta 75 lpcd, but the 'real consumption in many areas is more like 30–40 [lpcd]' (McIntosh, 2003: 68). A recent study of urban water supply in Accra, Ghana, suggested 41 lpcd when the primary water supply had to be carried to property, whereas piped water raised the consumption to 90 lpcd (Purshouse et al., 2015: 16). On the conservative side, therefore, we might be considering a domestic use in the range of 20–40 lpcd for the Islamic city. As noted, however, the overall urban consumption will be affected by non-domestic uses, and the total water demand is likely to have been much higher (see discussion below).

The water supply

The provisioning of water to urban communities often comprises the elements of raw water extraction, transportation to the city, water treatment, storage, distribution within the city, and waste water disposal.

Extraction

As we have already noted, the climatic and environmental context of the region makes it difficult to use rainfall for urban water supply or irrigation, other than as a seasonal supplement; rather, it was the Murghāb that was central to supplying water to the delta.

The extraction of water from the main river channel into canals would probably have been controlled by a dam, which Al-Muqaddasī stated was some 6 *farsakhs* (*c.* 34–8 km) south of Merv (Volin et al., 1939: 203).[6] This would place it a little further north than the location of the modern dam at Yolatan, somewhere in the region of modern-day Khuzeye-Beden. Al-Muqaddasī described the dam as having a round pond, from which canals branched out (Volin et al., 1939: 203). It is probable that a channel from this dam then branched closer to the city, to create the four canals in the immediate vicinity of Merv.

Delivery: the canal system

There are four canals named in Islamic sources (see Kennedy, 1999) (Figure 8.1). These are referred to by their Islamic names, although most probably pre-date the Islamic period in construction. From east to west, they were:

- The **Asadi al-Khorasani**: this flowed to the east of Erk Kala and Gyaur Kala, and could have been used to supply water to those cities (along with the Razik). It is likely that it was used for suburban activity to the east of the city, such as the extramural Buddhist monastic complex. It probably also supplied fields and settlements further north in the delta (see Williams, forthcoming). The existence of this canal may also have influenced the siting of the Islamic settlement camp of Shaim Kala, to the east of the old city.
- The **Razik**: this channel lay immediately to the west of Gyaur Kala, and was probably used to supply water to the city, before supplying fields and settlements further north. The channel is much more deeply incised than the other canals in the area (Figure 8.5), suggesting that it may have been one of the earliest channels. The siting of the Achaemenid and Seleucid cities, occupying the eastern bank of this 'river', would seem to support this, and it may even have been an ancient branch of the Murghāb, rather than a canal (see Williams, forthcoming).
- The **Madjān**: this flowed along a ridge of slightly higher ground to the west of the Razik (Figure 8.1), which was later to form the spine of the Islamic city of Sultan Kala (Figure 8.5, see below). It was initially used for irrigation, and possibly to supply settlements to the north. It was mentioned in written sources when the last Sassanid king, Yezdigerd III, was assassinated in a mill on the Madjān in AD 651, having been refused entry to the city (Volin et al., 1939: 96).
- The **Khormuzfarra**: this was the most westerly canal of the group, again probably originally constructed for irrigation and supplying towns to the north. This canal was crucial for supplying the elite building complexes and gardens to the west of the city of Sultan Kala, and to the industrial suburb (Williams, forthcoming).

It seems probable that part of this 'canal' system commenced as early as the sixth century BC, when the Achaemenids centralised urbanism in the delta. Indeed, the Razik canal/river channel to the west of Gyaur Kala was possibly part of enabling the process of more intensive occupation of the centre of the delta. The Madjān and Asadi al-Khorasani canals were certainly operating by the Sasanian period, in a system designed to provide agricultural irrigation to the delta, but that was also pivotal to supplying the urban populace, both at Merv and in the settlements further north.

The delivery of water to Sultan Kala

The Madjān was crucial to the development of the new Islamic city of Sultan Kala in the eighth century AD. Along its banks were constructed central administrative buildings (*dār al-imāra*), the Friday Mosque, and the central bazaars that were to form the heart of the new town (Williams, 2007: 45–7). In addition, many major water users, such as the large *hammams*, were situated along the main canal to ensure effective access to water (Figure 8.2).

Figure 8.2 The remains of a major bathhouse, constructed alongside the Madjān Canal, probably in the mid-eighth century AD. These baths were largely demolished during the construction of the town wall in the late eleventh century, but substantial elements were incorporated into the town wall and thus survived remarkably well. Source: author

Excavations across the Madjān revealed an irrigation channel (Figure 8.3), winding along a ridge of high ground (Figure 8.5) and running south to north, to the west of the Razik channel. This was a long-lived channel, which started as an earth-cut channel, with no apparent revetting, and then went through a long sequence of silting, cleaning and recutting. Silting in one of the later channels, before a final recut, has provisionally been dated to the seventh century AD by optically stimulated luminescence dating (Lisa Snape-Kennedy, personal communication[7]), which would confirm that the channel was certainly in use during the Sasanian period.

This channel was to become the primary water supply of the new city of Sultan Kala, *c.* AD 750. The earth-cut channel was replaced by substantial, and well-constructed, revetted fired-brick retaining walls (Figures 8.3 and 8.4). It was an open channel, *c.* 3.6 m wide at the surface, stepping down to the bed of the canal, which was 1.5 m wide, with

Figure 8.3 The Madjān Canal sequence: at the base, Sasanian irrigation ditches, recut on a number of occasions; above this the yellow fired-brick stepped walls of the Islamic canal; infilling this is tumbled rubble and debris from the Mongol sack of the city; and finally laminated deposits of water-laid silts of the stream that flowed through the post-sack ruins. Source: author

Figure 8.4 The Islamic-period Madjān Canal from above, showing the well-constructed retaining walls. Source: author

the overall canal *c*. 2.7 to 3 m deep. There is no evidence that the retaining walls were lined with mortar, or that there was any solid base to the channel. The canal was clearly very well maintained: despite its being in operation for *c*. 470 years there was no silting in its base (Figure 8.3), suggesting that it was regularly, and thoroughly, cleaned.

Distribution in the city

Water was delivered into the city from the central south-to-north Madjān Canal through a complex array of conduits, channels and water pipes. These relied on gravity to carry water from the main channel, almost exclusively running in an east–west direction down from the ridge upon which the Madjān was located (Figure 8.5).

- *Brick channels and ceramic water pipes*: the excavation across the Madjān revealed a complex water distribution system from the main channel (Figure 8.6) (Williams, forthcoming). The quality of the build, and the scale of replacement of these systems, clearly demonstrate the scale of maintenance that was undertaken to ensure that a reliable water delivery system was maintained for the city. Many of the ceramic water pipes were tapered to ensure a tight interlocking fit. It is not clear whether there was a functional difference between the two systems, which often lay side by side.

An open brick channel was observed leading to one of the district reservoirs (below), perhaps suggesting that there was a distinction between intermittent supply to reservoirs or tanks – filling them as needed – and constant water supply, through pipes, to specific buildings and fountains.

- *Brick conduits*: two were exposed by the South Turkmenistan Archaeological Inter-disciplinary Expedition of the Academy of Sciences of the Turkmen Soviet Socialist Republic (YuTAKE) work at the Kushmeihan Gate (Khodjaniasov, 1999). According to YuTAKE's records, one had a vault (0.8 m wide and 1.7 m high); the other had an inverted 'V'-shaped roof (0.6 m wide and 1.05 m high). These conduits were clearly capable of distributing much larger volumes of water than the brick channels and ceramic pipes discussed above, and perhaps they were designed to provide water for some of the larger district reservoirs (see below), or high-demand users, such as *hammams* (see below).

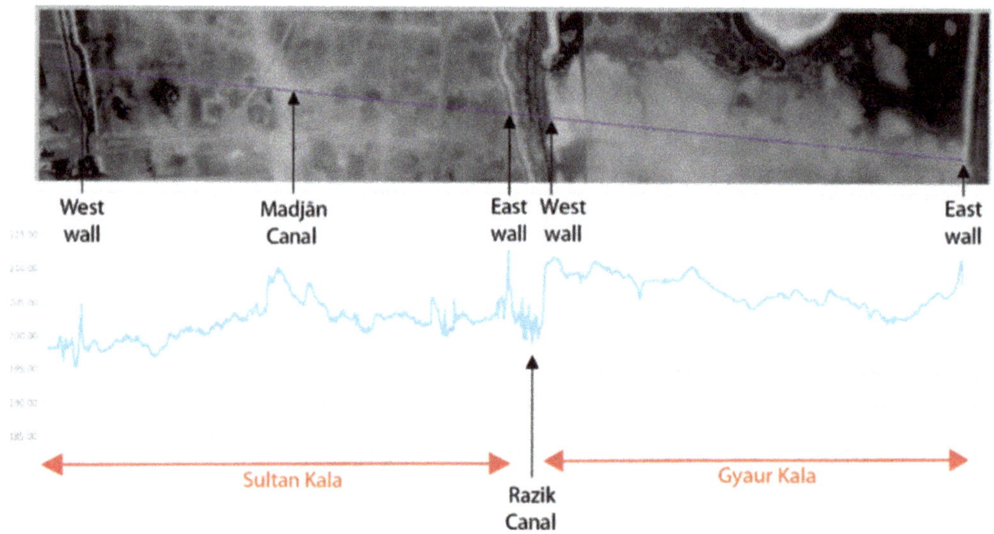

Figure 8.5 Cross-section of Sultan Kala and Gyaur Kala. The current surface of Sultan Kala slopes down from c. 210 m at the Madjān to c. 200 m at the western city wall, about 1 km away; to the east the slope is a little less pronounced, until it drops into the disturbed area of the Razik Canal, with current low points around 197 m. Note the considerably higher internal area of Gyaur Kala. Source: author

Figure 8.6 Excavations in the Madjān Canal, showing brick conduits and ceramic water pipes in the foreground. Source: author

Operation and maintenance

The tenth-century AD writer Al-Muqaddasī (translated in Volin et al., 1939 and by Collins in Al-Muqaddasī, 1994) stated that there were wooden boards, used to measure the water level, at the beginning of each canal. Each board was divided into units of one *shair* (3 cm): if the water came up to the 60-*shair* mark (1.8 m), a good harvest could be expected; if the water only moved up six divisions (0.18 m), it would be a lean year. Al-Istakhrī, also writing in the tenth century (*Masālik al-Mamālik*, translated by Kennedy and Moore, 1999: 123), stated that these wooden boards were also used in the town, for measuring water for each channel that supplied an individual street or quarter. It seems likely that the brick conduits and channels along the Madjān Canal were controlled by wooden sluices to regulate the flow of water from the main canal to various reservoirs and districts.

It is probable that some water filtration and removal of sediment would have taken place to the south of the town, before the water entered the city. In addition, a mortar-lined tank alongside the Madjān Canal in the northern part of the city (Williams, forthcoming), relined many times, may have been a settling tank for the removal of sediment from the canal water.

Al-Muqaddasī stated that there were often water shortages in Merv, so an efficient and fair water-distribution system was essential (Volin et al., 1939: 194). A special official (the *mīr-āb*) was in charge of the complex water-distribution system, and he in turn gave instructions for water distribution to those in charge of specific smaller waterways. An officer called the *dīwān al-kastabzūd* kept a record of all those entitled to a share in the water (Bosworth, 1969). Al-Istakhrī (in *Masālik al-Mamālik*) also stated that there were 10,000 people who were paid to work on the water supply (Kennedy and Moore, 1999: 123): probably an exaggeration, but indicative of the importance of water management to the delta's agriculture and cities.

Storage

Because of the very variable seasonal water flow in the main river (above), reserves would have been an important element in the management of the urban supply. Al-Muqaddasī stated that the people of Merv 'have covered and uncovered reservoirs complete with steps and gates, which are opened and connected up to the canals when it proves necessary' (Volin et al., 1939: 204).

District reservoirs

An unpublished survey by Terkesh Khodjaniasov (1987) stated that his team had identified 12 reservoirs in the northern half of the walled inner city, some 200–300 m apart, suggesting upwards of 24 for the inner city. During our aerial and field surveys, however, it has been impossible to argue convincingly which were large building complexes with substantial courtyards, and which might be reservoirs. However, Khodjaniasov excavated a trench across one reservoir, *c.* 25 m east of the Madjān. The excavation revealed a square pool, 20 m across and 3 m deep, with fired-brick retaining walls 0.4 m thick. The reservoir was filled by a narrow fired-brick channel from the Madjān. In front of the reservoir a square courtyard, 7.5 m across, with a fired-brick pavement, provided access to the reservoir.

This reservoir could have held a maximum of 1,200 cubic metres of water, or *c.* 1.2 million litres. At 40 lpcd, this would have been enough for 1,000 people for a month, although evaporation, especially in the hot summer months, from these uncovered tanks must have been significant. If there were 24 district reservoirs in the main city, assuming a similar size, they would have provided a capacity of nearly 29 million litres. However, that alone, with a population of 65,000, would only provide a reserve of 12 days at 40 lpcd, or 24 days at 20 lpcd.

Suburban reservoirs and tanks

A large square tank, with a fired-brick retaining wall, was explored by Khodjaniasov (1987) in the eastern part of the southern suburb. It was 54 m across and 2.5 m deep, giving a maximum capacity of 6.25 million litres. Khodjaniasov stated that early silt deposits in the reservoir contained material dated to the ninth and tenth centuries AD, perhaps suggesting a construction date in the eighth or ninth century. He also suggested that the reservoir was filled from the Razik Canal, some 170 m to the east, which seems likely, as the reservoir's construction pre-dates the southern city wall (constructed in the late eleventh century). Other possible reservoirs have been identified in the southern suburban area on the digital elevation model compiled from unmanned aerial vehicle (UAV) photography (Jorayev, forthcoming). Such reservoirs would have made a useful addition to the city water reserves, without drawing water off the central Madjān Canal. They would have been difficult to use for piped domestic water consumption in the main city, being much lower than the built-up residential areas to the north-west, but they would have been useful as a general reserve and would also have provided water to the large garden and orchard areas that lay at the periphery of the early Islamic town to the north (Williams, forthcoming).

The excavations of the eastern reservoir suggested that the tank had ceased to function in the eleventh or twelfth century AD (Khodjaniasov, 1987). This would have broadly coincided with the blocking of the canal access from the Razik by the construction of the southern suburb town wall in the late eleventh century (Brun et al., forthcoming). This suggests that demand had decreased, and, if they were servicing the gardens and orchards to the north, that the construction of the southern city wall to the north blocked access to that area. There is no evidence of a water-gate further to the north, which would have enabled the continued use of water from the Razik.[8]

Sardobas

A sardoba is a deep circular reservoir with a domed roof, located beneath ground level and entered via steps, fed either by channels or pipes from the canals, or by rainwater. The main advantage of sardobas is that the roof reduces evaporation, and a large amount of water can be stored at a cooler temperature. They are often found in association with public buildings and mosques.

Thus far, six sardobas have been identified at Merv: the best-preserved are at the original Friday Mosque in Gyaur Kala, at the suburban Mausoleum of Muhamad ibn Zaid, and by tombs in the southern walled suburb of Sultan Kala. The sardoba at the Mausoleum of Muhamad ibn Zaid is a good example of the type. Constructed in the early twelfth century AD (Pugachenkova, 1958: 303–10), the reservoir is 6.1 m in diameter, with floor and walls (to a height of 5.5 m) surviving, complete with a stepped entrance (Herrmann et al., 2002: 39). The floor and walls are of fired brick, bonded with a clay mortar externally, but on the interior with hard, waterproof mortar. Above the brick floor were layers of hard mortar, with a smooth surface. Two apertures survive, which connected the sardoba to supply pipes, perhaps filled from the nearby Khormuzfarra Canal. Assuming a water depth of 2 m, the sardoba could hold *c.* 230,000 litres.

Later modifications: The city wall

The construction of the city wall, in the late eleventh century AD, would have had a major impact on the water-management systems of the city. Where the main city wall crossed the Madjān Canal in the north, substantial modifications took place, both to defend the area and to ensure the continued use of the canal, maintaining its flow into the northern suburb. The area is called the Kushmeihan Gate. Terkesh Khodjaniasov undertook exploratory works here in 1999 (Khodjaniasov, 1999). These revealed a sluice gate constructed across the canal, comprised of two engaged brick pillars set either side of the channel, leaving a central gap of 0.8 m which could be closed by a wooden shutter. North of this, the canal was modified to pass underneath the city moat in a brick conduit. A complex series of outer works or defences were constructed to protect this vulnerable point. North of the moat, the canal re-emerged to flow through the north suburb, before exiting the northern walled suburb at the Poi Madjān Gate. Similar complex arrangements (seen on aerial photography) existed at the south gate of the main city. Some form of protection or control is also likely at the entrance/exit of the canal to the south and north walled suburbs, although aerial photography does not reveal the same scale of features, which suggests that these may not have been as heavily fortified (Brun et al., forthcoming).

The scale of the supply

If we assume that the Madjān Canal provided the core water supply of the city, with the Razik and Khormuzfarra supplying suburban areas and the gardens and orchards to the west and east of the city, then we can speculate on the scale of urban supply.

Using our excavations across the Madjān, we can *very crudely* explore stream flow using:

Quantity=$A*V*M$

where

A = Average cross-sectional area of the channel (channel *width* multiplied by water *depth*)
V = Estimate of the velocity of water flow. The canal had a low incline, so we would perhaps not expect it to be fast-flowing, although the flow rate is also a factor of the river discharge at the headwater dam
M = Mean water flow (velocity reduced by friction/resistance). Water at the surface travels faster than near the channel bottom and sides, because of friction or resistance (the lining impacts on this), wetted perimeter, etc.

Estimates:

- ***Area (A)***: the canal width is *c.* 1.5 m (slightly wider at the top). There are likely to have been significant fluctuations in the depth of water, but given the depth of the channel (*c.* 2.7 to *c.* 3 m), we might suggest that the maximum capacity was *c.* 1.5 m (A = 2.25), the minimum 0.5 m (A = 0.75).
- ***Velocity (V)*** (in metres per second = mps): there was no sediment in the canal, but that was probably the result of regular cleaning rather than an indication of water speed. This is very difficult to estimate for the Madjān: the flow might be as slow as 0.1 mps, to a moderate flow of perhaps 0.5 mps, an estimate based on analogies with modern low-incline streams and canal channels.[9]
- ***Friction (M)***: a resistance factor of 0.729 is used.[10] This is not directly comparable, but it reflects some of the complexity of water flow estimation.

The quantities of water are shown in Table 8.1

Table 8.1 Estimates of water flow and delivery

Width	Depth	Area (*A*)	Velocity mps (*V*)	Mean velocity (*F*)	m^3 per sec	m^3 per hour	Litres per hour	lpcd	Population	Hours
				Slow water flow (0.1 mps)						
1.50	1.50	2.25	0.10	0.729	0.16	590	590,490	40	65,000	4.40
1.50	1.00	1.50	0.10	0.729	0.11	394	393,660	40	65,000	6.60
1.50	0.50	0.75	0.10	0.729	0.05	197	196,830	40	65,000	13.21
				Medium water flow (0.2 mps)						
1.50	1.50	2.25	0.20	0.729	0.33	1181	1,180,980	40	65,000	2.20
1.50	1.00	1.50	0.20	0.729	0.22	787	787,320	40	65,000	3.30
1.50	0.50	0.75	0.20	0.729	0.11	394	393,660	40	65,000	6.60
				Faster water flow (0.5 mps)						
1.50	1.50	2.25	0.50	0.729	0.82	2952	2,952,450	40	65,000	0.88
1.50	1.00	1.50	0.50	0.729	0.55	1968	1,968,300	40	65,000	1.32
1.50	0.50	0.75	0.50	0.729	0.27	984	984,150	40	65,000	2.64

Scenarios might be:

- A peak flow for the rainy/meltwater season, with a depth of 1.5 m of water in the canal (A = 2.25) and a reasonably fast flow of water (V = 0.5 mps), would have provided 0.83 m^3 of water per second, or 2,952 m^3 per hour, equating to nearly 3 million litres per hour. Given a hypothetical population of 65,000 people in the main city,[11] requiring 40 lpcd (2.6 million litres), it would only take 53 minutes (0.88 hours) for that quantity to pass into the city. However, it is very unlikely, even at peak flow, that this volume was attained, or maintained.
- Perhaps more realistically, if we were to assume 1 m of water in the canal (A = 1.5) flowing at a moderate speed (V = 0.2 mps), the canal would supply nearly 800,000 litres per hour, and it would still only take about 3 hours 18 minutes to deliver enough for the 65,000 people.
- A minimum level, in the summer months, might be only 0.5 m water in the canal (A = 0.75), flowing slowly (V = 0.1 mps). This would supply around 200,000 litres per hour, and it would still only take about 13 hours 12 minutes to deliver the water for 65,000 people.

In practice, heavy water users (such as bathhouses, mosques, and industrial or craft processes), combined with the evaporation rates along the canal and especially in the large open reservoirs (see above), would have massively increased the demand on the urban water supply (even if gardens and orchards on the periphery were served by other means – see the discussion of reservoirs above). Doubling the estimate of the urban demand from 2.6 million litres to 5.2 million litres per day to allow for these additional demands would suggest that the Madjān would be stretched to cope in the middle of summer. As mentioned above, Al-Muqaddasī stated that there were often water shortages in Merv (Volin et al., 1939: 194), and management of the supply over the long hot summer must have been extremely challenging.

Waste water disposal

Waste water disposal probably relied on gravity in a similar way to the water supply: waste water would have drained to the east and west, down from the slight north–south ridge along which the Madjān Canal flowed and the main city. Some waste water, such as the discharge from bathhouses and the overflow from reservoirs and fountains, might have been used in these low-lying areas, thought to be garden and orchard spaces, especially in the east of the city (Williams, 2008). However, if sewage and waste water drained further east it would have entered the Razik Canal, and if to the west the Khormuzfarra Canal: were these used to carry away effluent water from the city? If these canals supplied towns in the north of the delta, this practice would have compromised the quality of the water that reached them. If, though, these canals were used during the Islamic period primarily to supply irrigation systems north of Merv, and not the settlements, then perhaps they could have carried effluent from Sultan Kala.

At present, there is no evidence of effluent flowing back into the Madjān Canal, although it has to be said that only a small area has been explored archaeologically. However, given that a number of very large building complexes were clustered along the canal in the northern walled suburb (Williams, forthcoming), it seems unlikely that the water quality would have been seriously compromised before the canal exited the city through the northern Poi Madjān Gate.

Water: Shaping the urban landscape

Michael Bonine (1979: 208) argued:

> Traditional Iranian cities have an orthogonal network of streets which ... did not develop from an outgrowth of streets around rectangular religious buildings or from the orientation of Iranian houses to maximize seasonal usage, but rather ... is due to irrigation systems. The orthogonal network of water channels corresponds to the slope of the land. Passageways follow these channels to reach various plots of cultivated land. Cities have expanded along the existing streets and water channels. The basic morphology of traditional Iranian cities was created by houses filling in adjacent rectangular fields and orchards.

The structuring of the city of Sultan Kala shows many similar attributes (Williams, 2007, 2008). The network of streets in Sultan Kala (Figure 8.7) had a strong east–west rhythm, lacking long north–south lines (except for the central Madjān Canal). The east–west routes are not straight, but they form clear segments of the cityscape, and it can be

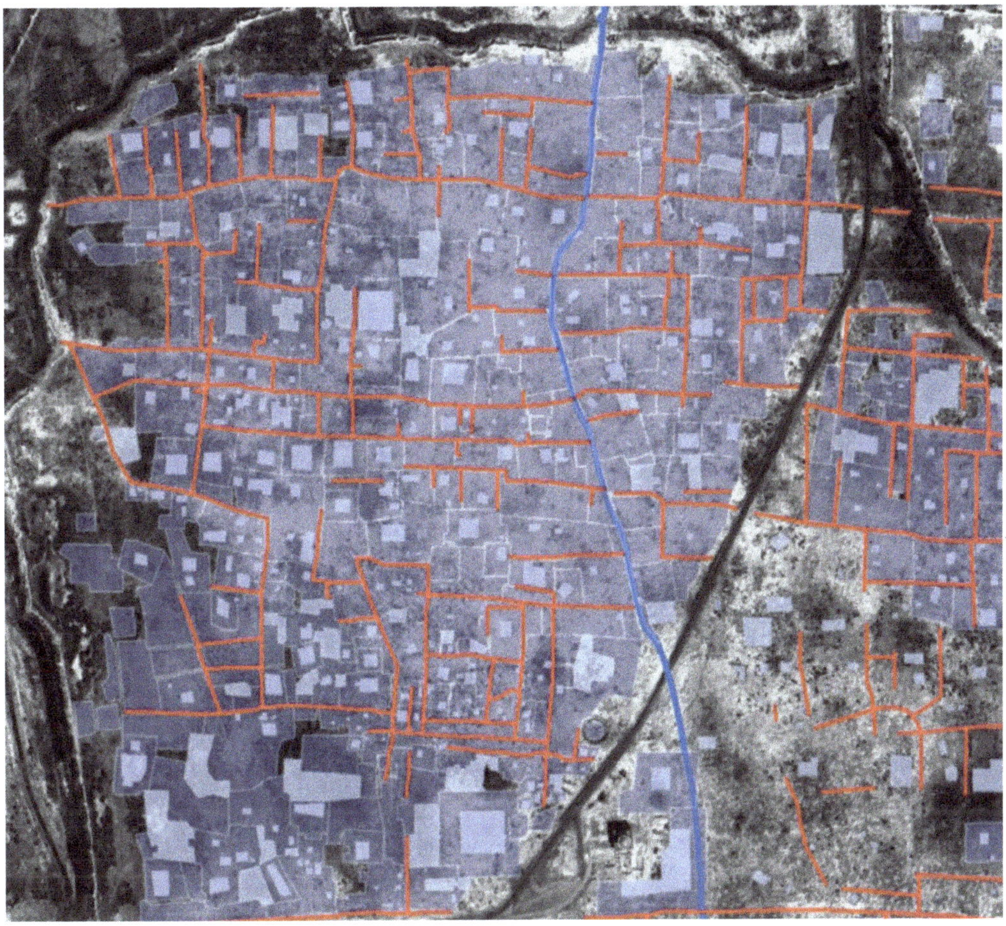

Figure 8.7 The north-western quarter of the Islamic city, showing digitisation (in progress at the time of writing): the central Madjān Canal (dark blue), major streets (red), minor streets (white), built-up land (blue) and open spaces/courtyards (light blue). North is at the top. Source: author

suggested that together these frame the development of land allocations and thus neighbourhoods. It is tempting to see this pattern as reflecting pre-existing field systems, like those Bonine saw in Iran. These may have dictated land allocation: elite groups, ethnic or religious communities, or trades and crafts (Southall, 1998: 226–28) may have been allocated land within this existing divided landscape.

Paul Wheatley (2001) cogently articulated the concept that Islamic cities are often characterised as functional, but only partially civil, the control and maintenance of shared local facilities contrasting with political elites' centralised facilities. At Sultan Kala it seems likely that the state directly intervened in the development of the new city, but in addition to providing centralised facilities – such as the administrative core (*dār al-imāra*), the congregational mosque and the central bazaars

(Al-Iṣṭakhrī, 1927: 258) – it most importantly provided the urban water supply. Converting a pre-existing irrigation channel into the substantial brick-lined Madjān Canal was a major civic undertaking, but it enabled water to lie at the very heart of the new urban development. The central canal system was a mechanism for meeting the spiritual and functional needs of the new Islamic town: it constituted a water supply capable of servicing the needs not just of the congregational mosque, but of all the district mosques, providing water not just to the major *hammams* constructed alongside the canal, but to all the district bathhouses that serviced each neighbourhood community, and not just to the elite, but to the district reservoirs that enabled neighbourhoods to function and thrive.

Water materially shaped the broad rhythm of the urban landscape, and shaped the daily lives of the people that occupied it.

Acknowledgements

Considerable thanks go to Dr Gai Jorayev for his work on the Unmanned Aerial Vehicle mapping and modelling of the Merv urban landscape, including the information for Figure 8.5. Thanks also go to UCL Qatar, and especially to Professor Thilo Rehren, the Director at the time, for the acquisition of the first UAV, which enabled so much of the broader survey to take place, and for his continued support of the project. Thanks also to Tish Prouse and his dedicated team of Turkmens who excavated the Madjān Canal section.

I am very grateful to our colleagues in Turkmenistan, especially Dr Mukhammed Mamedov, of the Ministry of Culture, and Rejeb Dzaparov and all the staff at the Ancient Merv Archaeological Park. Our team is too large to thank individually, which is a pity as the project would not have been possible without them! The work was supported by the excellent contributions of dedicated professional staff and students, especially from the Institute of Archaeology UCL, UCL Qatar, HWB, and CRAterre-EAG (Grenoble University).

Equally, the work would not have been possible without financial support from a variety of agencies, not least the World Monuments Fund, the Arts and Humanities Research Council, the British Academy, UNESCO, the British Embassy in Turkmenistan, the US Ambassadors Fund for Cultural Preservation, CyArk and UCL Qatar.

Notes

1. For a general review of the archaeology of the Murghāb Delta see the excellent work of the team from the University of Bologna (Gubaev et al., 1998; Salvatori et al., 2008).
2. Data from World Weather online, using Bairam Ali, the closest weather station to Ancient Merv: https://www.worldweatheronline.com/bairam-ali-weather-averages/mary/tm.aspx (accessed May 2017).
3. Measured at the hydrometric station of Takhta-Bazaar. Data calculated over 50 years.
4. The modern name of Antiochus Margiana is Gyaur Kala.
5. For example, see UNICEF (2016), 'Collecting water is often a colossal waste of time for women and girls', https://www.unicef.org/media/media_92690.html.
6. The length of a *farsakh* varied according to terrain and speed of travel. Ideas vary between 6.3 km and 5.67 km (Houtum-Schindler, 1888).
7. Doctoral work undertaken at Durham University, as part of the Persia and its Neighbours project.
8. This is complicated by the fact that the modern canal has destroyed much of the evidence of the Razik in the area. A branch of the modern canal 'breaks through' the Sultan Kala city wall, just to the north of the south-west corner of the walled town, and meanders inside the city before 'breaking out' again south of the junction of the city wall and the citadel of Shahriyar Ark. There is no evidence of structures to protect the entrance/exits, unlike on the Madjān Canal, and these are probably much later, post-abandonment, incursions.
9. Based on water flow velocities in irrigation canals in Iraq (De Araoz, 1962), and at the lower end of the concrete- and masonry-lined irrigation canals examined by Scobey (1939).
10. Based on a study of Iraqi irrigation canals (De Araoz 1962: 121).
11. Based on a population estimate of 80,000–100,000 for the entire city, including the suburbs, at its peak (Williams, forthcoming).

References

Al-Iṣtakhrī, Abu Isḥák al-Fárisí. 1927. *Kitāb al-Masālik wa l-mamālik*, trans. Michael Jan de Goeje. Leiden: Brill.

Al-Muqaddasī, Shams al-Dīn Abū ʿAbd Allāh Muḥammad ibn Aḥmad. 1994. *The Best Divisions for Knowledge of the Regions (Aḥsan al-taqāsīm fī Maʿrifat al-Aqālīm)*, trans. Basil A. Collins. [Doha]: Centre for Muslim Contribution to Civilization/Reading: Garnet.

Al-Ṭabarī, Abū Ja'far Muhammad ibn Jarīr. 1879–1901. *Tārīkh al-rusul wa-al-mulūk*, ed. Michael Jan de Goeje, trans. various. 15 vols. Leiden: Brill.

Badaev, Agadzhan G. 1994. 'Landscapes of Turkmenistan.' In Viktor Fet and Khabibulla I. Atamuradov (eds), *Biogeography and Ecology of Turkmenistan*, 5–22. Dordrecht: Kluwer.

Blair, Sheila and Jonathan Bloom (eds). 2009. *Rivers of Paradise: Water in Islamic Art and Culture*. New Haven, CT: Yale University Press.

Bonine, Michael E. 1979. 'The morphogenesis of Iranian cities.' *Annals of the Association of American Geographers* 69 (2), 208–24.

Bosworth, Clifford Edmund. 1969. 'Abū 'Abdallāh al-Khwārazmī on the technical terms of the secretary's art: A contribution to the administrative history of mediaeval Islam.' *Journal of the Economic and Social History of the Orient* 12, 113–64.

Brun, Pierre, David Gilbert and Tim Williams. In preparation. *The City Walls of Islamic Merv*.

Cattani, Maurizio, Barbara Cerasetti, Sandro Salvatori and Maurizio Tosi. 2008. 'The Murgab Delta in Central Asia 1990–2001: The GIS from research resource to a reasoning tool for the study of settlement change in long-term fluctuations.' In Sandro Salvatori, Maurizio Tosi and Barbara Cerasetti (eds), *The Bronze Age and Early Iron Age in the Margiana Lowlands: Facts and Methodological Proposals for a Redefinition of the Research Strategies*, 39–46. The Archaeological Map of the Murghab Delta Studies and Reports, 2. Oxford: Archaeopress.

Cerasetti, Barbara. 2002. 'A 5000-years history of settlement and irrigation in the Murghab Delta (Turkmenistan): An attempt of reconstruction of ancient deltaic system.' In Göran Burenhult and Johan Arvidsson (eds), *Archaeological Informatics: Pushing The Envelope. CAA 2001: Com-*

puter Applications and Quantitative Methods in Archaeology. Proceedings of the 29th Conference, Gotland, April 2001, 21–7. Oxford: Archaeopress.

Davies, Matthew. 2009. 'Wittfogel's dilemma: Heterarchy and ethnographic approaches to irrigation management in Eastern Africa and Mesopotamia.' *World Archaeology* 41 (1), 16–35.

De Araoz, Joaquin. 1962. 'Study of water flow velocities in irrigation canals in Iraq and their mathematical analysis.' *Bulletin of the World Health Organization* 27 (1), 99–123.

Deguilhem, Randi. 2008. 'The Waqf in the city.' In Salma K. Jayyusi, Renata Holod, Attilio Petruccioli and André Raymond (eds), *The City in the Islamic World*, 923–50. Leiden: Brill.

Dukhovny, Victor A. and Joop de Schutter. 2011. *Water in Central Asia: Past, Present, Future*. Leiden: CRC Press.

Genito, Bruno. 1998. 'The Achaemenids in the history of Central Asia.' In Abdyrakhman G. Gubaev, Gennadii Andreevich Koshelenko and Maurizio Tosi (eds), *The Archaeological Map of the Murghab Delta: Preliminary Reports 1990–95*, 149–58. Rome: Istituto Italiano per l'Africa e l'Oriente.

Gubaev, Abdyrakhman G., Gennadii Andreevich Koshelenko and Maurizio Tosi (eds). 1998. *The Archaeological Map of the Murghab Delta: Preliminary Reports 1990–95*. Rome: Istituto Italiano per l'Africa e l'Oriente.

Harrower, Michael J. 2009. 'Is the hydraulic hypothesis dead yet? Irrigation and social change in ancient Yemen.' *World Archaeology* 41 (1), 58–72.

Herrmann, Georgina. 1997. 'Early and medieval Merv: A tale of three cities.' *Proceedings of the British Academy* 94, 1–43.

Herrmann, Georgina, Helena Coffey, Stuart Laidlaw and Kakamurad Kurbansakhatov. 2002. *The Monuments of Merv: A Scanned Archive of Photographs and Plans*. London: Institute of Archaeology, University College London, and the British Institute of Persian Studies.

Houtum-Schindler, Albert. 1888. 'On the length of the Persian farsakh.' *Proceedings of the Royal Geographical Society and Monthly Record of Geography* 10 (9), 584–88.

Jorayev, Gaigysyz. In press. 'The uses of unmanned aerial vehicles in archaeology: Methodological considerations based on experiences at Ancient Merv, Turkmenistan.' *Advances in Archaeological Practice*.

Kamash, Zena. 2006. 'Water supply and management in the Near East 63 BC–AD 636.' Unpublished PhD thesis, University of Oxford.

Kennedy, Hugh. 1999. 'Medieval Merv: An historical overview.' In Georgina Herrmann (ed.), *Monuments of Merv: Traditional Buildings of the Karakum*, 25–44. London: Society of Antiquaries of London.

Kennedy, Hugh and Oliver Moore. 1999. 'Some ancient and medieval texts about Merv.' In Georgina Herrmann (ed.), *Monuments of Merv: Traditional Buildings of the Karakum*, 120–32. London: Society of Antiquaries of London.

Khodjaniasov, Terkesh. 1987. 'YuTAKE unpublished report: Report on the work of YUTAKE's XVIII Old-Merv Detachment. Autumn 1987.'

Khodjaniasov, Terkesh. 1999. 'Kushmeikhan Gate. YuTAKE unpublished report: Excavation work was carried out in September (13–21: IX) and October (15–19: X) 1999.'

Kottek, Markus, Jürgen Grieser, Cristoph Beck, Bruno Rudolf and Franz Rubel. 2006. 'World Map of the Köppen-Geiger climate classification updated.' *Meteorologische Zeitschrift* 15 (3), 259–63.

Maksimov, Sergei Pavlovich. 1914. *Obshchii otchet izyskanii na r. Murgabie, v tsieliakh izucheniia obezpechennosti orosheniia Murgabskago Gosudareva imieniia: 1907–1909 gg* (General summary of research on the River Murghab, with the aim of researching the irrigation sustainability of Murghab Ruler property: Years 1907–1909). St Petersburg: Tip. Ministerstva putei soobshchenii.

Manuel, Mark, Dale Lightfoot and Morteza Fattahi. 2017. 'The sustainability of ancient water control techniques in Iran: An overview.' *Water History* 10, 13–30.

McIntosh, Arthur C. 2003. *Asian Water Supplies: Reaching the Urban Poor*. London: Asian Development Bank.

Naff, Thomas. 2009. 'Islamic law and the politics of water.' In Joseph W. Dellapenna and Joyeeta Gupta (eds), *The Evolution of the Law and Politics of Water*, 37–52. Dordrecht: Springer.

Nesbitt, Mark and Sarah O'Hara. 2000. 'Irrigation agriculture in Central Asia: A long-term perspective from Turkmenistan.' In Graeme Barker and David Gilbertson (eds), *The Archaeology of Drylands: Living at the Margin*, 103–22. London: Routledge.

Orlovsky, Nikolai S. 1994. 'Climate of Turkmenistan.' In Viktor Fet and Khabibulla I. Atamuradov (eds), *Biogeography and Ecology of Turkmenistan*, pp. 23–48. Dordrecht: Kluwer.

Pugachenkova, Galina Anatol'evna. 1958. 'Puti razvitiia arkhitektury iuzhnogo turkmenistana pory rabovladeniia i feodalizma' (The development of the architecture of southern Turkmenistan in the periods of slave ownership and feudalism). *Trudy YuTAKE*, VI (whole volume).

Purshouse, Heather, Matthias Javorszky, Philip Antwi-Agyei, Andrew Sleigh and Barbara Evans. 2015. 'Modelling formal and informal domestic water consumption in Accra.' eprint, University of Leeds. http://eprints.whiterose.ac.uk/95008/. Accessed 14 June 2018.

Salvatori, Sandro, Maurizio Tosi and Barbara Cerasetti (eds). 2008. *The Bronze Age and Early Iron Age in the Margiana Lowlands: Facts and Methodological Proposals for a Redefinition of the Research Strategies*. Oxford: Archaeopress.

Sarianidi, Victor Ivanavich. 2002. 'The fortification and palace of northern Gonur.' *Iran* 40, 75–87.

Scobey, Fred C. 1939. 'Flow of water in irrigation and similar canals.' Washington, DC: Technical Bulletin 652, United States Department of Agriculture.

Skrine, Francis Henry and Edward Denison Ross. 1899. *The Heart of Asia: A History of Russian Turkestan and the Central Asian Khanates from the Earliest Times*. London: Methuen & Co.

Smith, Monica L. 2016. 'Urban infrastructure as materialized consensus.' *World Archaeology* 48 (1), 164–78.

Southall, Aidan. 1998. *The City in Time and Space*. Cambridge: Cambridge University Press.

Stride, Sebastian, Bernardo Rondelli and Simone Mantellini. 2009. 'Canals versus horses: Political power in the oasis of Samarkand.' *World Archaeology* 41, 73–87.

Volin, S. L., A. A. Romaskevich and A. Yakubovsky (eds). 1939. *Материалы по Истории Туркмен и Туркмении, том 1* (Materials on the history of Turkmen and Turkmenia, vol. 1). Moscow/Leningrad: Institute of Oriental Studies of the Academy of Sciences of the USSR/Institute of History under the CEC of the Turkmen SSR.

Watkins, Kevin. 2006. *Human Development Report 2006. Beyond scarcity: Power, poverty and the global water crisis*. New York: Palgrave Macmillan, for United Nations Development Programme (UNDP).

Wheatley, Paul. 2001. *The Places Where Men Pray Together: Cities in Islamic Lands, Seventh through the Tenth Centuries*. Chicago: University of Chicago Press.

Wilkinson, Tony and Louise Rayne. 2010. 'Hydraulic landscapes and imperial power in the Near East.' *Water History* 2 (2), 115–44.

Willems, Willem J. H. and Henck van Schaik (eds). 2015. *Water & Heritage: Material, Conceptual and Spiritual Connections*. Leiden: Sidestone Press.

Williams, Tim. 2007. 'The city of Sultan Kala, Merv, Turkmenistan: Communities, neighbourhoods and urban planning from the eighth to the thirteenth century.' In Amira K. Bennison and Alison L. Gascoigne (eds), *Cities in the Pre-Modern Islamic World: The Urban Impact of Religion, State and Society*, 42–62. Abingdon: Routledge.

Williams, Tim. 2008. 'The landscapes of Islamic Merv, Turkmenistan: Where to draw the line?' *Internet Archaeology* 25. http://dx.doi.org/10.11141/ia.25.1.

Williams, Tim. In preparation. *Medieval Merv: A Forgotten City on the Silk Roads*.

Wittfogel, Karl A. 1957. *Oriental Despotism: A Comparative Study of Total Power*. London: Yale University Press.

9

Water management across time: Dealing with too much or too little water in ancient Mesopotamia

Mark Altaweel

Abstract

Irrigation salinity has plagued agriculture in dry regions around the world. Simple mechanisms of fallowing allow fields to dry and remove salts from crop root zones. Drainage canals, or salt-tolerant plants that remove salts from soil layers, sometimes aid this process. This chapter explores how different fallowing strategies may have been applied to address concerns of over-irrigation that led to salinisation, while allowing farmers to produce sufficient crops so as not to under-irrigate. The chapter looks at southern Mesopotamia, for which historical and archaeological data provides information on irrigation societies from ancient to more recent periods. Results demonstrate optimal strategies for irrigation for different types of fields based on their drainage capacity. Fields that are well drained, often located along the levee crests, need shorter fallowing, such as one to two years, although longer fallow periods are required as salinisation increases. For more poorly drained areas, along levee slopes and basins, fallowing was generally longer, perhaps five years or more. Conditions of high salinity, however, may require shorter fallow periods, as fields become quickly saturated with salts.

Introduction

Managing water systems over variable time and over changing environmental conditions requires water system management to be adaptive and to be able to evolve so that given water use systems can be sustained. This is the case with irrigation systems, where in dry regions issues of salinisation are often present (Fritsch and Fitzpatrick, 1994). In ancient Mesopotamia salinisation has been identified as an issue: overuse of water resources can result in salinised fields, and underuse of irrigation in underproducing fields (Jacobsen and Adams, 1958; Jacobsen, 1982). In effect, while irrigation is necessary in southern Mesopotamia for crop production, overdependence on it can lead to diminished yields and salinised fields, while limiting it too much leads to low productivity. Scholars have also contested that mitigation of salts, in the form of sodium chloride as well as other mineral salts, would have been possible, and salinity may have not been a major concern in agriculture (Powell, 1985).

This chapter addresses cases in which irrigation agriculture is applied to a southern Mesopotamian case, using the cities and regions around Nippur and Uruk, to demonstrate how different adaptive strategies could limit salinity in cases where it might have become an impediment to irrigation agriculture. The intent is to demonstrate how a balanced use of irrigation could result in optimal strategies for different types of irrigated fields. The goal is also to demonstrate where irrigation salinity could become a serious problem, as drainage and fallowing may not allow adequate time for fields to recover. On the other hand, adaptive responses through optimal fallowing and drainage are discussed too. The object of this chapter is to demonstrate the types of strategies that could be undertaken to limit salinisation and also to adapt to environmental conditions that threaten productivity.

The chapter begins by giving historical background information about the region studied as well as about the process of salinisation; the applied methods, which include the modelling approach used, are then discussed. Within the methods, the data for the case is presented. The results of a simulation model are used to demonstrate strategies that might have been undertaken by agriculturalists in cases where salinity was a hindrance to agriculture. These include a demonstration of different fallowing strategies for three types of fields that have different qualities of drainage. Discussion of the implications of these results is presented, as well as a brief conclusion.

Background

Historical background

Periodic episodes of salinisation have been reported for southern Mesopotamia, where such salinisation has been blamed for the decline in settlement and episodes of declining agriculture spanning the Middle (*c.* 1600 BC) and Late (*c.* 1200 BC) Bronze Ages into the early second millennium AD (Butzer, 2012; Jacobsen and Adams, 1958). In fact, even throughout the twentieth century modern Iraq faced high salinity, resulting in major drainage canal projects that were intended to alleviate this (Al-Ansari et al., 2014). Although southern Mesopotamia in ancient periods was often home to some of the largest urban zones, such as Babylon and Uruk, and more recently Baghdad, overdependence on irrigation could lead to salinity in the region, because of its high temperatures and dry conditions. For evidence of salinisation, scholars have presented textual sources that record declining yields and a switch in the late third millennium BC to a greater use of barley, a more salt-tolerant grain. Although the third millennium BC is often seen as a period of widespread settlement in southern Mesopotamia, declining yields have been suggested as evidence of increased agricultural stress from salinisation (Jacobsen, 1982; Maekawa, 1974). Records of yield decline, by a third in places over the span of a few hundred years, with barley eventually making up nearly 98 per cent of the grain grown, form some of the relevant supporting data (Maekawa, 1973–4, 1984).

If we assume that some of the yield results mentioned above may reflect salinisation, we can also assume that human societies were not weak victims of their environment. Adaptive responses to increased salinity are likely in any period. The cultivation of barley is one possible example of such adaptation. However, other responses include fallowing, leaching of soils (using both natural and engineered means such as washing soils or purposely flooding fields), growing plants that absorb salts, and removing water to prevent salinity through drainage (Gibson, 1974; Jacobsen, 1982; Powell, 1985). Drainage canals are a possibility in ancient periods; they would have prevented both waterlogging and increased salinity. However, they would have required more extensive work and it is not clear if this occurred in ancient periods, or at least before late antiquity in the first millennium AD (Artzy and Hillel, 1988; Poyck, 1962). More recent studies of salinity in Iraq have shown that many mitigation strategies offer mostly temporary reprieve: soils can easily become filled with salts, or the salts are not easily removed from the soil profile once added. Often, the most effective method for removing salts from the root zone, where plant growth is most important, is simply to leave fields fallow so that salt naturally drains from the upper layers, even if it does not disappear entirely (Gibson, 1974). Short-term fallowing, however, may prove a temporary measure, as extended fallow periods may be needed to drain fields adequately. This includes multiple seasons of fallowing rather than a one-year-fallow/one-year-crop (biennial) fallow system, which is a more typical scenario in southern Mesopotamia. Balancing the right number of fallow years or cycles with rates of salinity therefore becomes a key problem for Mesopotamian agriculturalists.

Process of salinisation

Salinisation, or the addition of sodium chloride and other salt minerals to agricultural fields, is a common process in dry regions and in areas where poor drainage is present, there are high levels of salts in soils, over-irrigation is common, and a high water table exists (Chhabra, 1996; Smedema and Shiati, 2002). Capillary rise, through which the high evaporation rates common to hot regions often leave a crust of salt at the surface, and the use of salty irrigation water, are among the most common processes that add salt to plant root zones. Rainfall, if sufficient, can help to leach salts from fields. Areas that are low in slope or have poor drainage generally accumulate more salts.

In the past, fallowing was probably practised, as it allowed rain and natural drainage to remove salts from fields. Plants such as *Proserpina stephanis* and *Alhagi maurorum* were grown on fallow fields to help dry them out by diminishing capillary rise in the root zone and generally drying out subsoil layers. If salts are not removed in an agricultural yearly cycle from subsoil layers, it is possible that they will reappear the following season when irrigation is brought in and water rises through capillary action. Rapid evaporation, poor drainage, capillary action (i.e., a high water table) and a lack of sufficient rainfall to wash fields are all evident in southern Mesopotamia, making it highly susceptible to salinisation. While today salinity is a major problem in southern Iraq (i.e., southern Mesopotamia), it is assumed that similar processes occurred in the past, as the conditions for salinity present today were very likely true in different periods (Artzy and Hillel, 1988; Gibson, 1974; Jacobsen and Adams, 1958).

Methods

Applied modelling

To study the process of salinisation and methods of adaptation, a simulation model is created. This modelling method uses my earlier approach (Altaweel, 2013) to investigating salinisation, which builds on Prendergast's (1993) model of salinisation. Figure 9.1 highlights the model's agricultural steps, which integrate human intervention with a model that simulates salt change to the root zone due to irrigation. In effect, the steps represent a social-ecological modelling approach, in which human actions adapt and respond to environmental stimuli. Altaweel (2013) provides an appendix of the notation that summarises the model, which includes a site from which the code can be downloaded. The model is only summarised here, as it is available in the earlier work.

The basic model allows salt to accumulate in the root zone of agricultural areas: irrigation and rainfall add salts to fields. This accumulation is balanced by the fact that salts can be leached from soils, through human intervention or through natural washing or removal of salts. Although it is not known how human actors may have stimulated leaching, we can assume that some rate, even if minimal, occurred. Salt build-up occurs, and is measured through electric conductivity, in which decisiemens per metre (dS/m) is a common unit of measurement of salinity. Salts do not accumulate uniformly, as areas that have greater slope generally drain better. The leaching fraction and evaporation determine the rate of salts being added to the root zone. Yield is based on the effect of salt in the root zone and the type of agricultural crop (i.e., more or less resistant to salt); in applied modelling, barley, which is more resistant to salt, is used. Capillary rise is also a factor, as it brings salts to the surface and the root zone. The key decision to be made, then, is how long fields should remain fallow under conditions of irrigation, in which the water brought would lead to fields accumulating salt as it sat on fields and percolated into the root zone. Fallowing allows a period of time for salts to diminish from the root zone. However, extended fallowing can come at a cost to farmers, as they lose yield from the unproductive fields. In effect, farmers had to balance not only the amount of water, which promotes salinity, they applied to fields but also how long fields should remain in agricultural production: irrigating too much can lead to high salinity, and low irrigation results in low overall yields.

In a step-by-step manner, using Figure 9.1, the model applies an agent-based method (ABM; Bonabeau, 2002) in which agricultural agents apply rule-based agricultural mechanisms and the agents' decision focuses on fallowing activity. The first step (1) is for the farmer to decide to plant or leave a field fallow. A field left fallow becomes leached of salts during the agricultural year, while a field that is planted (2) will be irrigated. Rainfall (3) then falls (which can be shown by using a Markov chain on known rainfall patterns) during the agricultural year, bringing a small level of salts to fields as well as run-off. Irrigation (4) then occurs, which includes several sub-processes that affect root salinity. Salinity from irrigation and from rain is added to the root zone in these steps. A salinity result is produced, and fields are then harvested (5). The harvest value ranges from 0 to 1.0, where 0 represent no yield or a yield completely affected by salt, and 1.0 represents a yield not affected by salt. Farmers then decide if they will conduct extended fallowing (6), or long-term fallowing, the normal cycle being a biennial fallow. Fallowing decisions are affected by how much salt tolerance and fallow scaling a farmer should apply. Salt tolerance is how much yield reduction a farmer should accept. For instance, field yields of 0.5 and 0.4 apply to fields that give 50 and 40 per cent of their expected productivity because of the presence

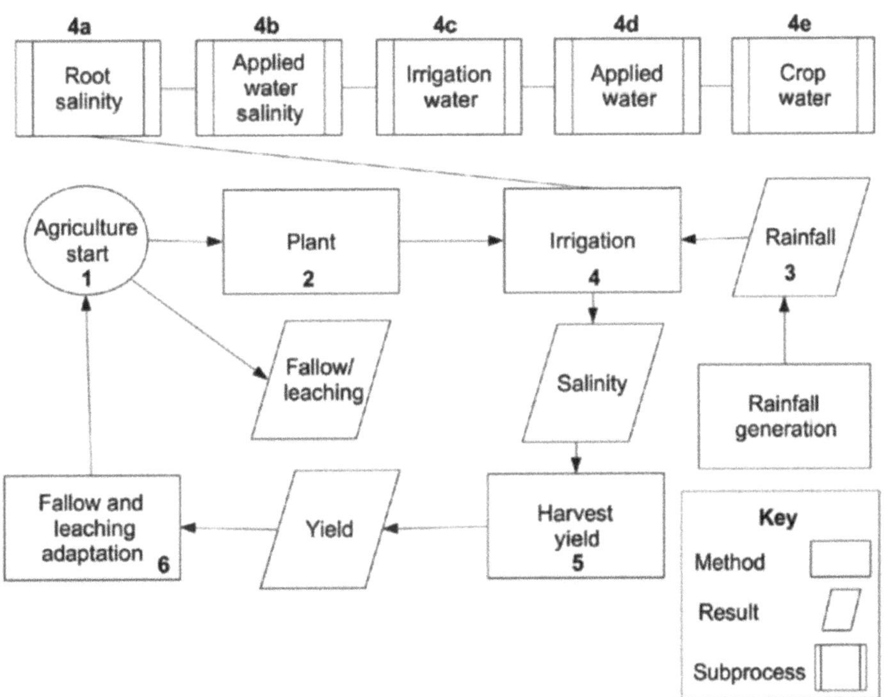

Figure 9.1 The steps applied in the model. Source: author

of salts. These values reflect what a farmer will tolerate in the productivity of fields. Fallow scaling regulates the response. If yields drop below a level tolerated by a farmer (e.g., 0.4, or 40 per cent, of the expected yield) because of the effects of salinisation, a fallow-scaling value regulates more or less years of additional fallowing. The greater the fallow-scaling value, the more time is given to fallowing, which allows fields to recover from salinity (for example through natural leaching or through built drainage channels).

Case study: Southern Mesopotamia

The data applied in the modelling presented below relates to the physical regions around Nippur and Uruk, major settlements in southern Mesopotamia during the Bronze Age (Finkbeiner, 1991; Gibson et al., 1992). The area was surveyed by Adams (1981). Sites such as Nippur and Uruk are long-lived settlements that were occupied from at least the fifth to fourth millennium BC to the early or mid first millennium AD.

Satellite data, specifically Shuttle Radar Topography Mission (SRTM; http://asterweb.jpl.nasa.gov/, accessed 3 February 2017) and Advanced Spaceborne Thermal Emission and Reflection Radiometer (ASTER; http://asterweb.jpl.nasa.gov/, accessed 3 February 2017), has been released that can be used for terrain elevation data. This data is useful in the model discussed below, as it highlights where canal levees are evident in relation to Bronze Age sites such as Nippur and Uruk (Figure 9.2a). It also defines terrain slope, showing where potential field types might be located. Slope data and imagery distinguish three irrigated areas, classified as levee crest (LC), levee slope (LS) and basin (B) fields. Fields that are LC are relatively well-drained areas located along the banks of canals; lower clay content and coarser sediments such as silt and sand are found in such fields. Fields that are LS are less well drained, with poorer leaching of salt, and have higher clay content. They are intermediate in their location along slopes of levees; silts enable some drainage and leaching to occur. Fields that are B are the worst for drainage and leaching because of a high percentage of clay, which means these fields generally have very low slope.

Data on settlement location was obtained from Hritz (2005, 2010). For soil data, Buringh (1960) and Powers (1954) are used to reconstruct soil profiles from the region. These profiles affect leaching factors, depth of soils, and the level of the water table. The soils are generally saline-alkali in composition, although salinity is affected also by the composition of

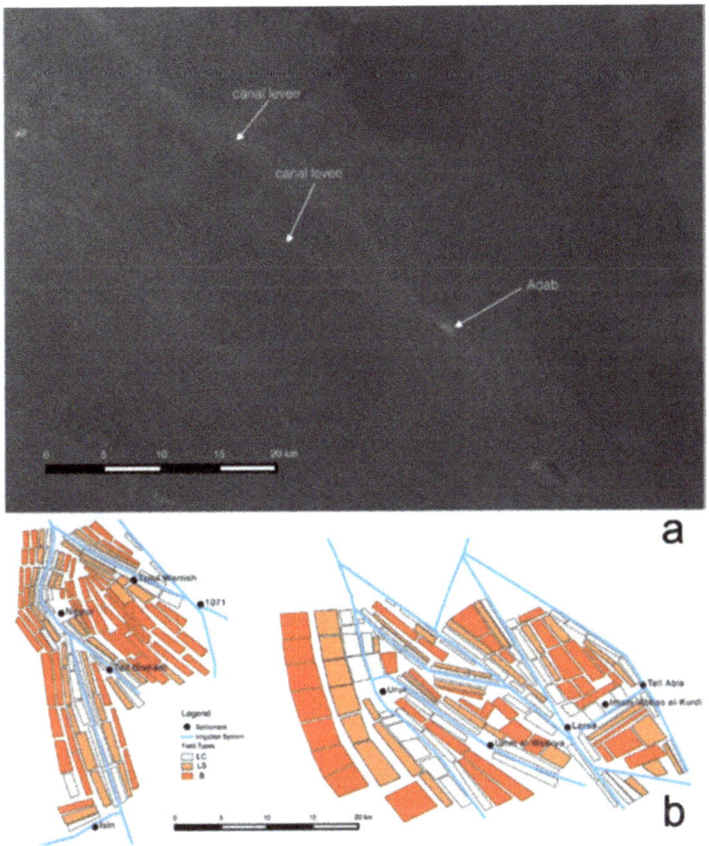

Figure 9.2 a. Canal levees, shown as light-coloured streaks, are determined using SRTM and ASTER elevation data, where slight elevation contrasts make them visible in the surrounding terrain. The ancient city of Adab (mostly a Bronze Age city) is indicated. b. Fields simulated in the region around the ancient sites of Nippur and Uruk. The field types simulated are levee crest (LC), levee slope (LS) and basin (B). Source: author

the clays, silts and sands present. The thickness of soil layers, along with electric conductivity (EC), for the water table is also present in other studies used here (see Jorenush and Sepaskhah, 2003). Capillary movement of water is a significant factor in southern Iraq (Barica, 1972; Goudie, 203) because the water table is high. Dry conditions, in which irrigation would be depended on for agriculture, can be assumed for many periods; data can be obtained from rainfall stations in southern Iraq for the period between 1930 and 1955 (NOAA, 2017). Rainfall patterns can be varied using a Markov chain algorithm, although generally conditions stay fairly stable where rainfall falls mainly from autumn to spring and summers are dry. Evaporation is an important factor; studies by Al-Khafaf et al. (1989) can be used to estimate what the rates are for the hot and dry conditions that would be found in southern Iraq/Mesopotamia.

Table 9.1 Types of model inputs used in scenarios and their data sources

Data input	Data source
Pan evaporation/coefficient (E_p)	Al-Khafaf et al., 1989
Empirical coefficient (*K*)	Al-Nakshabandi and Kijne, 1974
Rainfall salinity (C_r)	Prendergast, 1993
Irrigation salinity (C_w)	Kiani and Mirlatifi, 2012; Prendergast, 1993
Threshold salinity (*A*)	Barrett-Lennard, 2002; FAO, 2017
Crop coefficient (K_c)	Araya et al., 2011
Soil typology	Powers, 1954; Buringh, 1960
Fallow seasons (*FA*)	Jacobsen and Adams, 1958
Yield (*Y*)	Barrett-Lennard, 2002; FAO, 2017
Soil layer (*d*)	Barica, 1972; Dieleman, 1977
Water table conductivity (EC_{wt})	Jorenush and Sepaskhah, 2003
Yield response factor (K_y)	Doorenbos and Kassam, 1979
Leaching fraction (*LF*)	van Hoorn, 1981; Lyle et al., 1986
Capillary rise (*J*)	Goudie, 2003; Jorenush and Sepaskhah, 2003
Landscape and settlements	Adams and Nissen, 1972; Adams, 1981; Hritz, 2005; USGS, 2017
Rainfall (*R*)	NOAA, 2017
Percentage yield reduction (*B*)	FAO, 2017
Salt tolerance (*ST*)	
Fallow season scaling (*T*)	Gibson, 1974; Poyck, 1962
Leaching efficiency (E_l)	van Hoorn, 1981

Certainly some yield loss could be tolerated; however, allowing too much salt to be introduced could do long-term damage to fields. Extended fallowing outside the normal biennial crop cycle was probably applied as a primary means of mitigating salinisation (Gibson, 1974; Poyck, 1962). For modelling scenarios, barley is the crop used, as it is more salt-tolerant (Jacobsen, 1982). Planting occurs in the autumn, and irrigation primarily in the spring, when more water is available. Other variables in the model are the salinity threshold and the percentage yield reduction related to barley, which can be found in FAO tables (Tanji and Kielen, 2002: Annex 1). The model variables used, and their information sources, are listed in Table 9.1; they are similar to Altaweel (2013: Table 1). Figure 9.2b indicates the settlements and fields modelled in scenarios that use physical data gathered from the cited sources.

Results

Table 9.2 lists the variables and inputs, including their ranges where relevant, used in scenarios. Simulations were executed for 100 years, with multiple runs for parameter settings averaged in the results given.

Scenario one

The first scenario investigates conditions in which irrigation salinity is low, at a value of 1.0 dS/m (Al-Khafaf et al., 1990). The scenario demonstrates the effects of moderate salinity levels on agricultural output under conditions of no fallowing and biennial fallow. While in this case salinity conditions are low, salts do build up in soils without the aid of any fallowing. Figure 9.3 shows how field conditions change over time as fallowing is removed altogether from modelling runs. In effect, all field types are affected by salinisation (Figure 9.3a), although LC fields are the least affected and the influence of salinity generally occurs later. Overall,

Table 9.2 Input values for parameters. From left to right, LC-, LS- and B-type fields' input are indicated in columns where three data inputs are present.

Variable	Value	σ
E_p	1.1 m	0.2 m
K	0.6	
C_r	0.008 dS/m	
C_w	1.0 dS/m	
A	8.0 dS/m	
K_c	0.83	0.075
FA	1	
Y	1	
d	5	
ECwt	2.0 dS/m	
K_y	1	0.05
LF	0.20/0.175/0.10 m	0.05/0.04/0.03 m
J	0.35/0.50/0.70 m	0.075/0.125/0.175 m
R	see NOAA 2012 tables	
B	5.0 % per dS/m-1	
ST	0.1–1.0	
T	1–60	
E_l	0.40/0.35/0.30	

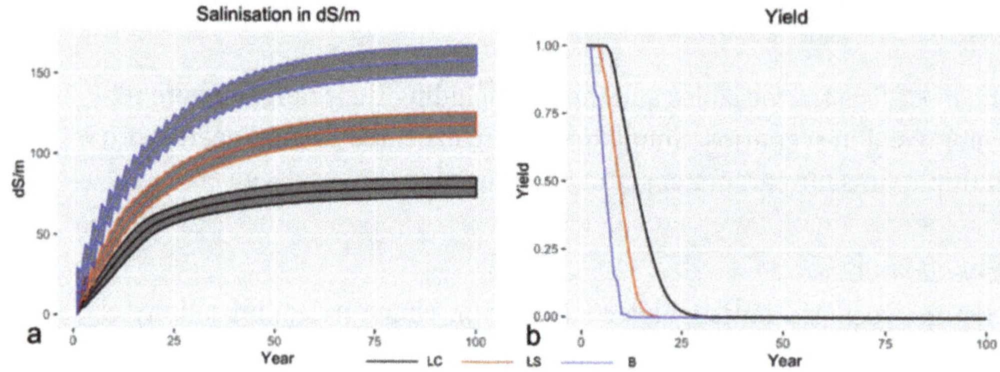

Figure 9.3 a. Salinisation and b. yield for fields. Shaded areas indicate one standard deviation from the mean result. Source: author

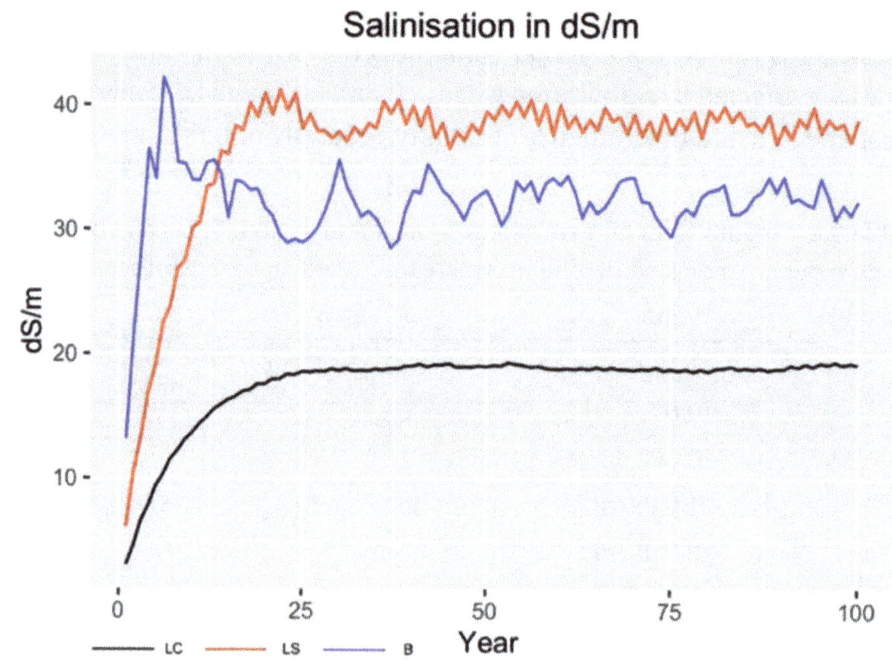

Figure 9.4 Fallowing helps to stabilise salinity in fields, as demonstrated here. Source: author

however, yields drop rapidly (Figure 9.3b), and after between 5 and 20 years all field types are almost completely affected by salinisation. In effect, yield reaches 0 (or is 100 per cent affected by salinisation) during that time because fields cannot fully drain salts as yearly irrigation is added.

Once biennial fallowing is enabled, however, salinity becomes more balanced or stable (Figure 9.4). In cases where irrigation salinity is low, minimal fallowing may be needed. Fallowing can bring about a salt balance in fields, since it gives adequate time for the natural removal of salts: salt might be present but overall salinity is lower and yields are more moderately affected. Scenario two demonstrates how such yields are affected and what the optimal responses are to enable improved overall yield results.

Scenario two

As the first scenario demonstrated, absence of fallow adjustments lead to overly saline fields, while biennial fallowing helps lead to a salt balance. This scenario addresses how farmers could adapt to optimise their production and at what levels of irrigation they can produce the best yields for given fields. Regular biennial fallowing is now implemented along with extended fallowing to optimise fields. Farmers can add additional fallow years to allow irrigation water to drain further from fields. However, farmers have to decide how much water to add or allow on their fields over a given time, because over- or under-fallowing can result in lost production. Figure 9.5a–c indicate what values of salt tolerance and fallow scaling allow the highest total yield, that is, yield over time, at the end of 100-year simulations. Figure 9.5d–f show the fallow years required for different salt tolerance and fallow-scaling values in simulations. Results stabilise (that is, an equilibrium is reached) by the end of simulations, as demonstrated in Figure 9.4, allowing for the effects of the parameters to be measured and indicated in Figure 9.5.

From these results, it is evident that different types of fields require different adaptive responses. Levee crest fields, that is, those that are better drained, have the best yields when salt tolerance is at 0.35, 0.45, 0.75 and 0.85 (Figure 9.5a), and fallow scaling, which controls how long fallow periods should be, ranges between 5 and 10. What this means is that some salinisation should be tolerated, specifically at levels that lead to yield reduction to 0.35, 0.45, 0.75 and 0.85 levels. However, this means that shorter fallow periods are better (Figures 9.5d, 9.6). In fact, fallow was only about one year in the best yield results, which is close to the normal biannual cycle with no additional years of fallowing. For LS fields, a different pattern is evident, as these fields are not as well drained. In these fields, a lower tolerance of salinity is required, and the best results were at 0.95 for salt tolerance (Figure 9.5b). Rather than too much salt, only minimal salt should be allowed. Fallow scaling is mostly low, at ranges of 5–10, although some of the better results also had high fallow scaling (60–65). This means that responses should generally have low levels of tolerance of the salt that affects yields (at 0.95 or 95 per cent

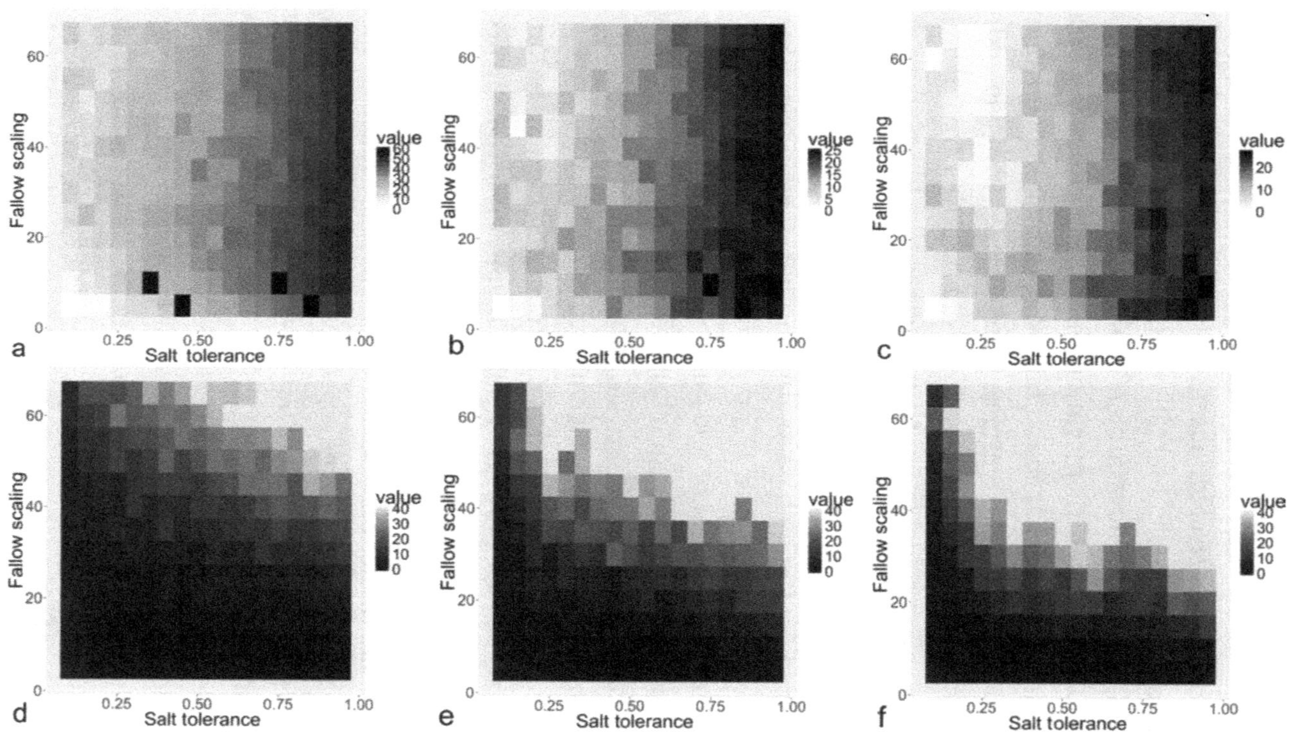

Figure 9.5 Results demonstrating yield (a–c) and fallow years (d–f) for different field types (LC (a and d), LS (b and e), and B (c and f) fields). Source: author

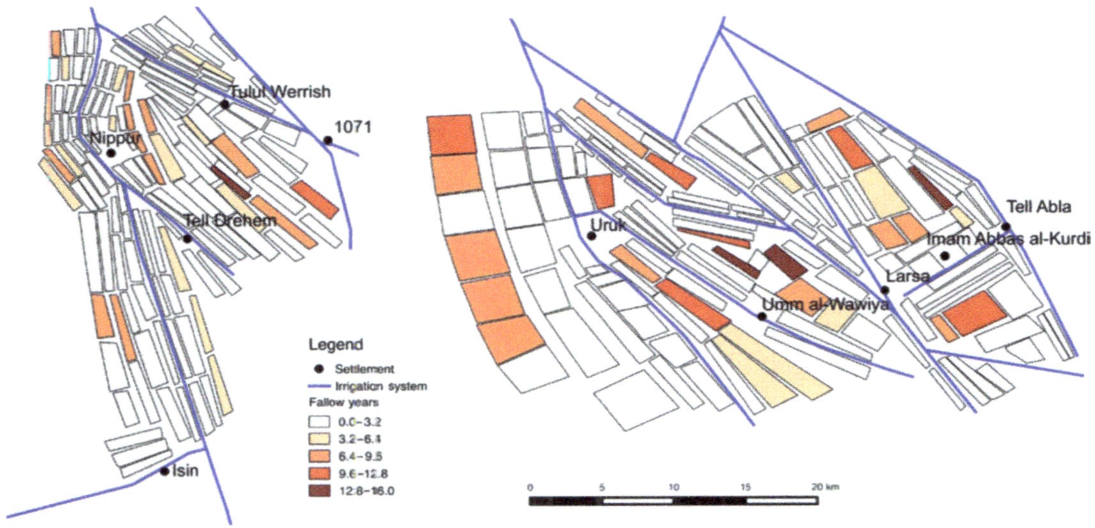

Figure 9.6 Fallow-year values for different field systems in the modelled regions. Results reflect the best adaptive salt tolerance and fallow-scaling values for LC fields; other field types are also shown. In this case, B fields show 3.2 or more fallow years. Source: author

of expected yield), but the responses for fallowing can vary from short to longer periods. Short fallow has the benefit of putting fields back into production more quickly, while longer fallow allows higher-producing fields to recover for each year of production. Overall, the number of fallow years that gave the best results ranged from 1.02 to 3.4 (Figure 9.5e). In other words, some extra fallowing is required to improve overall yields from the normal biennial cycle. For B fields, which were the worst-draining, salt tolerance was comparable to LS fields, at 0.9–0.95 (Figure 9.5c), for the best overall yields, with fallow-scaling values also being comparable to LS fields. As for overall fallow years, the results showed that 1.7–4 years of fallow led to the best yields (Figure 9.5f).

Increasing the salinisation to 4.0 dS/m begins to affect which strategies and adaptations were more suitable for farmers who wished to increase their yields. As LC fields became more saline, and salinity increased more rapidly, salt tolerance at 0.95 and fallow scaling between 40 and 65 now became preferable (Figure 9.7a). In other words, lower tolerance of salt and longer fallowing were preferred. In this case, all the best yields required more than 4 years of fallow (Figure 9.7d). The effect of increased salinisation meant that LS fields were more similar to LC fields, where salt tolerance at 0.95 was required, although longer or shorter fallow periods led to comparable results, where fallow scaling ranged between 10 and 60 for the best yield results (Figure 9.7b). Fallow years increased to more than 5 years in the best yield results observed (Figure 9.7e). For B fields (Figure 9.7c), the results varied more. Salt tolerance at 0.95, that is, yields being 0.95 affected by salts, as with other field types, could be a beneficial strategy. However, lower salt tolerance, at 0.3 and 0.1, led to greater yield. Mostly, lower fallow scaling (5–10), that is, the mechanism controlling fallow times, was needed to create better yields for this field type. In effect, fallow periods were shorter (Figure 9.7f) for the higher-yielding salt tolerance and fallow-scaling values. The best resulting yields ranged from 1.36 and 2.23 fallow years. The reason for this is that in the basins fields became salty very quickly, which meant that long fallow periods gave little advantage. Longer fallow periods would allow the salts to drain, but the yield during that time would be very low. Tolerating higher levels of salt, and having shorter fallow periods, was better for these poorly-drained fields.

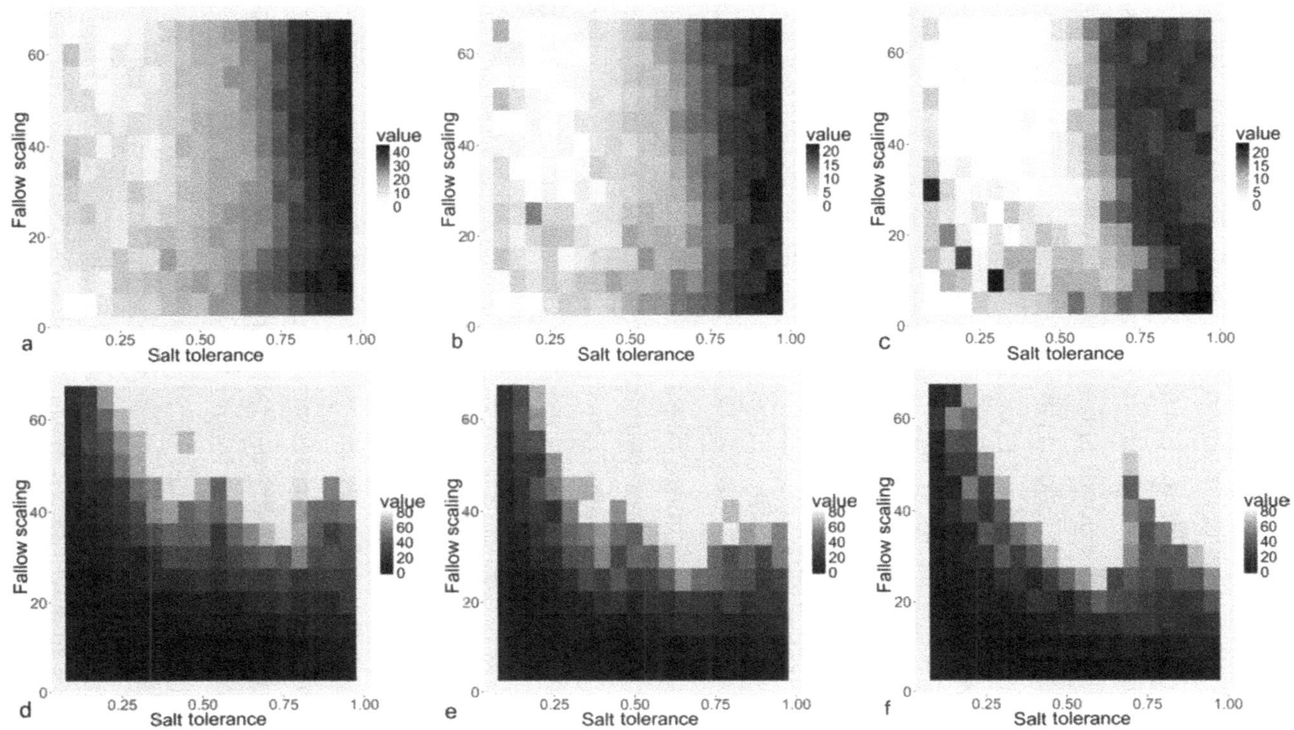

Figure 9.7 Results demonstrating yield (a–c) and fallow years (d–f) for different field types (LC (a and d), LS (b and e) and B (c and f) fields) when irrigation salinity is increased to 4 dS/m. Source: author

Discussion

The scenarios modelled above indicate that progressive salinisation can be a significant problem for agricultural fields even at low irrigation salinity (1.0 dS/m), without any human intervention. Adaptive changes, including leaching and fallowing fields, can have beneficial results. For relatively well-drained fields, such as LC fields, biannual fallow is generally sufficient, particularly if some level of tolerance of salt in fields is accepted (e.g., 0.35–0.85). However, for fields that are less well or poorly drained (LS and B), progressive salinisation can build even in cases where multi-year fallow is evident and when salinity is relatively low. For optimal yields, multi-year fallow is needed for fields that are poorly drained, ranging from 1 to 4 years (1.02–3.4 for LS and 1.7–4 for B), with light tolerance of salt on fields (e.g., 0.9–0.95).

Where salinity is increased to more moderate and greater levels, such as where the climate becomes drier (e.g., 4.0 dS/m), irrigation salinity becomes a longer-term problem even for well-drained fields such as LC. A different strategy might be required, which tolerates only very limited levels of salt, at around 0.95, and follows a more vigorous fallowing regime of up to 4 years. For LS fields a similar strategy might be required, but 5 or more years might be necessary for better yields. Interestingly, for the worst-drained fields (B), shorter fallow periods (1.36–2.23 years) are needed to limit higher levels of salinity because salinity occurs more rapidly, negating the benefits of long-term fallow. In effect, more frequent cropping and irrigation to optimise these poorly-drained fields are a better strategy.

Overall, in scenarios of low salinity, well-drained fields should be allowed to tolerate moderate levels of salt. In the absence of fieldwork or empirical evidence from the past, this chapter suggests that this may have been the practice in cases such as Bronze Age Mesopotamia. For less well-drained fields, multi-year fallowing was required. If the salinity in irrigation water increases, longer fallowing is needed in even the best-drained fields. On the other hand, poorly-drained fields are so determinately affected that the best strategy is to limit fallow years and crop, thus irrigate, more frequently. Given that B fields often showed high salinity, the outputs suggest that such places may have been avoided for agriculture. These regions could have been useful for providing reeds, natural food sources (e.g., waterfowl) and other resources (Renger, 2007), whereas irrigation agriculture may have been too costly in relation to its benefits.

Conclusion

The chapter demonstrates issues of salinisation and methods for mitigating its effects on different agricultural fields in Mesopotamia. While even mild levels of salinisation can have detrimental effects on crop yields, mitigation strategies, including biennial and extended fallowing, can minimise its effects. The results have helped to show how frequently field types could be irrigated in relation to the amount of time needed for fallowing and the leaching of field salts. They suggest that moderate levels of salinisation can be tolerated by settlements that depend on irrigation agriculture, so long as many of the fields that settlements depend on are relatively well drained. However, more substantial irrigation salinity, at levels of 4.0 dS/m or more, are likely to put greater stress on agriculturalists, as many fields, including well-drained ones, will need more extended fallowing,.

Balancing water input into fields is not an easy task, as drainage and the salinity levels of irrigation itself affect the optimal fallow and irrigation periods. Future avenues of research could investigate how irrigation practices in different regions compare to those in Mesopotamia, where climate and field types both strongly affect salinisation. Field testing results, and investigation of whether archaeological or textual evidence indicates the use of mitigation strategies, could be effective in demonstrating how significant a problem salinisation was. What is presented here is a guide to what we should expect from field systems if salinisation was a major threat, as past research has suggested. These results could be applied to Bronze Age Mesopotamia, where salinity may have been a long-term problem. Results demonstrate conditions in which salinity becomes a greater problem, such as increased aridity and the presence of salt in the water used for irrigation.

Acknowledgements

This research was supported financially by a Grant-in-Aid for Scientific Research from the Ministry of Education, Culture, Sports, Science and Technology of Japan (Project No.23310190, 'Ecohistory of Salinisation and Aridification in Iraq').

References

Adams, R. McC. 1981. *Heartland of Cities: Surveys of Ancient Settlement and Land Use on the Central Floodplain of the Euphrates*. Chicago: University of Chicago Press.

Adams, R. M. and H. J. Nissen. 1972. *The Uruk Countryside: The Natural Setting of Urban Societies*. Chicago: University of Chicago Press.

Al-Ansari, N., A. A. Ali and S. Knutsson. 2014. 'Present conditions and future challenges of water resources problems in Iraq.' *Journal of Water Resource and Protection* 6, 1066–98.

Al-Khafaf, S., A. Andan and N. M. Al-Asadi. 1990. 'Dynamics of root and shoot growth of barley under various levels of salinity and water stress.' *Agricultural Water Management* 18 (1), 63–75.

Al-Khafaf, S., F. A. Sharhan, P. J. Wierenga and A. D. Iyada. 1989. 'Some empirical relations for the prediction of soil evaporation, transpiration and root water uptake under field conditions.' *Agricultural Water Management* 16 (4), 323–35.

Al-Nakshabandi, G. A. and J. W. Kijne. 1974. 'Potential evapotranspiration in central Iraq using the Penman method with modified wind function.' *Journal of Hydrology* 23 (3–4), 319–28.

Altaweel, Mark. 2013. 'Simulating the effects of salinization on irrigation agriculture in southern Mesopotamia.' In T. J. Wilkinson, McG. Gibson and M. Widell (eds), *Models of Mesopotamian Landscapes: How Small-Scale Process Contribute to the Growth of Early Civilizations*, 219–39. Oxford: Archaeopress.

Araya, A., L. Stroosnijder, G. Girmay and S. D. Keesstra. 2011. 'Crop coefficient, yield response to water stress and water productivity of teff (*Eragrostis tef* (Zucc.).' *Agricultural Water Management* 98 (5): 775–83. doi:10.1016/j.agwat.2010.12.001.

Artzy, Michal and Daniel Hillel. 1988. 'A defense of the theory of progressive soil salinization in ancient southern Mesopotamia.' *Geoarchaeology* 3 (3), 235–8.

Barica, Jan. 1972. 'Salinization of groundwater in arid zones.' *Water Research* 6 (8), 925–33.

Barrett-Lennard, E. G. 2002. 'Restoration of saline land through revegetation.' *Agricultural Water Management* 53 (1–3), 213–26.

Bonabeau, Eric. 2002. 'Agent-based modeling: Methods and techniques for simulating human systems.' *Proceedings of the National Association of Sciences* 99 (Supp. 3), 7280–7.

Buringh, P. 1960. *Soils and Soil Conditions in Iraq*. Baghdad: Ministry of Agriculture.

Butzer, Karl W. 2012. 'Collapse, environment, and society.' *Proceedings of the National Association of Sciences* 109 (10), 3632–9.

Chhabra, Ranbir. 1996. *Soil Salinity and Water Quality*. Rotterdam: Balkema.

Dieleman, P. J. (ed.). 1977. *Reclamation of Salt Affected Soils in Iraq: Soil Hydrology and Agricultural Studies*. Wageningen: International Institute for Land Reclamation and Improvement.

Doorenbos, J. and A. H. Kassam. 1979. 'Yield response to water.' FAO Irrigation and Drainage Paper 33. Rome: FAO.

FAO. 2017. Threshold Salinity. http://www.fao.org/docrep/005/y4263e/y4263e0e.htm.

Finkbeiner, Uwe. 1991. *Uruk: Kampagne 35–37, 1982–1984: Die archäologische Oberflächenuntersuchung*. Ausgrabungen in Uruk-Warka. Endberichte, 4. Mainz: Zabern.

Fritsch, E. and R. W. Fitzpatrick. 1994. 'Interpretation of soil features produced by ancient and modern processes in degraded landscapes. 1. A new method for constructing conceptual soil-water-landscape models.' *Australian Journal of Soil Research* 32 (5), 889–907.

Gibson, McGuire. 1974. 'Violation of fallow and engineered disaster in Mesopotamian civilization.' In T. E. Downing and M. Gibson (eds), *Irrigation's Impact on Society*, 7–19. Tucson: University of Arizona Press.

Gibson, McGuire, Judith A. Franke, Miguel Civil, Michael L. Bates, Joachim Boessneck, Karl W. Butzer, and Ted A. Rathbun and Elizabeth Frick Mallin. 1992. *Excavations at Nippur: Twelfth Season*. 2nd printing. Oriental Institute Communications, 23. Chicago: Oriental Institute of the University of Chicago.

Goudie, Andrew S. 2003. 'Enhanced salinization.' In A. S. Alsharhan and W. W. Wood (eds), *Water Resources Perspectives: Evaluation, Management and Policy*, 287–94. Amsterdam: Elsevier.

Hritz, Carrie. 2005. 'Landscape and settlement in southern Mesopotamia: A geo-archaeological analysis.' Unpublished PhD thesis, University of Chicago.

Hritz, Carrie. 2010. 'Tracing settlement patterns and channel systems in southern Mesopotamia using remote sensing.' *Journal of Field Archaeology* 35 (2), 184–203.

Jacobsen, Thorkild. 1982. *Salinity and Irrigation Agriculture in Antiquity Diyala Basin Archaeological Projects: Report on Essential Results, 1957–58*. Bibliotheca Mesopotamica, 14. Malibu: Undena Publications.

Jacobsen, Thorkild and Robert M. Adams. 1958. 'Salt and silt in ancient Mesopotamian agriculture.' *Science* 128 (3334), 1251–8.

Jorenush, M. H. and A. R Sepaskhah. 2003. 'Modelling capillary rise and soil salinity for shallow saline water table under irrigated and non-irrigated conditions.' *Agricultural Water Management* 61 (2), 125 41.

Kiani, A. R. and S. M. Mirlatifi. 2012. 'Effect of different quantities of supplemental irrigation and its salinity on yield and water use of winter wheat (*Triticum aestivum*).' *Irrigation and Drainage* 61 (1), 89–98.

Lyle, C. W., A. H. Mehanni and A. P. Repsys. 1986. 'Leaching rates under a perennial pasture irrigated with saline water.' *Irrigation Science* 7 (4), 277–86.

Maekawa, K. 1973–4. 'The development of the é-MÍ in Lagash during Early Dynastic III.' *Mesopotamia* 8–9, 77–144.

Maekawa, K. 1974. 'Agricultural production in ancient Sumer: Chiefly from Lagash materials.' *Zinbun* 13, 1–60.

Maekawa, K. 1984. 'Cereal cultivation in the Ur III period.' *Bulletin on Sumerian Agriculture* 1, 73–96.

NOAA (National Oceanic and Atmospheric Administration). 2017. Environmental Data Rescue Program. http://docs.lib.noaa.gov/rescue/data_rescue_iraq.html. Accessed 9 February 2017.

Powell, Marvin A. 1985. 'Salt, seed, and yields in Sumerian agriculture: A critique of the theory of progressive salinization.' *Zeitschrift für Assyriologie und Vorderasiatische Archäologie* 75 (1), 7–38.

Powers, W. L. 1954. 'Soil and land-use capabilities in Iraq: A preliminary report.' *Geographical Review* 44 (3), 373–80.

Poyck, A. P. G. 1962. *Farm Studies in Iraq: An Agro-economic Study of the Agriculture in the Hilla-Diwaniya Area in Iraq*. Mededelingen van de Landbouwhogeschool te Wageningen, Nederland, 62–1. Wageningen: H. Veenman & Zonen.

Prendergast, J. B. 1993. 'A model of crop yield response to irrigation water salinity: Theory, testing and application.' *Irrigation Science* 13 (4), 157–64.

Renger, Johannes. 2007. 'Economy of ancient Mesopotamia: A general outline.' In Gwendolyn Leick (ed.), *The Babylonian World*, 187–97. New York: Routledge.

Smedema, Lambert K. and Karim Shiati. 2002. 'Irrigation and salinity: A perspective review of the salinity hazards or irrigation development in the arid zone.' *Irrigation and Drainage Systems* 16 (2), 161–74.

Tanji, Kenneth K. and Neeltje C. Kielen. 2002. 'Annex 1. Crop salt tolerance data.' In *Agricultural Drainage Water Management in Arid and Semi-Arid Areas*. FAO Irrigation and Drainage Paper 61. Rome: Food and Agriculture Organization. http://www.fao.org/DOCREP/005/Y4263E/y4263e0e.htm. Accessed 15 June 2018.

USGS. 2017. Earth Explorer. https://earthexplorer.usgs.gov/.

van Hoorn, J. W. 1981. 'Salt movement, leaching efficiency, and leaching requirement.' *Agricultural Water Management* 4 (4), 409–28.

10
Framing urban water sustainability: Analysing infrastructure controversies in London

Sarah Bell

Abstract

Water infrastructure embodies social and cultural values, delivers a vital natural resource for everyday consumption and provides a physical buffer between city-dwellers and changeable natural environments. At global, national and local scales, water policy and debate are framed according to assumptions about nature, water, technology and culture. Analysis of urban water sustainability literature reveals five distinct but overlapping framings: sustainable development, ecological modernisation, socio-technical systems, urban political ecology and radical ecology. Applying these frameworks to an analysis of controversies in London's water infrastructure related to desalination and the construction of an intercepting sewer tunnel shows the underlying values and knowledge in environmental debates. Decisions about water infrastructure are consistent with ecological modernisation theories and policies, but other framings are evident within debates and draw attention to alternative technology options and wider consequences for the environment and society.

Introduction

Water is an essential element of cities, shaping culture, urban form, public health and environmental quality. Water infrastructures of drainage, supply and waste disposal are among the most ancient urban technical systems. Infrastructures are designed, built and managed according to physical laws and technical expertise, and they are subject to political, social and cultural choices. Culture, technology, nature and politics are all entwined in analysis, discussion and debate about urban water systems.

Urban water infrastructure can be conceived of as an assemblage of technologies, institutions, hydrological resources and ecosystems. Critical philosopher of technology Andrew Feenberg (1993) links the social purpose and the technical form of technologies:

> [S]ocial purposes are embodied in the technology and are not therefore mere extrinsic ends to which a neutral tool might be put. The embodiment of specific purposes is achieved through the 'fit' of the technology and its social environment.

The technical form of urban water infrastructure is shaped in relationship to its social and natural environment. Questions about water infrastructure therefore cross technical and political boundaries. Who deserves access to water and sanitation, and at what price? What is an appropriate use of water? What is the role of the state, the private sector and citizens in water services? Who pays for water infrastructure? How much water should be allocated to maintaining natural ecosystems? What quality of water should be returned to the environment?

Water infrastructure embodies social and cultural values, just as it delivers a vital natural resource and provides a physical buffer between city-dwellers and changeable natural environments (Gandy, 2004). In this, water reflects wider environmental discourse and politics (Dryzek, 2013; Myerson and Rydin, 1996). Since the 1970s water has been a key element of international deliberations about sustainable development and sustainable cities. At the national and local scales, water policy and debate are framed according to assumptions about nature, water, technology and culture. Analysis of urban water sustainability literature reveals five distinct but overlapping framings: sustainable development, ecological modernisation, socio-technical systems, urban political ecology and radical ecology. Here, a summary of the key elements of each of these frameworks provides a general overview, which is then applied to two recent technical controversies in London's water infrastructure: combined sewer overflows and desalination. The use of the five frameworks to describe specific cases shows how contrasting analysis of the same problem leads to different priorities in solutions. It shows gaps and contradictions, as well as complementarity, between different framings. Framing urban water sustainability through different discourses and theories can expose fundamental political differences, but it may also be the basis of more transparent and informed deliberation about future water systems.

Frameworks for sustainability

Sustainability is a contested concept. How to protect and restore the environment while maintaining human well-being and development is a long-standing question for political deliberation, science and engineering (Dobson, 2000). The role of technology in achieving sustainability is also widely debated, some promoting technology as the answer to environmental problems and others pointing to industrial technologies as the root cause of the global ecological crisis (Davison, 2001).

A range of political and analytical approaches to addressing environmental problems have emerged since the 1960s (Dryzek, 2013). Five distinct frameworks can be seen to operate in professional, academic and policy discussions about urban water systems. Sustainable development originates from international negotiations aimed at ensuring continued economic and social development within ecological and resources limits (United Nations World Commission on Environment and Development, 1987). Ecological modernist approaches seek to use technology and the market to reduce environmental impacts without undermining economic growth or disrupting modern lifestyles (Mol and Sonnenfeld, 2000). Socio-technical-systems approaches highlight relationships between technology, society and the environment to reveal barriers and opportunities for sustainable transitions (Brown and Farrelly, 2009). Political ecology analyses socio-environmental problems to show the relationships between political power, capital accumulation and environmental degradation, highlighting uneven distribution of environmental costs and benefits under the dominant capitalist model of development (Swyngedouw et al., 2002). Radical ecology calls for a restructuring of human society based on ecological principles and deep respect for nature (Zimmerman, 1987).

These five frameworks reflect the dynamic nature of environmental discourse, including debate about urban water infrastructure. There are overlaps, complementarity and contradictions between different framings of how to solve environmental problems. The frameworks provide analytical lenses through which to see different alignments of the technology of urban water infrastructure within its social and natural environments. This helps to reveal the political and ethical choices to be made in selecting technologies and designing policies to achieve urban water sustainability.

Sustainable development

Sustainable development was famously defined in the UNCED *Our Common Future* report of the Brundtland Commission as 'development that meets the needs of the present without compromising the ability of future generations to meet their own needs' (United Nations World Commission on Environment and Development, 1987: 8). This definition reflects efforts to address environmental problems such as loss of biodiversity, pollution and desertification without undermining the processes of global development that are the basis for alleviating poverty and delivering basic needs such as food, health care and education. It represents a pragmatic trade-off between environmentalist calls to constrain population and economic growth and the dominant model of industrialisation as the basis for development. Sustainable development was conceived of as a universal objective and promoted through Agenda 21 and other international agreements reached at the UNCED Conference on Environment and Development in Rio di Janeiro in 1992 (United Nations, 1992). While achieving development without undermining the quality of the environment and resources in the future remains globally relevant, more recent discussions have focused on the Global South, governments of the Global North generally reducing emphasis on sustainable development as a policy framework.

Water has been an important element of global discussions about sustainable development since the 1977 UN Conference on Water in Mar del Plata, Argentina (Falkenmark, 1977). Universal access to clean water and safe sanitation has long been an objective of development, and was affirmed as a human right by the United Nations General Assembly (2010). Management of water resources for agriculture and industry as well as for municipal supply led to the development of the concept of integrated water resources management, which has underpinned global policy approaches to water since Agenda 21. Universal access to water and sanitation, within the framework of integrated water resource management, is one of the Sustainable Development Goals agreed by the United Nations (2015). Sustainable Development Goal 17 is to '[e]nsure availability and sustainable management of water and sanitation for all'.

The preceding Millennium Development Goals – to reduce by half the proportion of people without access between 2000 and 2015 – were achieved for water but not for sanitation (United Nations World Water Assessment Programme, 2015). In 2015, 800 million people lacked access to water and 2.5 billion lacked access to sanitation. Provision is better in cities than in rural areas, but 18 per cent of people living in cities, that is, 700 million urban residents, don't have access to improved sanitation.

Sustainable development as a framework for urban water sustainability emphasises the importance of universal access, particularly for the urban poor. It directs international efforts towards developing infrastructure and technologies to achieve universal provision, including financing and appropriate governance structures. Water and sanitation are the foundation for good public health, upon which other goals such as education and economic development can take place. Sustainable development recognises the need for integrated management of water, within the limits of local hydrological and ecological systems, and confirms the importance of economic growth to achieving development and environmental objectives.

Ecological modernisation

Ecological modernisation theory developed in Germany, the Netherlands and the United Kingdom in the 1980s and 1990s as a counterpoint to the environmental politics that pitted environmental protection against modernisation and industrialisation (Huber, 2005; Mol and Sonnenfeld, 2000; Hajer, 1995). Rather than conceiving of modernisation as the root cause of environmental problems, we frame it as the basis of solutions. Environmental harm is the outcome of incomplete modernisation, and so modern institutions such as the market, government, and science and technology need to be reformed, rather than abandoned, in order to better account for the environment (Mol, 1996). Technology has a particularly important role in ecological modernisation, and the role of government is largely to create the conditions in which markets and industries can innovate to improve resource efficiency and solve environmental problems, rather than to impose strict regulation to reduce environmental impacts (Huber, 2005). Environmental modernisation is evident in policies such as emissions trading, subsidies for environmental technologies, industrial ecology and behaviour change, and green consumer campaigns.

A similar approach is evident in the United States in the movement for resource and energy efficiency. The book *Natural Capitalism* outlines four key principles for improving resource efficiency and restoring the natural environment (Hawken et al., 2000):

- radical resource productivity to dramatically improve the efficiency of technologies and manufacturing,
- biomimicry to design technologies inspired by natural systems,
- a service and flow economy decoupling economic growth from resource extraction, and
- investment in natural capital to restore ecosystems.

In 2015 the US-based Breakthrough Institute published the *Ecomodernist Manifesto*, which emphasises the role of technological innovation in decoupling human development from environmental impacts and resource use (Asafu-Adjaye et al., 2015). Technologies such as desalination, nuclear power and agricultural intensification are promoted as means of delivering continued benefits to humanity, including poverty reduction and improved standards of living, while reducing human impact on the environment and leaving more space for non-human nature.

In contrast to the 2015 manifesto, European developments in ecological modernisation theory have responded to criticisms of technological optimism by giving greater attention to the need for institutional reform and to the demand for resources as well as efficiency in supply (Mol and Sonnenfeld, 2000). Sustainable consumption complements earlier efforts to devise sustainable production systems, addressing the role of individual behaviour, choices and lifestyles. Sustainable consumption may include technological innovation, but also new business models, and incentives for consumers to reduce the environmental impacts and resource use.

Ecological modernisation frames urban water sustainability as a problem of efficiency and a technical challenge for developing new sources of water. Demand management programmes, particularly those that emphasise economic incentives for reducing consumption and improving water use efficiency, are consistent with ecological modernisation theories. Desalination has also been framed as the ultimate, modern, technical solution to water scarcity, particularly when powered by renewable energy. Within an ecological modernisation framing, water sustainability can be achieved without major reorganisation of infrastructure and culture, but requires improvements in efficiency and technical innovation to develop new water supply.

Socio-technical systems

Theories of socio-technical systems come from systems science and engineering and the social studies of science and technology. Methods such as soft systems methodology and system dynamics provide frameworks for incorporating human and social systems, along with natural and technical systems, into models to allow for deliberation about current and future scenarios (Checkland, 1999). These approaches have their origins in systems engineering, systems theory and cybernetics, starting from biophysical and technical systems and broadening the approach to incorporate social and human elements. Socio-technical theories also incorporate the work of sociologists, philosophers and anthropologists who study the role and function of science, technology and innovation in society, and as social phenomena in themselves. They include actor–network theory, large technical systems theory, the social construction of technology, social practice theories and transitions theory.

Actor–network theory analyses social, technical and natural entities in the same terms (Latour, 1993). Rather than dividing the world into social and technical, or natural and cultural, elements to be understood using different theories and methods, actor–network theory analyses the material relationships between human and non-human networks. Such analysis reveals the role that non-human actors play in shaping society, and the social values and interests that are embodied in technologies and in human interaction with natural systems.

Large technical systems theories explain the co-evolution of infrastructures, associated technical norms, standards and institutions, and political and social change. Thomas Hughes's study of power networks traces the evolution of electricity supply from local production and distribution to complex, interconnected systems operating on the national and continental scales (Hughes, 1993). While technology is central to large technical systems theories, it is only one element of a much broader analysis of networks and systems that underpin modern, industrial society.

The social construction of technology draws attention to the social groups involved in the development of technologies and the process of stabilising particular social and political values and contexts (Bijker, 2009). The history of key technical developments, such as the bicycle and electricity, shows that technologies took many different forms in their early days (Bijker, 1997). The stabilisation of a particular technical form is not just the outcome of technical efficiency and optimisation but reflects the relative power and influence of different social groups. What appears to be a neutral, technical object is the outcome of social processes of design and development choices.

Social practice theories emphasise the interactions between technologies, infrastructures, cultures and everyday life (Shove, 2003). Consumption of resources such as water is enabled and constrained by particular forms of infrastructural provision. Technologies, such as automatic washing machines, fit into wider infrastructural systems of water supply and sewerage. The availability of water in turn makes laundry a convenient, daily practice which drives up social and cultural expectations regarding the cleanliness and freshness of clothing. Cultural norms about body odour and definitions of hygiene are therefore shaped by technologies and infrastructure. Water consumption is associated with everyday practices that co-evolve with cultures and technologies, rather than as a direct or conscious individual decision about resources or the environment.

Socio-technical transitions theories analyse the emergence, diffusion and stabilisation of innovation within institutional and social contexts. The multi-level theory consists of niches in which new technologies and practices first emerge in a specific local context, regimes of professional and policy institutions that set standards and technical norms, and in the broader landscape of the economy, culture and politics in which socio-technical systems are situated (Geels, 2011). Transition to sustainable socio-technical systems requires reform across all three levels, and the success or failure of particular innovations can be explained by their alignment or incompatibility with dominant regimes and landscapes.

A socio-technical systems framing of urban water sustainability draws attention to the relationships between technical, social, cultural and political factors. As technical systems co-evolve with social and cultural norms and institutions, the achievement of sustainability requires simultaneous analysis and action that can account for different social groups, technical options and institutional forms.

Political ecology

Urban political ecology recognises that society and environment are co-constructed, in processes of socio-environmental change (Swyngedouw et al., 2002). Cities are not separate from 'nature', just as rural landscapes and 'wilderness' areas are shaped by social processes and cultural meanings (Swyngedouw, 2009). Political ecology draws attention to the uneven distribution of the costs and benefits of socio-environmental change, linking social and economic inequality to environmental problems and infrastructure decisions. In particular, it analyses the role of capital and neoliberal policies in driving urban ecological processes, with unequal consequences. Processes of environmental harm are linked to social and economic exploitation and disenfranchisement (Loftus, 2009).

In highlighting the political nature of socio-environmental problems and their solutions, urban political ecology rejects depoliticised representations of sustainable development (Swyngedouw, 2010). Sustainability and justice cannot be achieved without addressing the power of dominant actors and discourses, which requires a fundamentally political analysis to inform activism and policy alternatives.

The role of the private sector in water infrastructure is a particular concern of urban political ecologists (Bakker, 2010). Linking the provision of basic services to private profit taking undermines the goal of universal provision of water as a human right, and leads to forms of infrastructure and service contracts that maximise return on investment rather than benefits to society and the environment. 'The financialisation of the water sector' refers to the growing role of water infrastructure as a vehicle for international investment (Loftus and March, 2016). Infrastructure decisions tend towards capital-intensive options which increase the value of assets and the opportunities for returns from complex capital financing arrangements. Government regulation and policy can provide constraints and incentives for infrastructure finance, but financial drivers may override the goals of environmental protection and social benefit in decision making.

Political ecology is critical of neoliberal policy on water infrastructure, including privatisation and commodification. Demand management based on water metering, and pricing without adequate regulation, can have a disproportionate impact on low-income households, while high-income households are associated with high per capita consumption (Gandy, 2003). It also shifts responsibility for resource management away from infrastructure managers and onto individual consumers, undermining collective responsibility for the universal provision of basic water services and environmental protection.

While political ecology is critical of the privatisation of water infrastructure, its analysis of power in relation to water as a socio-environmental problem in cities reveals the limitations of public and communal approaches to provision (Bakker, 2010). Municipal and communal ownership of infrastructure tends to reflect the interests of powerful actors, to the detriment of the environment and vulnerable citizens, and is prone to problems arising from mismanagement and poor technical and economic capacity. Analysing water infrastructure provision as a socio-environmental process highlights the political nature of decisions and discourse, on the local, regional and global scales.

Radical ecology

Since the 1960s environmental activists and scholars have considered the environmental crisis in modern, industrial society to be the outcome of a fundamentally exploitative and destructive relationship with nature. In contrast to reformist approaches, radical ecology looks for the root cause of environmental problems in the basic structures of Western society and culture. Deep ecologists call for an ecocentric world view, which places the needs of non-human nature at the centre of human culture (Naess, 1986). Social ecologists point out the connection between the exploitation of nature and the exploitation of people under capitalism (Bookchin, 2005). Ecological feminists specifically link the oppression of women and of nature in Western culture.

Deep ecology is most closely associated with movements to protect 'wilderness' areas from development, including anti-logging and anti-dams campaigns (Sessions, 1998). Deep ecology was first defined by Norwegian philosopher Arne Naess in 1972, who contrasted it with 'shallow ecology', which characterised reformist approaches that focused on pollution reduction, resource depletion and other environmental issues from an anthropocentric, or human-centred, viewpoint (Naess, 1984). Deep ecology is the deep questioning of human relationships to the natural world, and leads to proposals for bioregional communities as the basis for human development, with deep connections to local landscapes, ecological processes and non-human nature. Deep ecology has been criticised within and outside the environmental movement for focusing exclusively on human relationships with nature, without addressing inequality and patterns of domination within society.

Social ecologists and ecological feminists more explicitly address connections between domination of nature and structures of power within human society and culture. Social ecologist Murray Bookchin (2005) proposes that the ecological crisis is the outcome of the hierarchical structure of modern capitalism, which requires reorganisation of society into more decentralised, self-organising communities. Ecological feminists, including Karen Warren, Val Plumwood and Carolyn Merchant, have analysed the specific association of women and nature in the exploitative structures of Western culture. Women are classically associated with nature, while men represent culture. Women and nature are therefore both subject to domination by masculine culture, so that ecological politics must also be feminist (Warren, 1990). In moving away from hierarchical power structures based on domination and submission, ecological feminist responses to the ecological crisis emphasise negotiation of relationships with the 'other', accommodating difference (Plumwood, 2003).

Radical ecologists therefore emphasise the value of water for nature in cities and their catchments. Water as a natural material, a force in shaping landscapes and fundamental to all ecological processes, is an important element for understanding human relationships with the natural world. Modern construction of dams, treatment works, pipe networks and flood defences represents efforts to control and dominate water. More sustainable approaches 'make space for water' and recognise its value to human well-being as part of natural systems that can be integrated into urban landscapes.

Water controversies in London

With a growing population, limited resources and ageing infrastructure, London, like many major cities, faces considerable challenges in achieving water sustainability. London receives an annual average rainfall of 640 mm. The current population of 8.9 million people is forecast to grow to 11 million by 2050 (Greater London Authority, 2017). Without the development of new water sources or significant reduction in per capita water use, demand is forecast to outstrip supply of water by 2020 (Thames Water, 2015). More than 20 per cent of the water supplied to London is lost through leakage in the distribution network, which includes pipes that are more than 150 years old in some areas (Carrington, 2017). London's basic sewer system was also constructed in the nineteenth century, and sewer overflows into the River Thames are a major source of pollution.

London's water infrastructure is owned by a private company, Thames Water, and is regulated by the Office for Water (Ofwat), the Environment Agency (EA) and the Drinking Water Inspectorate. The privatised water sector in the UK brings benefits and constraints in achieving sustainability. Privatised water companies are able to raise capital for investment in major new projects, including environmental improvements. However, investment must be shown to deliver value for money to water customers. Integration of the water infrastructure with wider goals for urban sustainability is more challenging than for cities where the water utility is municipally owned and operated (Dolowitz et al., 2018).

In recent years, two water infrastructure projects have generated controversy in London: the desalination plant at Beckton, which was opened in 2011, and the Tideway Tunnel, which at the time of writing is under construction as a solution to the problem of combined sewer overflows. Analysis of these projects using the five frameworks for urban water sustainability shows that debates about infrastructure reflect different values and conceptions of the nature of environmental problems and solutions.

Desalination

The Thames Gateway Water Treatment Works, also known as the Beckton desalination plant, was completed in 2011. With the exception of periods of operation for testing and maintenance, the plant has never supplied water to London. It was originally planned in response to low reservoir levels following dry winters in 2004–2006, but more recently has been justified as a 'resilience measure' to provide a backup water source in times of extreme drought (Loftus and March, 2016).

Controversy over the plant centred on the refusal of the Greater London Authority under Mayor Ken Livingstone in 2006 to grant planning permission for its construction. The permission was refused on environmental grounds, and a judgement that Thames Water had not adequately considered other options for preventing London's future water shortages, including water recycling and leak reduction. When Boris Johnson was elected Mayor in 2008, one of his first actions was to approve the desalination plant, and construction began soon after.

Desalination is an energy-intensive and expensive source of clean water, with the potential to cause local environmental harm by entraining marine life in the intake and discharging highly saline water into receiving environments. The Beckton plant won a Sustainability Award from the Global Water Awards in 2009 for its environmental protection measures and use of renewable energy (Global Water Intel, 2009). The plant is located in the Thames Estuary, and is designed to treat brackish rather than saline water, which reduces the energy required for treatment compared to seawater desalination, which is more common in other parts of the world. The operating conditions set by the Environment Agency for the Beckton desalination plant require that a drought is declared and specific low-flow conditions are met in the River Thames (3,000 million litres per day (Ml/d) or less for 10 consecutive days at Teddington Weir; Thames Water, 2015). These conditions have not been met since the plant opened in 2011, and so the plant has yet to operate as designed.

Sustainability was a core concept in the controversy about the Beckton desalination plant. Mayor Ken Livingstone argued that the plant was unsustainable, particularly compared to less energy-intensive alternatives such as water reuse. The global water industry recognised the plant's sustainability credentials with an award. This shows the contested nature of sustainability, and its discursive flexibility in debates about water technologies and infrastructures.

From the point of view of sustainable development, the cost, energy-intensity and local environmental impacts of the Beckton desalination plant must be considered in balancing the social need for a secure water supply. Assessment of the plant against these basic elements of sustainable development, particularly in comparison with alternative strategies for drought response and water resource management, shows it to be a costly and resource-intensive solution to the risk of future water scarcity. Concerns about the environmental and energy impacts of the plant can be seen in Mayor Livingstone's objections, and in the operating conditions imposed by the Environmental Agency to ensure that it is only used as a source of water in times of extreme drought, not as an element of normal water supply to London.

Desalination is often presented as the ultimate technical solution to water scarcity, producing freshwater from the vast oceans, which account for 97 per cent of the water on Earth (Shiklomanov, 2000). The Ecomodernist Manifesto promotes desalination as one of a suite of technologies that will decouple human development from environmental impacts (Asafu-Adjaye et al., 2015). Using renewable energy to power desalination, as is the case in the Beckton plant, is seen as further evidence of the capacity of technology to solve environmental problems and avert resource shortages (Ghaffour et al., 2015). Although more costly than conventional water resources, desalination follows the basic law of substitution for scarce resources, refining ever more contaminated reserves as higher-quality sources of the resource are exhausted. Desalination as a resilience measure aims to minimise the disruption of society caused by natural events. Social functions, including water consumption, are maintained as technology is utilised to solve the problem of environmental uncertainty and greater demand for water from growing populations.

The socio-technical systems concept of 'lock-in' refers to the stability of existing technologies, practices and institutions, creating conditions in which particular forms of innovation are favoured over those that require greater disruption. Anique Hommels uses the concept of the 'obduracy' of urban infrastructure and form to describe the endurance of dominant systems and approaches (Hommels, 2005). As a supply-side solution to water scarcity, owned and operated by the incumbent utility, the desalination plant represents a relatively minor change to London's water infrastructure. It fits within existing economic and environmental regulatory processes for the water industry, and avoids the need for dramatic changes to water consumption or more socially challenging alternatives such as potable reuse or distributed supplies such as greywater recycling and rainwater harvesting. Although more energy-intensive and expensive than some alternatives, desalination is favoured because it is most compatible with the existing socio-technical form of water infrastructure in the city.

Alex Loftus and Hug March have analysed the Beckton desalination plant as an outcome of the financialisation of the water sector in the UK (Loftus and March, 2016). They show that the plant's credentials as an ecological modernisation and industrial ecology solution to drought in London are only possible because of a vast network of international capital investment. The availability of investors as diverse as an Australian bank, a Canadian teachers' pension fund and Chinese and Middle Eastern sovereign wealth funds to provide capital for infrastructure projects provides important context for the decision making about water in London. The desalination plant is not merely a response to localised resource shortages, but a global investment opportunity and market for international engineering consultants and technology providers. The project was constructed by a consortium of the UK-based Interserve and Atkins Water, and the Spanish water company Acciona Agua, using membrane technologies from the Dutch firm Norit and the US-based Hydronautics. A political ecology analysis of the Beckton desalination plant highlights the role of international investors and firms, and the logic of capital growth through infrastructure expansion as key drivers for desalination as the preferred option for addressing the risk of water scarcity during drought.

The technical promise of desalination to provide a limitless source of water for human use runs counter to radical ecological goals of living within local ecological and hydrological systems. While radical ecologists promote alternative patterns of human settlement and society based on living in partnership with nature, desalination expands the boundaries of the control of nature using advanced technology. Through desalination, all water on Earth becomes a potential resource for human use, perpetuating a culture of exploitation that radical ecologists claim is the root cause of the environmental crisis. As a supply-side option it reduces incentives for citizens in London to adapt their lifestyles and live within local environmental variability and limits. While ecological modernisation measures such as the use of renewable energy and techniques to minimise the environmental impacts of the abstraction of the raw water and the discharge of the brine by-product, industrial-scale desalination is fundamentally incompatible with radical ecological approaches to sustainability.

The international recognition of the sustainability of the Beckton desalination plant reflects an ecological modernisation framework, which is consistent with engineering and industrialist culture and discourse. Criticisms of the desalination plant are also grounded in alternative framings of sustainability, pointing out the expense and energy intensity of the technology, the financial benefits to international investors and suppliers, and the distraction from a more fundamental reorganisation of London society and culture to change its water use to live within local hydrological limits.

Combined sewer overflows

Combined sewer overflows (CSOs) are a significant environmental problem for a number of cities, particularly those with sewerage infrastructure built during the nineteenth century (Dolowitz et al., 2018). Combined sewers receive both waste water from buildings and surface water run-off from roofs, streets and hard surfaces. In order to prevent sewers flooding streets and buildings, these systems are designed to overflow into local rivers during heavy rainstorms. In London the sewers were originally designed by the Metropolitan Board of Works in the mid-nineteenth century to overflow into the Thames on average four times per year (Bazalgette, 1865). By the beginning of the twenty-first century overflows were occurring 50 times per year on average. The increased frequency of overflows is the result of the reduced permeability of urban surfaces, which is due to paving and building over open space, and the increased baseload flow of waste water from an increased population. In 2012 the European Court of Justice ruled that the UK was in breach of the European Union Urban Waste Water Treatment Directive, with respect to combined sewer overflows in London.

In 2000 Thames Water commissioned the Thames Tideway Strategic Study to evaluate the options 'to protect the Thames Tideway from the adverse effects of wastewater discharges' (Thames Tideway Strategic Study, 2005: 5). The study was overseen by a steering committee chaired by independent engineer Chris Binnie and included members representing the EA, the Department for Environment, Food and Rural Affairs (DEFRA), the Greater London Authority and Thames Water, with an observer from Ofwat. The committee reported in 2004, recommending the construction of a 35 km tunnel from Hammersmith in west London to the Crossness Sewerage Treatment Works in the east, and a separate tunnel to receive CSOs from the River Lee. The study investigated alternative options, including sustainable drainage systems (SuDS), which prevent inflow of surface water to the sewers by increasing infiltration and storage across the city. SuDS features include ponds, swales, rain gardens, green roofs and rainwater harvesting (Woods-Ballard et al., 2007). The report concluded that SuDS were not a suitable solution to CSOs in London because of the highly urbanised nature of the city, the impermeability of its clay soils, excessive costs and a lack of natural receiving waters for surface water run-off. The route originally recommended for the Tideway Tunnel was subsequently revised to include the Lee Tunnel, reduce the overall length of the Tideway Tunnel to 30 km, and discharge at Beckton, rather than Crossness treatment works.

The proposed Tideway Tunnel was the subject of considerable controversy, particularly during public consultation over the Development Consent Order for the project to be approved by a government minister as a nationally significant infrastructure project. In 2011 the Thames Tunnel Commission was formed by local authorities likely to be impacted by its construction (Thames Tunnel Commission, 2015). The commission investigated SuDS as an alternative to the interceptor tunnel, or as a complement to a smaller tunnel. They questioned the water quality standards set by the Thames Tideway Strategic Study as unnecessarily strict, thereby effectively ruling out SuDS as a potential solution, despite their wider environmental benefits. The commission also pointed out that SuDS were more difficult to implement than an interceptor tunnel in London, despite the environmental benefits, because planning and financing structures were more favourable to large infrastructure projects than to distributed interventions that controlled surface water at source. Other criticisms of the project included a report by Chris Binnie, the original chair of the Thames Tideway Strategic Study, questioning the increasing cost estimates for the tunnel, from £1.7 billion in the original study to £4.1 billion by 2014, and claiming that developments in SuDS techniques and upgrades to the sewer infrastructure since 2004, including the construction of the Lee Tunnel, meant that the tunnel was no longer required (Griffiths, 2014).

The Tideway Tunnel received its Development Consent Order in August 2014. In June 2015 a new company, Bazalgette Tunnel Ltd (operating as 'Tideway'), was formed to construct, own and operate the tunnel. Major contracts for the construction were agreed in 2016. Construction is underway and is expected to be completed by 2023. SuDS are now promoted through local planning requirements and the Greater London Authority's Sustainable Drainage Action Plan, primarily for the wider benefits associated with reducing run-off and increasing urban greening, rather than as the solution to CSOs in London (Greater London Authority, 2016).

The framing of different arguments in the debate about the Tideway Tunnel in London reflects wider debates about urban water sustainability. While stakeholder interests, such as profitable operation for the water company and avoiding the disruption caused by construction for riverside local governments, are important in determining particular positions within the argument, wider debates about the suitability of the tunnel or SuDS to solve the problem of CSOs in London demonstrate the intersection between technology and values in environmental decisions.

As an environmental protection measure to restore the health of the River Thames in the context of growing population and urbanisation in London, the tunnel is consistent with a sustainable development framing of urban water sustainability. The tunnel was largely supported by recreational river users, fishers and environmental organisations. It promises that London can continue to grow, without adversely impacting the local environment. The Thames Tideway Strategic Study addressed environmental, economic and social factors, including public health. The consultation and enquiries undertaken as part of the Development Consent Order process also addressed economic, social and environmental factors, albeit within a planning framework in the UK in which the principles of sustainable development are of less significance than maintaining economic growth.

Sustainable development principles were also invoked by opponents to the tunnel and proponents of SuDS (Thames Tunnel Commission, 2015). Local councils and residents argued that the tunnel would cause unfair disruption to local communities. It was argued that SuDS could deliver a wider range of environmental benefits and local employment opportunities than the tunnel, representing a more sustainable option.

The tunnel is a large technical solution to a persistent environmental problem, and is therefore consistent with ecological modernisation policies and theories. The idea of an interceptor tunnel itself is not innovative, but detailed modelling and design of the tunnel and associated surface facilities show the use of innovative, technical tools in demonstrating the sustainability of the project. The use of private finance and private ownership is also consistent with ecological modernisation, with the state acting as a regulator to ensure economic efficiency and environmental outcomes.

London's sewerage system was designed and built as a series of interceptor sewers in the nineteenth century by Sir Joseph Bazalgette and colleagues at the Metropolitan Board of Works (Bazalgette, 1865; Halliday, 2001). The Tideway Tunnel is a continuation of the infrastructural logic laid down in the strategy for managing waste water and surface water in London more than 150 years ago. The tunnel can therefore be considered an outcome of infrastructural lock-in, with engineers, owners and regulators driven towards the solution that is least disruptive to existing arrangements within the city. With the creation of a new private company to construct and own the tunnel within the established regulatory frameworks of the water sector in England, it can be seen to be consistent with long-established processes of infrastructural provision in the city. By contrast, the legislative, planning and ownership structures for SuDS are less stable in England, requiring more complex arrangements and oversight. The socio-technical and institutional landscape of London is therefore more conducive to large centralised infrastructure solutions to water-management problems than decentralised systems such as SuDS.

The formation of Bazalgette Tunnel Ltd shows the value of the project to international investors and the capacity of the private sector to raise capital to deliver infrastructure projects. From a political ecology point of view this is an indication of the financialisation of the water sector (Loftus and March, 2016). Bazalgette Tunnel Ltd is owned by a consortium including German insurance company Allianz, the multinational Amber Infrastructure Group, UK-based fund manager Dalmore Capital, Dutch fund manager DIF, International Public Partnerships and Swiss Life Asset Managers. The £4.1 billion investment with guaranteed income provided by regulated water bills paid by London customers, and investment risks guaranteed by UK Treasury, provides a stable return to international capital fund managers. Water infrastructure is therefore as important as an investment vehicle as it is as a solution to an environmental problem, with significant financial benefits accrued to international investors.

A radical ecology approach to CSOs in London emphasises SuDS as the means to bring nature into the city, manage surface water locally and reduce impacts on the River Thames. From this perspective the tunnel is a continuation of nineteenth-century engineering models based on domination and control of water and nature for human benefit, while SuDS involves working within local hydrological systems to create wider benefits for environmental and human health and well-being. SuDS is part of a strategy of 'making space for water' in cities, while the interceptor sewer maintains separation of people from water, far below ground.

Conclusions

Water infrastructure is largely taken for granted in modern cities, yet it is increasingly subject to controversy. Water networks must adapt to changing populations, environmental conditions, cultures and politics.

Debates about the sustainability of current and future water infrastructures reflect wider environmental discourse, politics and culture.

Desalination and combined sewer overflows in London show the dominance of ecological modernisation as the key framework for urban water sustainability. This emphasises technological innovation and the role of the market in delivering solutions to environmental problems with minimal disruption to existing lifestyles and institutions. However, controversy over the Beckton desalination plant and the Tideway Tunnel demonstrates the existence of alternative framings of the role of technology and culture in determining the sustainability of water infrastructure in the city. Political ecology analysis focuses on the financialisation of the water sector as a driver for large capital investment in centralised infrastructure solutions over decentralised technologies. The concept of socio-technical lock-in explains the power of existing infrastructures to determine the viability of proposals for future development, including the stability of institutions and cultures as well as the physical obduracy of the pipes and treatment works themselves. Sustainable development has been deployed in favour of and against the proposed projects, depending on where the balance of trade-offs between ecological, economic and social costs and benefits is drawn. Radical ecology perspectives on water infrastructure in London despair at the continuation of nineteenth-century models of engineering based on domination of water for human benefit, and at the loss of opportunities to enhance natural solutions and live within local hydrological limits.

Infrastructures are more than mere technical systems for delivering materials and services to society. The form of infrastructure reflects the social world it inhabits, and in turn shapes cultural and political possibilities. Decisions about infrastructure reflect complex interactions between technology, culture and nature, which may be understood through theoretical and political frameworks that reflect different knowledge and values. The recent history of water infrastructure in London reveals that these complex relationships are far from settled, and pose ongoing challenges for democratic decisions and hydrological sustainability.

References

Asafu-Adjaye, John, Linus Blomqvist, Stewart Brand, Barry Brook, Ruth DeFries, Erle Ellis et al. 2015. 'An ecomodernist manifesto.' www.ecomodernism.org. Accessed 4 December 2018.

Bakker, Karen J. 2010. *Privatizing Water: Governance Failure and the World's Urban Water Crisis*. Ithaca, NY: Cornell University Press.

Bazalgette, Joseph William. 1865. *On the Main Drainage of London: And the Interception of the Sewage from the River Thames*. London: Clowes and Sons.

Bijker, Wiebe E. 1997. *Of Bicycles, Bakelites, and Bulbs: Toward a Theory of Sociotechnical Change*. Inside Technology. Cambridge, MA: MIT Press.

Bijker, Wiebe E. 2009. 'Social construction of technology.' In Jan Kyrre Berg Olsen, Stig Andur Pedersen and Vincent F. Hendricks (eds), *A Companion to the Philosophy of Technology*, 88–94. Oxford: Wiley-Blackwell.

Bookchin, Murray. 2005. *The Ecology of Freedom: The Emergence and Dissolution of Hierarchy*. Oakland, CA: AK Press.

Brown, R. R. and M. A. Farrelly. 2009. 'Delivering sustainable urban water management: A review of the hurdles we face.' *Water Science and Technology* 59 (5), 839–46.

Carrington, Damian. 2017. 'Water companies losing vast amounts through leakage, as drought fears rise.' *The Guardian*, 11 May. https://www.theguardian.com/environment/2017/may/11/water-companies-losing-vast-amounts-through-leakage-raising-drought-fears. Accessed 4 December 2017.

Checkland, Peter. 1999. *Systems Thinking, Systems Practice: A 30-Year Retrospective*. New York: John Wiley & Sons.

Davison, Aidan. 2001. *Technology and the Contested Meanings of Sustainability*. Albany: State University of New York Press.

Dobson, Andrew. 2000. *Green Political Thought*. 3rd edn. London: Routledge.

Dolowitz, David Peter, Sarah Bell and Melissa Keeley. 2018. 'Retrofitting urban drainage infrastructure: Green or grey?' *Urban Water Journal* 15 (1), 83–91.

Dryzek, John S. 2013. *The Politics of the Earth: Environmental Discourses*. 3rd edn. Oxford: Oxford University Press.

Falkenmark, Malin. 1977. 'UN Water Conference: Agreement on goals and action plan.' *Ambio* 6 (4), 222–7.

Feenberg, Andrew. 1993. *Critical Theory of Technology*. Oxford: Oxford University Press.

Gandy, Matthew. 2003. *Concrete and Clay: Reworking Nature in New York City*. Urban and Industrial Environments. Cambridge, MA: MIT Press.

Gandy, Matthew. 2004. 'Rethinking urban metabolism: Water, space and the modern city.' *City* 8 (3), 363–79.

Geels, Frank W. 2011. 'The multi-level perspective on sustainability transitions: Responses to seven criticisms.' *Environmental Innovation and Societal Transitions* 1 (1), 24–40.

Ghaffour, Noreddine, Jochen Bundschuh, Hacene Mahmoudi and Mattheus F. A. Goosen. 2015. 'Renewable energy-driven desalination technologies: A comprehensive review on challenges and potential applications of integrated systems.' *Desalination* 356, 94–114.

Global Water Intel. 2009. 'The 2009 Global Water Awards: Sustainability award.' https://www.globalwaterintel.com/global-water-intelligence-magazine/10/2/uncategorized/2009-global-water-awardssustainability-award. Accessed 4 December 2017.

Greater London Authority. 2016. 'London sustainable drainage action plan.' https://www.london.gov.uk/sites/default/files/lsdap_final.pdf. London: Greater London Authority. Accessed 16 June 2018.

Greater London Authority. 2017. 'Population projections.' https://www.london.gov.uk//what-we-do/research-and-analysis/people-and-communities/population-projections. Accessed 4 December 2017.

Griffiths, Ian. 2014. 'Planned London super sewer branded waste of time and taxpayer money.' *The Guardian*, 27 November. https://www.theguardian.com/environment/2014/nov/27/london-super-sewer-branded-waste-time-money. Accessed 4 December 2017.

Hajer, M. 1995. *The Politics of Environmental Discourse: Ecological Modernization and the Policy Process*. Oxford: Clarendon Press.

Halliday, Stephen. 2001 *The Great Stink of London: Sir Joseph Bazalgette and the Cleansing of the Victorian Metropolis*. Stroud: History Press.

Hawken, Paul, Amory B. Lovins and L. Hunter Lovins. 2000. *Natural Capitalism: Creating the Next Industrial Revolution*. New York: Little, Brown and Co.

Hommels, Anique. 2005. 'Studying obduracy in the city: Toward a productive fusion between technology studies and urban studies.' *Science, Technology, & Human Values* 30 (3), 323–51.

Huber, Joseph. 2005. 'Technological environmental innovations.' Forschungsberichte des Instituts für Soziologie 2005-1. Universität Halle-Wittenberg. https://www.ssoar.info/ssoar/bitstream/handle/document/11467/ssoar-2005-huber-technological_environmental_innovations.pdf?sequence=1. Accessed 2 July 2018.

Hughes, Thomas Parke. 1993. *Networks of Power: Electrification in Western Society, 1880–1930*. Softshell Books edn. Baltimore, MD: Johns Hopkins University Press.

Latour, Bruno. 1993. *We Have Never Been Modern*. Cambridge, MA: Harvard University Press.

Loftus, Alex. 2009. 'Rethinking political ecologies of water.' *Third World Quarterly* 30 (5), 953–68.

Loftus, Alex and Hug March. 2016. 'Financializing desalination: Rethinking the returns of big infrastructure.' *International Journal of Urban and Regional Research* 40 (1), 46–61.

Mol, Arthur P. J. 1996. 'Ecological modernisation and institutional reflexivity: Environmental reform in the late modern age.' *Environmental Politics* 5 (2), 302–23.

Mol, Arthur and David Sonnenfeld. 2000. *Ecological Modernisation around the World: Perspectives and Critical Debates*. London and Portland, OR: Frank Cass.

Myerson, George and Yvonne Rydin. 1996. *The Language of Environment: A New Rhetoric*. London: UCL Press.

Naess, Arne. 1984. 'Identification as a source of deep ecological attitudes.' In Michael Tobias (ed.), *Deep Ecology*, 256–70. San Marcos, CA: Avant Books.

Naess, Arne. 1986. 'The deep ecological movement: Some philosophical aspects.' *Philosophical Inquiry* 8, 10–31.

Plumwood, Val. 2003. *Feminism and the Mastery of Nature*. London: Routledge.

Sessions, George. 1998. 'Introduction: Deep ecology.' In J. Baird Callicott and Michael E. Zimmerman (eds), *Environmental Philosophy: From Animal Rights to Radical Ecology*, 165–82. 2nd edn. Upper Saddle River, NJ: Prentice Hall.

Shiklomanov, Igor A. 2000. 'Appraisal and assessment of world water resources.' *Water International* 25 (1), 11–32.

Shove, Elizabeth. 2003. *Comfort, Cleanliness and Convenience: The Social Organization of Normality*. New Technologies/New Cultures Series. Oxford: Berg.

Swyngedouw, Erik. 2009. 'The political economy and political ecology of the hydro-social cycle.' *Journal of Contemporary Water Research & Education* 142 (1), 56–60.

Swyngedouw, Erik. 2010. 'Apocalypse forever?' *Theory, Culture & Society* 27 (2–3), 213–32.

Swyngedouw, E., M. Kaïka and Esteban Castro. 2002. 'Urban water: A political-ecology perspective.' *Built Environment* 28 (2), 124–37.

Thames Tideway Strategic Study Steering Group. 2005. 'Thames Tideway Strategic Study: Steering Group Report.' https://infrastructure.planninginspectorate.gov.uk/wp-content/ipc/uploads/projects/WW010001/WW010001-001256-8.1.2_Thames_Tideway_Strategic_Study_Steering_Group_Report.pdf. Accessed 2 July 2018.

Thames Tunnel Commission. 2015. 'Response to the invitation to submit evidence.' https://www.ccwater.org.uk/wp-content/uploads/2014/02/Evidence-to-the-Thames-Tunnel-Commission-August-2011.pdf. Accessed 4 December 2017.

Thames Water. 2015. 'Final water resources management plan 2015–2040: Main report. Section 3: Current and future demand for water.' Reading: Thames Water. Accessed 4 December 2017.

United Nations. 1992. 'United Nations Conference on Environment and Development. United Nations Conference on Sustainable Development, Rio de Janeiro, Brazil, 3 to 14 June 1992: Agenda 21.' https://sustainabledevelopment.un.org/content/documents/Agenda21.pdf. Accessed 4 December 2017.

United Nations. 2015. 'Sustainable development knowledge platform.' Department of Economic and Social Affairs. https://sustainabledevelopment.un.org/?menu=1300. Accessed 4 December 2017.

United Nations General Assembly. 2010. 'Resolution adopted by the General Assembly: 64/292. The human right to water and sanitation.' New York: United Nations. http://www.un.org/es/comun/docs/?symbol=A/RES/64/292&lang=E. Accessed 17 June 2018.

United Nations World Commission on Environment and Development (WCED). 1987. 'Towards sustainable development.' In Gro Harlem Brundtland, *Our Common Future*, 43–65. Oxford/New York: Oxford University Press.

United Nations World Water Assessment Programme. 2015. *Water for a Sustainable World: The United Nations World Water Development Report 2015*. Paris: UNESCO.

Warren, Karen J. 1990. 'The power and the promise of ecological feminism.' *Environmental Ethics* 12 (2), 125–46.

Woods-Ballard, B., R. Kellagher, P. Martin, C. Jefferies, R. Bray and P. Shaffer. 2007. *The SuDS Manual*. London: CIRIA.

Zimmerman, Michael E. 1987. 'Feminism, deep ecology, and environmental ethics.' *Environmental Ethics* 9 (1), 21–44.

III
Water and societies

11

Early Indian Buddhism, water and rice: Collective responses to socio-ecological stress: Relevance for global environmental discourse and Anthropocene studies

Julia Shaw

Abstract

This chapter focuses on early forms of community engagement with water and environmental control in ancient India, as responses to environmental and climate-change challenges to human health and well-being, and the relevance of this material to global debates within contemporary environmental humanities-based studies on the one hand, and environmental, public and 'planetary' health discourse on the other. A key argument is that despite environmental archaeology's recent engagement with Anthropocene studies, its traditional emphasis on the practical and technological responses to environmental stress and climate change tends to overlook the religio-philosophical and epistemological roots of historically specific human–environmental relationships. Conversely, within the environmental humanities, ancient traditions of religious-philosophical knowledge are often cited as an inspiration for global environmental ethics. Yet the question of whether such traditions in early Indian contexts support attitudes towards nature and its resources in 'eco-friendly' ways has seen little input from archaeology. Drawing on landscape data from central India, and with a particular emphasis on the Buddhist tradition, this chapter assesses the degree to which archaeological evidence can aid scholarly understanding of early

Indian attitudes towards animals, plants, food production, and land and water use. It examines early Buddhism's relationship with land and water management, and new forms of food production as responses to social and environmental stress on the one hand, and as agents of new cultural attitudes towards food and the body on the other. Drawing on 'devolved' examples of religion-based institutional management of land and water resources, it argues that water control was central to the gradual and long-term process of monumentalisation of early Buddhist monasticism and its entanglement with its broader socio-ecological environment.

Introduction

This chapter focuses on early forms of community engagement with water and environmental control in ancient India as responses to environmental and climate-change challenges to human health and well-being, and the relevance of this material to global debates within contemporary environmental humanities and environmental health circles. It calls for those studying environmental events past and present to give greater thought to the religio-philosophical and epistemological roots of the historically specific human–environmental relationships that underlie our current environmental and climate-change crisis, and to question how differing attitudes towards the relationship between humans and non-humans may produce distinct environmental trajectories and responses to extreme events. It also questions the relevance of diachronic approaches to collective responses to environmental stress to current debates within environmental, and 'planetary' health agendas (Watts et al., 2017; Whitmee et al., 2015), and public health circles that emphasise the element of community responsibility in tackling specific health challenges (Deprez and Thomas, 2016). Finally it stresses the need for greater integration between environmental ethics, Anthropocene and climate change studies, and recent environmental health discourse based on developments within epigenetics (Shaw, 2016a), as echoed more recently within bioethics circles (Lee, 2017; Macer, 2017).

Within contemporary environmental discourse, ancient Indian traditions of religious-philosophical knowledge have been drawn upon as resources for informing global environmental ethics. Yet the question as to whether they support attitudes towards nature and its resources in 'eco-friendly' ways has seen little input from archaeology. This chapter will assess the degree to which archaeological evidence can aid scholarly understanding of early Buddhist attitudes towards animals, plants, food production, and land and water use, and contribute to broader discourse within Environmental Humanities and Anthropocene studies (for a comparative assessment of Buddhism and Hinduism in this respect, see Shaw, 2016a). In particular it questions how early Buddhist communities responded to new environmental challenges from the mid first millennium BC to the early centuries AD, in the face of widespread reactions to established Vedic worldviews, rising urbanism, and changing socio-political structures. I examine the relevance of 'devolved' examples of religiously based institutional management of land and water resources to scholarship and to activism, focusing on adaptive responses to present-day environmental and climate change, and its impact on human health, well-being and suffering. I mediate between two polarised approaches to Buddhism and ecology, one of which views Buddhism as 'eco-*dharma*' (*dharma* in this context referring to Buddha's teachings) partly on the basis of its supposed preoccupation with non-violence (*ahiṃsā*), and the well-being and (alleviation of) suffering (*dukkha*) of non-humans; a second drawing on philosophical-theological arguments to discredit environmentally engaged models of Buddhism (Schmithausen, 1997). I argue that the latter critique perpetuates the canonical model of passive monks removed from worldly concerns, apparently innocent of archaeological evidence of socially engaged forms of monastic governmentality and 'landlordism' from the late centuries BC. Some scholars (e.g. Elverskog, 2014) *have* referenced this material, but ironically enough only as a means to *discount* the eco-*dharma* argument. In doing so they misconstrue modern environmentalism as a rejection of agriculture altogether, rather than recognising it as a quest for greater human:non-human equilibrium, and by focusing on the non-human dimension of suffering alone, they overlook the potential convergences between modern and ancient versions of environmental ethics.

I will examine early religious institutions that acted as agents of knowledge, by influencing archaeologically attested changes in land and water use, food culture and land tenure in early historic India, as illustrated by the changing dynamics between irrigated rice and non-irrigated wheat (and millet) crops, especially as urbanism and related phenomena spread westwards from the Gangetic valley. To date, Indian archaeobotany has focused on the Neolithic origins of domestication (Fuller, 2005; Kingwell-Banham et al., 2015), with less emphasis on later periods, or on how religio-philosophical developments influenced or responded to changes in food production and consumption (for vegetarianism, for example, see Shaw, 2016a). The key case study here is the Sanchi Survey Project (SSP), a multiphase landscape-based assessment of the socio-ecological basis of Buddhist propagation from the late centuries BC (Shaw, 2007) (Figure 11.1). Using the distribution of Buddhist monasteries, habitational settlements and water-resource systems around the early monastic complex at Sanchi, I focus on early Buddhism's relationship with land and water management and new forms of food production as responses to social and environmental stress, and as expressions of new cultural attitudes towards food and the body. I contend that the *saṅgha* (monastic order)'s engagement with environmental control (a relationship that is reformulated within Hindu contexts in later years: Shaw, 2016a) was – like its posited involvement with

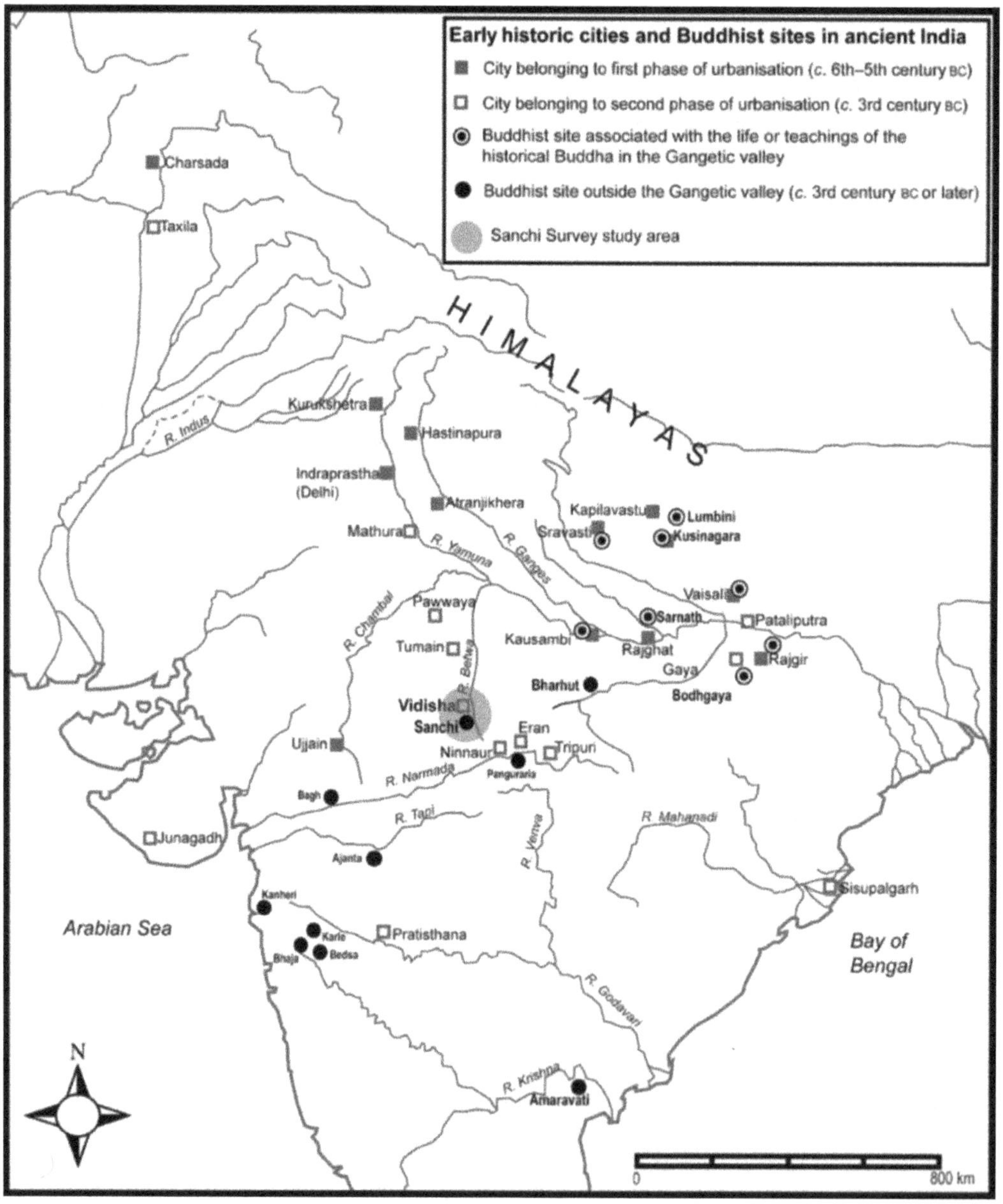

Figure 11.1 Major Buddhist and urban sites in early historic India.
Source: author

medicine (Zysk, 1998) and banking (Schopen, 1994) – both an instrument of lay patronage, and closely related to Buddhism's deeper preoccupation with human suffering (*dukkha*) and the means of its alleviation: one of the Buddha's key teachings was that humans suffer if they do not live correctly (Schlieter, 2014). Here I focus specifically on the human:non-human:environment dimension of this emphasis on 'correct action', and the gradual and long-term process of monumentalisation and human entanglement with the socio-ecological environment that underpinned the development of sedentary monasticism in the Sanchi area; central to this process was the Buddhist monastery's involvement with water and water control.

Archaeology, religion and Anthropocene studies

A growing field of religion and ecology has focused on the potential of environmentally oriented theologies to inform current and future 'green' policy making (Jenkins et al., 2016). By contrast, despite several agenda-setting papers on archaeological conservation and landscape ethics (Dalglish, 2012), there is a paucity of archaeologically oriented studies of the ethical dimensions of Anthropocene discourse, and, in particular, the religion–ecology–environmental change interface (see, however, Shaw, 2016c). Similarly, with recent exceptions (Riede and Klevnäs, 2016), archaeology has a very scant presence within Environmental Humanities research forums. Moreover, it has only recently begun to respond to earth scientists' framing of the Anthropocene as a post-Industrial Revolution phenomenon (Crutzen and Stoermer, 2000), with a flurry of position statements that seek to highlight archaeology's relevance to collaborative climate change research (e.g. Ellis et al., 2016; Murphy and Fuller, 2017). A key argument in such discourse is that the Anthropocene should be regarded as a much older, Neolithic event (Braje, 2016; Dalby, 2016; Edgeworth, 2014; Lane, 2015; Randall, 2016), as represented by evidence of crop and animal domestication from the Mid-Holocene transition 6,000 years ago (Fuller, 2007; Hodder, 2012: 75–6), and even earlier, from 8000 BC, in South and East Asia, and the Near East (Boivin et al., 2016; Morrison, 2015). Emerging evidence of the long-term history of forest exploitation also challenges modern environmentalist narratives of deforestation as a largely post-industrial phenomenon (Clement et al., 2015; Evans, 2016; Morrison and Lycett, 2014).

In South Asia, pre-modern climate-based models of urban decline and 'collapse' have focused on the posited weakening of the summer monsoon in 2.1 ky BC / 2100 BC and its supposed impact on the transformation of Harappan urbanism after *c.* 1900 BC (Dixit et al., 2014). However, there has been little questioning of how the posited impact of such events on land use and settlement distribution extended to the religio-philosophical realm and to those traditions concerned with human health and well-being. This lacuna is in keeping with the general exclusion of the non-physical sciences from the aforementioned collaborative frameworks envisaged by environmental archaeologists (Ellis et al., 2016; Murphy and Fuller, 2017). Additional uncertainties persist for the ensuing period, from the composition of the Rig Veda to the re-emergence of urbanism and complex political organisation in the Gangetic valley around a thousand years later, when the religio-philosophical traditions discussed here, including Buddhism, began to take shape. Traditional explanations for this 'second urbanisation' include metallurgical innovation, with the iron plough and axe facilitating the expansion of intensive agriculture into previously inaccessible, forested, areas.

Such models have been challenged by improved survey methods, revised iron chronologies (Tewari, 2003), and revisionist theories regarding the history of forest exploitation in South Asia (Morrison and Lycett, 2014). Moreover, the precise relationship between urbanisation and environmental history, especially as urban polities spread westwards from the Gangetic valley from about the third century BC, is still an unsettled matter, as is the interface between environmental change, the transformations of material culture, and religion-based intellectual and practical responses as manifested in orthodox and heterodox challenges from the mid first millennium BC, to older Vedic rituals and worldviews (Shaw, 2016a). Buddhism, for example, is viewed variously as a catalyst for and an outcome of urbanism, whereby it provides a means of tackling the suffering associated with urbanism's more negative attributes such as poverty, illness and pollution (Bailey and Mabbett, 2003; Shaw, 2013a: 88–90). Here there are obvious parallels with the mixed outcomes of rapid urban development in the modern day. One of the many ironies here is that Buddhism flourished in times of upheaval, partly because monks could mediate between different political and economic forces at times of turbulent social change (Shaw, 2013a: 89).

Archaeology, environmental ethics and epigenetics

Despite the crucial deep-temporal perspective offered by archaeology's engagement with Anthropocene studies, it is important not to underplay the particular gravity and uniqueness of our current environmental crisis, whose close link with the petrochemical and agro-pharmaceutical industries, and associated use of synthetic, and often toxic, chemicals on an unprecedented scale, set it apart from pre-industrial examples of human:nature entanglements. The tendency within popular media-oriented environmental activism tropes, as well as Anthropocene studies, is to emphasise anthropogenic environmental stress and climate change at the expense of their consequent human disease and suffering, which stem as much from individual contributory factors, such as synthetic chemical use, as from the direct effects of climate change itself. However, given developments in environmental medicine that demonstrate how our synthetically altered environment is changing human and non-human animals at an intergenerational level through epigenetic, genetic and endocrine disruption (DellaValle, 2016; Parry and Dupré, 2010; Dupré, 2016; Genuis, 2012; Mostafalou and Abdollahi, 2013), the message of Rachel Carson's (1962) ecological canon is as urgent as ever. Nevertheless, current environmental ethics discourse has been slow to respond to such developments (Lee, 2017; Macer, 2017; Shaw, 2016a). Archaeology's recent emphasis on 'entanglement' (Hodder, 2012), which views human-constructed environments largely through the prism of agricultural 'domestication' and forest exploitation, could benefit from the multidirectional perspective offered by the epigenetic model, while environmental archaeologists too need to give consideration to the ethics of their own professional (Riede et al., 2016) and personal (Shaw, 2016b) actions. My position is that highlighting the points of convergence between the 'entanglement' theme in the social sciences and new directions within environmental health, including new sustainable development models (United Nations, 2015), and related medical initiatives (Watts et al., 2017; Whitmee et al., 2015) vastly increases the scope for bringing 'green' agendas into mainstream political activism.

A similarly one-sided interpretation of environmentalism as being concerned with 'nature' as an entity removed from humans underscores much of the scholarly debate about the prominence, or not, of environmental ethics in early Buddhism. Thus, the relevance of the Buddhist doctrine of suffering (*dukkha*) to the environmental movement is usually linked to its concern for animal welfare, rather than to the human fallout of environmental stress. By emphasising a human-centric approach to suffering and the means of its alleviation, as in examples of early Indian Buddhist engagement with environmental control, one might better draw on convergences between past, present and future environmental ethics, and encourage a more holistic and intergenerational approach to human:non-human care (Shaw, 2016a). This perspective sits well with recent calls from public health theorists who stress the element of community responsibility as a tool for tackling current health challenges (Deprez and Thomas, 2016), and arguments within contemporary and future disaster management discourse that place engagement with religion high on the agenda for tackling climate change and extreme environmental events (Chester, 2005; Hulme, 2010).

This new emphasis is significant against the recognition of religion-as-'worldview' which in many cultural contexts is the primary modulator of empirical knowledge about humans' place in the world, and for codifying frameworks of purity or cleanliness versus pollution or dirt, or of harmful versus safe human:non-human relationships (Shaw, 2013b, 2016a) . This may be contrasted with secular contexts in which scientifically driven government legislation is often the last word for determining beliefs about climate change, environmental health, disease aetiology, and related consumption and lifestyle choices that impact on global climate patterns (Holm et al., 2015), although certain environmentalist positions have themselves been described as forms of 'secular religion' (Latour, 2013; Shaw, 2016b).

Archaeologies of religion and 'nature' in ancient India

In response to the current environmental crisis, ancient Indian traditions of religious-philosophical knowledge are often presented as forms of pre-modern ecological utopia whose expectation of care for the environment contrasts with modern realities. Archaeology, with its uniquely deep temporal, and broad spatial, perspective, is well placed for testing such utopian theologies, particularly those that perpetuate notions of the primordial sacred forest which, disrupted only by colonial or industrial developers, is presented as the epitome of 'nature' in contrast to agro-urban zones as manifestations of 'culture' (Morrison and Lycett, 2014: 150). Advances in archaeobotany, geoarchaeology and remote sensing have highlighted the socially constructed element of the 'wild', with large areas of previously designated 'virgin' forests in Cambodia (Evans, 2016) and the Amazon (Clement et al., 2015), for example, now known to have undergone agro-urban development in ancient times.

Similarly, archaeology has helped to challenge the impression, central to the rejuvenation-based conservation models that have arisen in opposition to 'big dam' World Bank development agendas (Agarwal and Narain, 1997), of the unquestionably benign and sustainable nature of pre-modern water-management systems, particularly those associated with early temple and monastic administrations. While offering 'devolved' alternatives to Wittfogelian centralised models of water management, many such dams (those in South India extending up to 3 km in length) may themselves be regarded as 'big dams' (Morrison, 2010). And despite their sacred status and associations with religious authority, such dams and their associated reservoirs are not necessarily disconnected from state power, especially given the mutual interconnection between imperial rule and land-owning Hindu deities (Willis, 2009). These gods residing in temples are still the biggest land-owning entities in India (Sontheimer, 1964), a legal privilege invested through imperial powers from the mid first millennium AD onwards (Willis, 2009), arguably in competition with the earlier Buddhist-based models of land and water management discussed below (Shaw, 2007). As Morrison (2010: 192) states, Indian reservoirs 'were always politically and religiously charged features', bound up with motives of power and profit.

However, while the varied impact on those living upstream and downstream means that dam building rarely proceeds without opposition, the 'sustainable' reputation of many pre-modern dam traditions reflects to some degree their tendency to follow highly localised design models (Sutcliffe et al., 2011), and codified rules of community reciprocity over access to water supplies (Agarwal and Narain, 1997). It is important not to discount the potency of individual ecological concepts within specific religious traditions, and to consider the long-term histories of intergenerational human:non-human:environment engagement, out of which individual religio-philosophical, socio-ecological and agrarian models (and subsequent problems or imbalances) emerge.

Buddhism and ecological ethics

To this end, let us now examine the debates regarding early Buddhist 'environmentalism', which, roughly speaking, fall into two main camps, the so-called 'eco-apologists' (Swearer, 2006) promoting the idea of 'eco-*dharma*' (Dorje, 2006; Harvey, 2007; Oliver, 2004), and the 'eco-critics' arguing that the Buddha's original teachings have been distorted and misappropriated by Western environmentalism (Harris, 1997: 388, 395; Schmithausen, 1997; Strain, 2016; Swearer, 2006). The 'eco-*dharma*' argument draws principally on early Buddhism's commitment to non-violence (*ahiṃsā*) and the alleviation of suffering (*dukkha*), and to later Mahāyāna traditions of the Buddha of Compassion, and the expectation of care towards the natural world. Further, much has been made of Arne Næss's (2003: 271) self-professed alignment with the later (and largely Chinese) Buddhist doctrine of the Origination in Dependence as an overt influence on the development of his philosophy of 'deep ecology' (Dorje, 2006: 1095), resulting in what critics (e.g., Schmithausen, 1997) view as anachronistic and historically inaccurate associations between Western environmentalism and Buddhist thought in general (Strain, 2016: 197; Shaw, 2016a). Schmithausen (1997) argues that, far from being actively concerned with the environment, early Buddhists were impressed not so much by the 'beauty' of nature as by its negative aspects. Early Buddhists followed a form of 'passive environmentalism', seeking not to transform or subjugate nature but to transcend it spiritually through detachment.

A similarly 'passive' stance is highlighted in Buddhist attitudes towards vegetarianism and diet in general. Although compassion (*karuṇā*) towards animals, and retribution for animal-killing, in this and the next life, are recurring themes in early Buddhist texts (Stewart, 2014), the Buddha permitted monks to eat meat as long as the animal was not considered to have been killed for this purpose (Mcdermott, 1989: 274; Ruegg, 1980: 234–5). Early Buddhists are seen therefore as being less concerned with the physical or energetic impact of meat consumption on the eater's body or spirit, let alone the well-being of the 'eaten', than with questions of morality and self-restraint (Mcdermott, 1989; Ruegg, 1980: 235). This assumed 'passive environmental action', which may not always benefit the eater – as illustrated by the story of Buddha's final, fatal, meal (Wasson and O'Flaherty, 1982) – contradicts the Hindu *tridoṣa*-based framework which associates certain diets with different body types determined by caste (Smith, 1990; Shaw, 2016a), with specific ingredients and with modes of preparation and cooking that pose major threats to the eater's purity. We may also view such assumed passivity as a kind of 'opting out' of ethical debates: by being required to eat whatever is put in their begging bowl, monks are exempted from orthodox purity:pollution-based consequences of dietary decisions. It is important, however, to reassess critically anthropological models of 'ritual' purity and pollution (e.g., Morrison, 2012) that have persisted since Douglas's (1966) seminal work, and which fail to distinguish between polluting actions or substances which require ritual purification as prescribed in Sanskrit texts, and actions bracketed off as 'ritual' in the broader anthropological sense of habitualised action (see Shaw, 2016a).

Monasteries as gardens: Transcendence or control of nature?

A similar level of detachment is promoted by recent scholarship on early Buddhist monasteries as ascetic inversions of urban expressions of sophistication, power and wealth, such as the courtly garden, itself an epitome of control and artifice (Ali, 2003; Schopen, 2006). Within such frameworks, sculptural depictions of plant imagery at early Buddhist sites are viewed not as real plants, but as idealised utopias, or '*dharma* spaces' (Brown, 2009), representing monks' transcendence of, rather than engagement with, 'nature'. Similarly, sculptural depictions of women, traditionally viewed as 'nature' spirits (*yakṣī*) are, according to such models, viewed as courtesans, whose role in well-known romantic tales set in palace gardens made them ideal mnemonic symbols of monks' transcendence of worldly pleasures (Shimada, 2012).

However, while Schopen (2006: 498–505) suggests that such monastic gardens were places from which to view 'nature' from a safe distance, remarking on the beautiful 'views' of the bucolic countryside from early monastic sites, the point is that, in the Sanchi area at least, the landscape data discussed below paints a rather different kind of 'view', one characterised by cultivated, 'managed' agrarian and hydraulic landscapes, interspersed by semi-urban habitational sites (Shaw, 2007, 2015). In fact, the control and harnessing of 'nature', and particularly water, is what the Buddhist monastery excelled at, and the garden was the ideal visual medium for illustrating such skills. As well as the monastery-governed hydraulic systems described below, examples include the ostentatious display of water-storage facilities at early rock-cut complexes in the Deccan (Shaw, 2004). I have argued that these features were instruments of lay patronage, symbolising not the *saṅgha*'s transcendence of the 'natural' world, but rather its ability to 'tame' and harness natural resources, to harvest and store water in regions of climatic uncertainty (Shaw, 2004, 2007, 2013a). While local populations were otherwise dependent on rain-making cults for timely rainfall and both drought and flood control, mediated through the propitiation of dangerous serpent deities (Shaw, 2004, 2016a; Cohen, 1998), they were now assured reliable and timely water supplies via new frameworks of technological and administrative knowledge. Water shortage is a primary cause of human suffering, especially in regions where 90 per cent of the annual rainfall occurs in two to three months, and the *saṅgha*'s ability to alleviate this suffering was made explicit through outward symbols of its engagement with environmental control.

This angle can be extended to the aforementioned sculpted plant imagery and 'nature spirits', as symbols of the *saṅgha*'s ability to 'live well' with 'nature', harness its resources, and overcome its more dangerous and unpredictable elements. Much of the plant imagery depicted in early Buddhist relief sculpture, for example, depicts aquatic species such as lotus along with fish, turtles or snakes. For the SSP area, I argue that such imagery reflects the 'watery' landscape out of which the monuments on Sanchi hill emerge (Shaw and Sutcliffe, 2001, 2003, 2005; Shaw et al., 2007). This hydraulic landscape, comprised of the reservoirs discussed below, forms the basis of an emerging 'Buddhist economics' (better known in later South East Asian contexts: Green, 1992; Harvey, 2000: 215–19; Pryor, 1990; Schumacher, 1973) that sustains monks as a non-producing section of society, and provides a practical means for alleviating the human suffering (*dukkha*) that forms the focus of the Buddha's earliest teachings. The 'swampy' vegetation depicted in early Buddhist art accords with the irrigated rice-growing environment in the Gangetic valley from where the earliest Buddhist missions spread outwards in the late centuries BC. A key argument, supported by preliminary pollen sequences from excavated reservoir deposits in the Sanchi area, is that the monastic culture in central India was also predicated on a predominantly rice-growing economy (Shaw et al., 2007). Similarly, as argued convincingly by Cohen (1998), *yakṣī* sculptures (and their serpent deity counterparts, *nāgas*), were possibly etic and generalised attempts to represent the 'local', with specific iconographies reflecting older ritual-ecological realities in the Gangetic valley heartland (Shaw, 2004; also Dalton, 2004). Symbolically too, Sanchi hill and related sites can be viewed as rising from primordial waters in ways that mirror cosmogonic references to Mount Meru at many sacred sites across South and South East Asia. As argued below, the earliest monasteries here were established only after long periods of pre-monumental engagement with comparatively 'wild', peripheral zones of the settled landscape. One might justifiably view such imagery, therefore, as mnemonic indicators of monks' transformation (rather than transcendence) of nature on the one hand, and their transcendence of conventional modes of urban-based production and consumption on the other.

This discussion is relevant to debates about Buddhism's 'middle way' approach to asceticism and agro-urban culture, which contrasts with the alleged 'anti-civilizational' stance of its Hindu ascetic counterparts (Olivelle, 2006; Shaw, 2016a). Benavides (2005: 82) regards Buddhism as neither a rejection nor an affirmation of urban society, but rather a 'commentary' on new attitudes to labour, consumption and wealth. It also implied new attitudes towards purity and pollution: while monks are prohibited from engaging in labour, the lay donations (*dāna*) on which they depend are generated by the polluting work of others. Through a reworking of Hindu purification rituals (Shaw, 2016a), the donor is cleansed of resultant pollution through gift-giving rituals which fuel the development of institutionalised monasticism (Schopen, 1994; Shaw, 2013a): while the Buddha and his earliest followers found support from the merchant and royal classes of Magadha in about the fifth century BC, Buddhism's spread out of the Gangetic valley was linked to the patronage of the Mauryan Emperor Asoka several centuries later. However, it was not until the post-Mauryan period (second to first centuries BC) that the most widespread process of Buddhist construction took place, this time fuelled predominantly by collective patronage (Shaw, 2013a). Following a logic similar to that of modern 'crowdfunding' campaigns (Agrawal et al., 2014; Shaw, in press), single buildings were funded through multiple donations, with individual architectural components bearing the names of individual donors. Epigraphs recording land, village and labour grants appear in the early to mid first millennium AD, which, according to Schopen (1994), coincided with the eventual development of sedentary monasticism.

As argued elsewhere (Shaw, 2011, 2013a), however, this problematic chronological framework overlooks abundant material and textual evidence of monastic landlordism from the second century BC in Sri Lanka (Coningham et al., 2007; Coningham and Gunawardhana, 2013; Gilliland et al., 2013; Gunawardana, 1971) and central India (Shaw and Sutcliffe 2001, 2003, 2005). Not only has this material been largely ignored by scholarship on the history and chronology of institutionalised monasticism, it has also had scant presence within recent 'Buddhist environmentalism' discourse. A rare exception is Elverskog (2014), who, ironically enough, draws on this material to *contest* the eco-*dharma* argument, holding that an agriculturally engaged monastery was at odds with the ideals of environmentalism. Such a stance obviously betrays a vision of environmentalism as an outright rejection of land exploitation rather than as a quest for sustainable human:non-human relationships. It also shows similarities to Eurocentric (as well as, ironically, post-colonial) approaches to Indian nature conservation that place the 'solution' to forest 'degradation' in wilderness 'reclamation', in contrast with the mainstream model of forest 'management' in Europe (see Morrison and Lycett, 2014: 150). Elverskog's stance also has echoes of canonical scholarship which presents forest monasticism, divorced from the forces of production, as the 'original' and higher ascetic path in distinction from the more socially engaged model of urban monasticism (Gombrich, 1988). On the other hand, the eco-*dharma* debate, with its focus on the non-injury of animals, is also flawed because it fails to acknowledge the *saṅgha*'s engagement with environmental control as a reflection of its concerns with human suffering, its causes, and its means of alleviation, which I argue should remain at the forefront of our understanding of Buddhist forms of environmental control (Shaw, 2016a).

Case study: The Sanchi Survey Project

Since its inception in 1998, the Sanchi Survey Project (SSP) has sought, through several phases of survey and excavation (Shaw, 2004, 2007, 2011, 2013a, 2013c, 2015; Shaw and Sutcliffe 2001, 2003, 2005; Shaw et al., 2007; Sutcliffe et al., 2011), to relate the westward spread of Buddhism and other religious traditions from the Gangetic valley to central India during the late centuries BC to aspects of socio-economic and ecological history, including urbanisation, state formation and new forms of land use, food production and environmental control (Figure 11.1). The study covers approximately 750 km^2 around the UNESCO World Heritage, Buddhist site of Sanchi, and the ancient city of Vidisha several kilometres further north, with documented site-types ranging from settlements and monastic sites to land use and 'non-site' data (Figure 11.2). Sanchi's earliest constructional history is connected with royal Mauryan patronage, and to a second major phase of propagation during the post-Mauryan period (Phase II), fuelled by collective, largely non-royal patronage (Shaw, 2007, 2011).

A central hypothesis which the study sought to test is that, while monks would have initially relied on local begging rounds for their daily nutritional needs, as monastic communities grew in size, a more integrated patronage system, based on an agricultural surplus rather than a subsistence-based economy, was necessary (Bailey and Mabbett 2003: 70–2). By Phase II, the size and scale of Buddhist monasteries (Shaw, 2011) and the relative configuration of settlements and dams are sufficiently similar to those underlying Sri Lankan monastic landlordism (Gunawardana, 1971) to warrant the assumption of a similar three-way exchange system between land-owners, monasteries and the agricultural laity (Shaw and Sutcliffe, 2001, 2003). The Sri Lankan textual and epigraphical evidence suggests that this system was mediated through the monastic administration of land and water resources, while local communities handled the physical running of the dams and irrigation

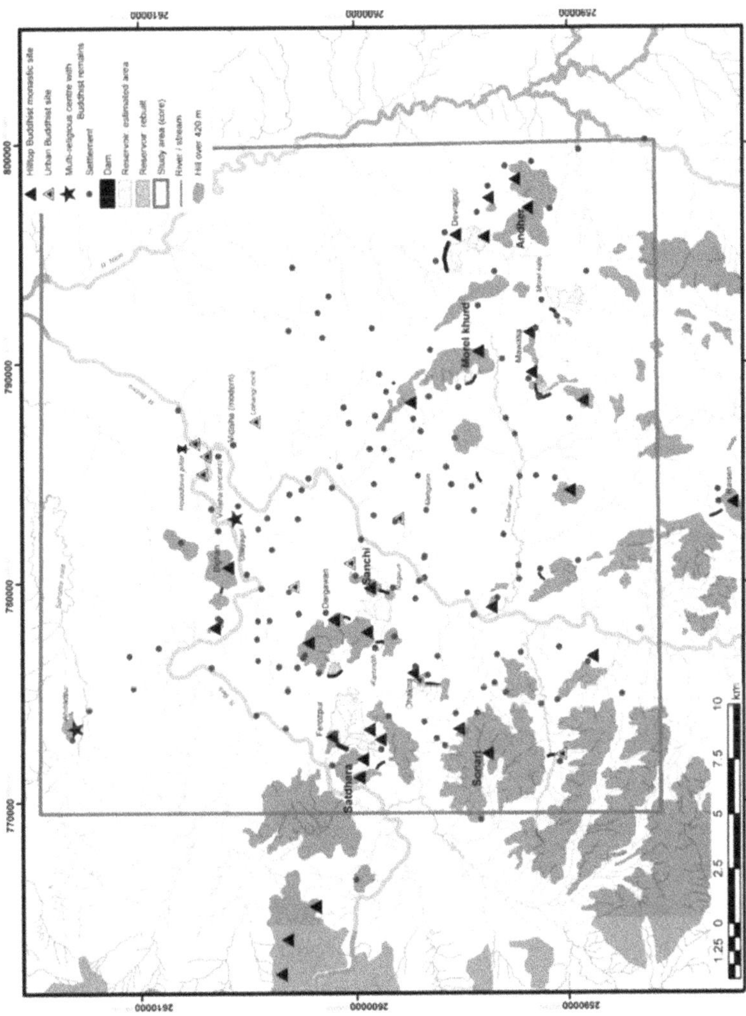

Figure 11.2 Sanchi Survey Project study area: site distribution. Source: author

facilities that they provided. Recent survey and excavation work in Sri Lanka has supported the idea of a 'theocratic' hydraulic landscape in Anuradhapura's hinterland (Coningham et al., 2007; Coningham and Gunawardhana, 2013), with comparative studies in eastern India (Sen et al., 2014: 67), Bihar (Rajani, 2016: 6), and the north-west (Olivieri et al., 2006: 131–3). Similar relationships are inferred for Thotlakonda in Andhra Pradesh, where ceramic assemblages suggest on-site food preparation and storage by non-monastic staff, rather than individual begging rounds (Fogelin, 2006: 152–3, 165).

A number of supporting propositions have been offered based on the SSP landscape data (Shaw, 2007, 2013c). First, Chalcolithic settlement distribution and ceramic evidence attest to limited sedentary occupation within the low-lying areas from at least the second millennium BC, with increased settlement density during the middle to late first millennium BC (Shaw, 2015: 394–5) (Figure 11.3). We may infer that the incoming *saṅgha* did not choose completely unsettled areas, and that the establishment of large monumental monastic complexes was a gradual process. Secondly, the distribution of painted rock-shelters and prehistoric tools attests to peripatetic, possibly hunter-gatherer, occupation in the hilly zones from at least the Chalcolithic period (Figure 11.4). Associations between such non-agricultural zones and later, property-renouncing ascetic groups, show that it is possible that, as a 'non-producing' entity, the incoming *saṅgha* had limited choice over which places it could occupy. Many prehistoric rock-shelters surviving around the edges of these complexes were adapted for monastic use, possibly as an intermediate stage between peripatetic and sedentary monasticism as represented by Phase II monasteries; similar 'monastic shelters' distributed widely throughout central India (Shaw, 2007: 37, 117) bear a close resemblance to the drip-ledged shelters (*lena)* of Sri Lanka, whose third-century BC donative inscriptions represent the earliest form of lay-monastic patronage in that region (Coningham, 1995) (Figure 11.5). As in Sri Lanka and South East Asia, some of the SSP shelters were undoubtedly occupied into later periods as part of a two-tiered urban/forest model of monasticism. Future excavation and rock-art analysis is expected to clarify their chronological and historical relationship to structural monasteries and institutional monasticism, while palynological and geoarchaeological analyses will shed light on how Buddhist propagation related to wider patterns of forest clearance and land tenure. This is important given the *saṅgha*'s alleged pioneering role in encouraging population shifts into new areas (Ray,

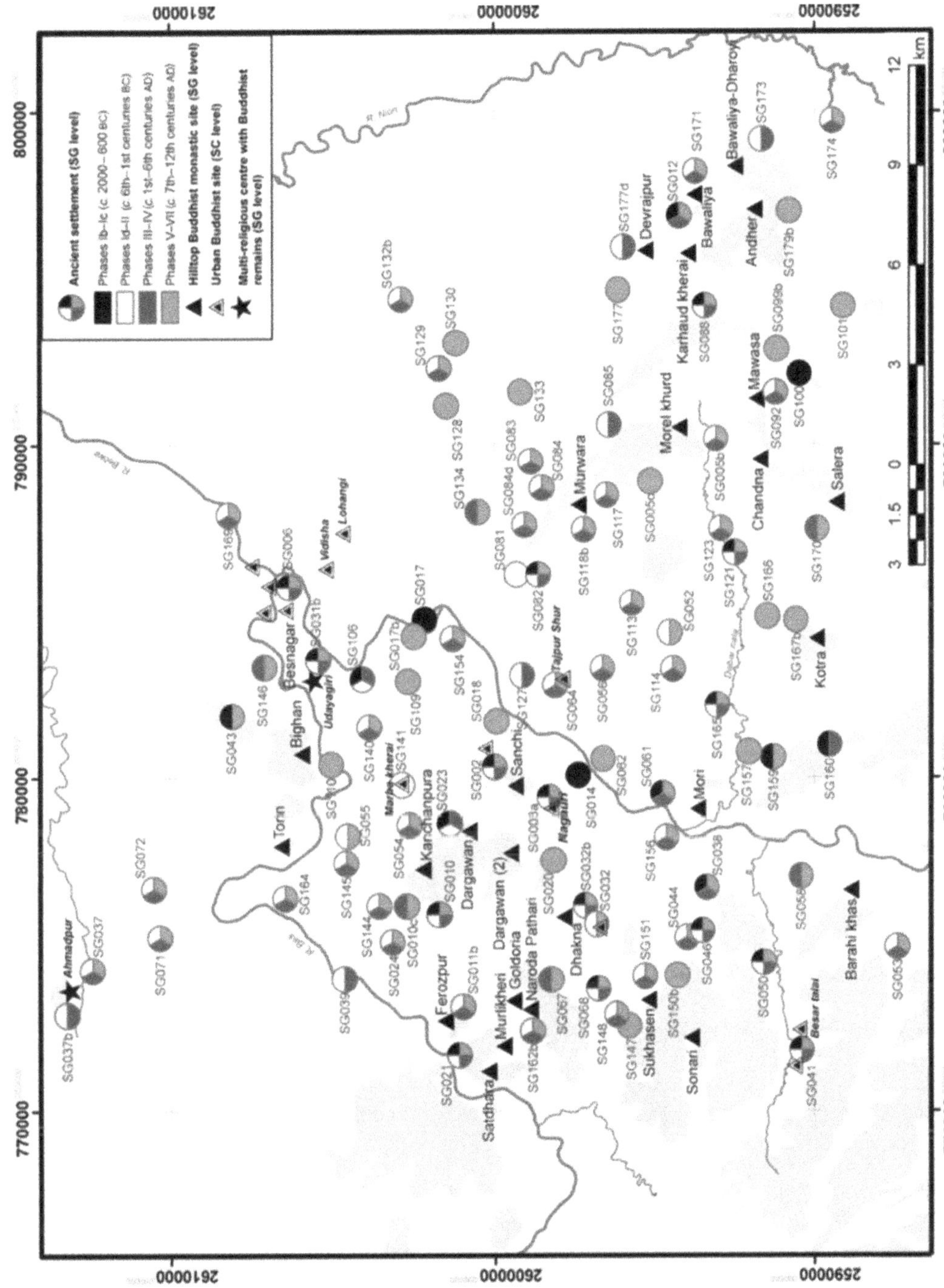

Figure 11.3 SSP settlement chronology. Source: author

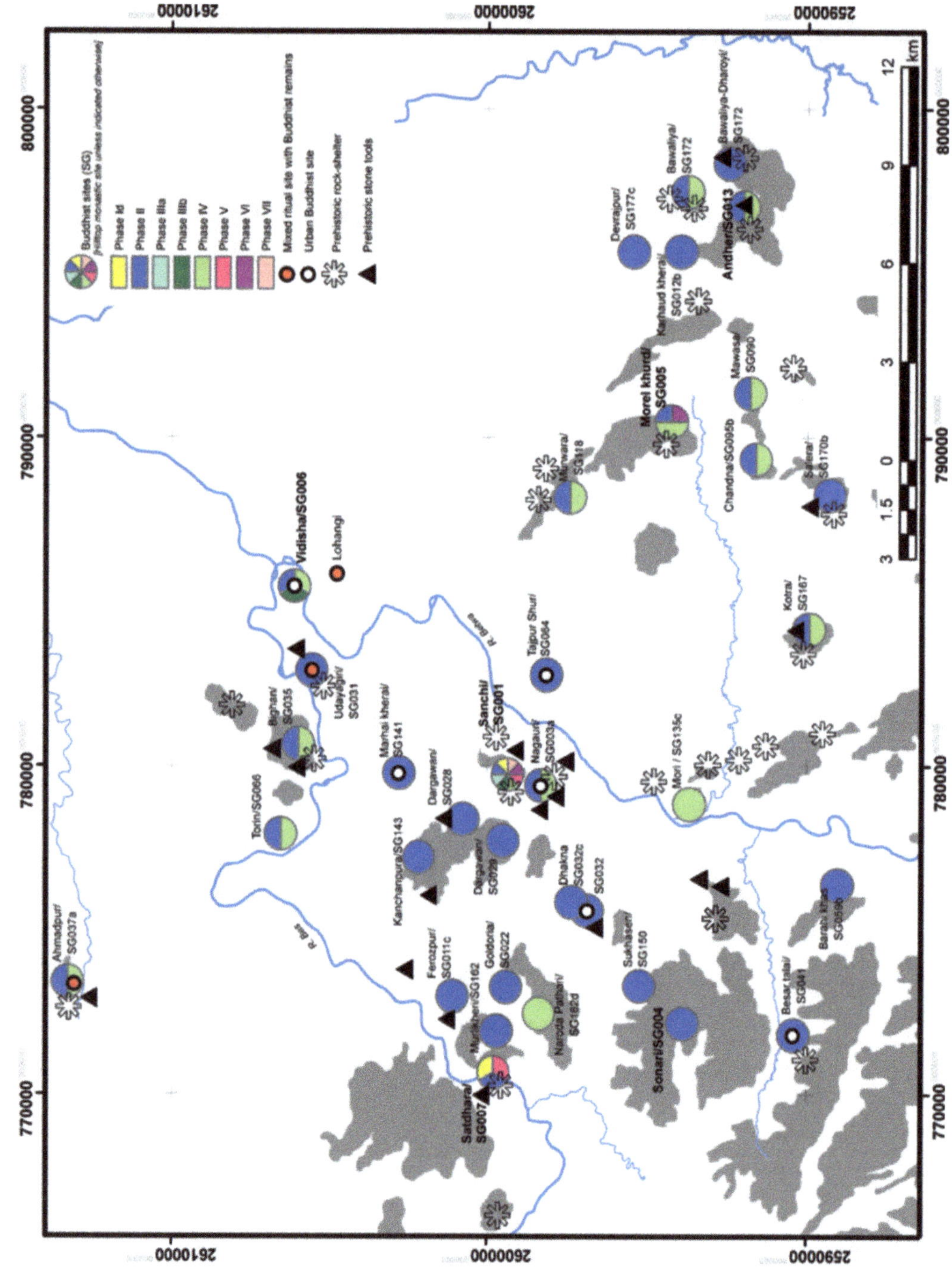

Figure 11.4 'Monastic shelter' distribution. Source: author

1994: 5), and more recently posited comparisons with medieval examples of Hindu and Islamic 'monastic governmentality' (Chatterjee, 2015).

Many of the SSP 'monastic shelters' contain Chalcolithic paintings of wild animals in hunting, post-butchery or hide-preparation scenes, or with animals' internal organs and skeletal structure visible (Neumayer, 2011). This rich and only partially documented repertoire has yet to be incorporated into aforementioned debates about Buddhist attitudes to 'nature', or about the broader history of anatomical knowledge which hitherto has been presented as a unilinear trajectory from Vedic sacrifice-based observations to the eventual formal codifications of the classical Ayurveda medical tradition, and possibly involving along the way travelling Buddhist monks who, unlike their Hindu counterparts, were not restricted by orthodox prohibitions of contact with polluting bodily substances (Zysk, 1986, 1998). Human knowledge of, and interaction with, animals and animal anatomy outwith the Sanskrit textual tradition does not figure in this narrative, an obvious shortcoming which needs to be addressed through collaborative textual and archaeological research (Shaw, 2016a). Buddhist forest monks would have had as close dealings with the hunting communities that created this art as they did with local farmers, and indeed, as today, there would have been considerable overlap between these two lifeways. When such monastic 'locales' do become monumentalised, the element of fortification is prominent, with key structures raised on high platforms and surrounded by high boundary walls (Shaw, 2007: 110–45). These features evidently provided security against potential hostile human action, but also against wild animals, a frequent motif in early Buddhist art alongside the aforementioned plant imagery, possibly as reminders of the more dangerous aspects of 'nature' with which monks were closely familiar. Tigers, bears and leopards are still everyday hazards in many parts of central India, and the older, peripatetic, monsoon retreat-based tradition, out of which sedentary monasticism grew, would have given monks direct exposure to nature in its most wild, unpredictable and often dangerous forms, in ways that presented real threats to human well-being and longevity. Of relevance here, for example, is the 'Bhaya-bherava Sutta' ('Discourse on fear and dread') from the *Sutta Piṭaka* of the Pāli Canon (*Majjhima Nikāya* v. 4) in which the Buddha teaches meditation practices to jungle-dwelling ascetics in order to overcome their fear of wild animals (Nāṇamoli and Bodhi, 1995: 102–7), with many other examples demonstrating similarly the degree of fear with which certain wild or uninhabited places were associated (Shaw, in press).

Reservoirs, rice production and monastic landlordism

Finally, our model of early monastic governmentality in central India draws heavily on a group of 17 stone-faced, earthen dams, distributed across the study area in close proximity to monastic sites (Shaw and Sutcliffe, 2001, 2003, 2005; Shaw et al., 2007; Sutcliffe et al., 2011) (Figure 11.2). Surviving to heights of up to 6 m, and lengths of up to 1,400 m, they supported reservoirs with areas of up to 3 km^2, and volumes of up to 3 million m^3. Some, such as the main dam below Sanchi hill, were built on gradually sloping terrain, providing inundation reservoirs for upstream irrigation; others extended across deeper valleys for downstream irrigation, with evidence of spillways and sluice gates (Figure 11.5). Using associated material and OSL/TL dating of dam and reservoir sediments (Shaw et al., 2007), we dated the earliest construction phase to about the third to second centuries BC, in keeping with the Mauryan and post-Mauryan phases at Sanchi and neighbouring Buddhist sites.

Additional hypotheses about reservoir design and function, and associated land use, were tested through analyses of surface remains and present-day local hydrology. Individual reservoir storage capacity of up to 3 million m^3 far exceeds the needs of modern agriculture, which, until the introduction of tube-well irrigation in the 1990s, was dominated by rain-fed wheat cultivation (Shaw and Sutcliffe, 2001, 2003, 2005). The near absence of irrigation in recent times is heavily influenced by the local, highly moisture-retaining black-cotton soils which store sufficient water from monsoon recharge to support winter wheat crops throughout the growing season. Low population density following several recorded waves of famine-related emigration from the fourteenth century onwards is another factor; in the aftermath of the 1899–1900 drought (*Imp. Gaz.*, 1908: vol. ix, 374–5), it was suggested that any major population resurgence would require agricultural re-intensification, including the adoption of irrigation (Watt, 1889–93: vi, 151). This raises obvious questions regarding land use during the late centuries BC, when site distribution in the Vidisha hinterland attests to considerably higher settlement density than today, with 29 out of 133 settlements occupying levels two to four of a six-tiered hierarchy (Shaw, 2007: 226; cf., however, Hawkes, 2014, who misinterprets such discussions as promotion of an exclusively 'urban' model of local monasticism), and with large monasteries, now abandoned, spread over every hilltop. While the aforementioned drought-related migrations influenced low irrigation use in recent times, the earlier decline in fortunes of Buddhism, from around the tenth

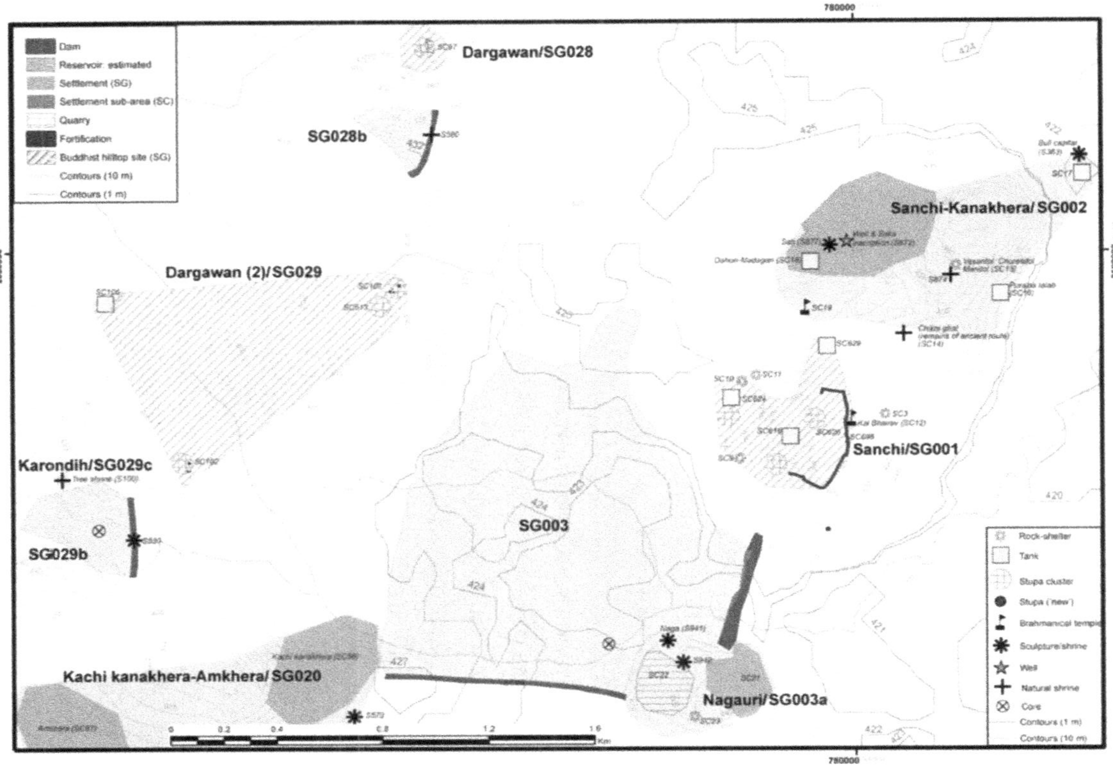

Figure 11.5 Sanchi hill: ritual monuments, settlements and water-resource systems. Source: author

century, was probably more significant. Some water bodies were evidently used in later periods for bathing and domestic supplies, but the decline of Buddhism was a key catalyst for their ceasing to function as irrigation resources. The abandonment of monasteries would have led to serious economic downturn, lowering food production requirements, and rendering the upkeep and repair of the reservoirs unviable (for later Buddhist history, and alternatives to the homogeneous model of post-Gupta Buddhist decline in India, see Willis, 2013; Skilling, 2014; Shaw, 2015). Similar explanations, together with climate fluctuation (Lucero et al., 2015: 1148; Gilliland et al., 2013: 1026–7), are posited for the disuse of reservoirs in Dry-Zone Sri Lanka, with further analogies in South India where Chola and Pallava dam inscriptions recorded terms and conditions for reservoir administration and desiltation. Once these sources of patronage dried up, so did the reservoirs (Davison-Jenkins, 1997: 93; Venkayya, 1906). The hypothesis that the Sanchi dams were already out of action by the earliest recorded drought and consequent depopulation of the fourteenth century – and possibly even earlier, given recent evidence of rainfall variation of the monsoon during the early second millennium AD (Jung et al., 2004; Gunnell et al., 2007) – is supported by nineteenth-century famine records which demonstrate that those areas with continued use of traditional irrigation systems had better immunity to the effects of severe drought (Agarwal and Narain, 1997: 182–3).

Our principal argument is that the Sanchi dams were designed not for wheat cultivation, the current staple in the area, but for newly introduced rice, the staple of the Gangetic valley since at least the second millennium BC, whose westward and southward spread, according to hitherto untested hypotheses, did not occur until early historical periods (Fuller, 2005). Although, today, supplementary tube-well irrigation of up to 50 mm can be generated without canal transfer, extensive canalisation is necessary for extensive water distribution; larger growing areas are needed for wheat than for rice, which requires concentrated areas with minimum irrigation depths of about 800 mm (Shaw, 2007: 250). The high cost of dam construction and maintenance makes more sense for rice cultivation, largely because of radically increased depth and intensity of irrigation but also because of the high water-storage capacity of local soils. By contrast, dam and canal construction is less cost-effective for increasing wheat yields than bringing new land into cultivation, especially when the availability of cultivable land was not a limiting factor in the area.

From historical water-balance records a total reservoir storage capacity of 19.5 $m^3 \times 10^6$, and an irrigation capacity of 24 km^2, have been estimated across the study area (Shaw and Sutcliffe, 2001, 2005).

This illustrates the intensive nature of rice irrigation, which requires significant water storage to supply relatively small but highly productive growing areas to meet high food needs, although the variable irrigation requirements of pre-modern rice varieties are an important consideration here (Deb, 2017). An estimated average annual rice crop of approximately 2400–3600 t/year is significantly higher than for wheat. For example, late nineteenth-century records show that the annual yield for unirrigated wheat across the whole Central Provinces (a British Indian province with legal standing between 1864 and 1936, that incorporated parts of present-day Madhya Pradesh, Maharashtra and Chhattisgarh) was just 90 to 135 kg/ha (Watt, 1889–93: iv, 153). While this comparison is an over-simplification, it demonstrates the impact of introduced irrigated rice on food production as a response to rising urban and non-producing populations (Lucero et al., 2015: 1143–6, for similar links in Dry-Zone Sri Lanka), our suggestion being that it formed part of a cultural package that accompanied the spread of new religio-cultural traditions from the Gangetic valley.

Our account of locally cultivated rice is closely related to the highly seasonal nature of local cultivation and climate, with over 90 per cent of total rainfall (annual average 1,300 mm) occurring between June and September. This pronounced seasonality was a key factor in the development of irrigation technologies in ancient India, with water-storage facilities providing both monsoon flood control and insurance against drought; it also led to the prevalence, across much of monsoon-dependent India, of a double-cropping cycle, rice being planted in summer and wheat in winter (Sutcliffe et al., 2011). We have argued, however, that local conditions made central India better suited to a complementary cropping system, with the Sanchi reservoirs providing intensive irrigation for rice cultivation, while supplementing rain-fed cultivation of both summer *kharif* and winter *rabi* crops in surrounding areas. This contrasts with the double-cropping cycle suggested for medieval reservoirs of the western Deccan (Morrison, 1995: 212), as well as the traditional *ahar* dams of southern Bihar which provide downstream irrigation of summer *kharif* rice crops, and upstream irrigation of winter *rabi* crops (wheat, barley and sometimes rice) within the reservoir beds themselves (Agarwal and Narain, 1997: 86–98; Sutcliffe et al., 2011: 281–2). Designed in response to local sandy soil conditions which cannot support an unirrigated winter crop, the *ahar* design would make little sense for central India's clay-rich, moisture-retaining soils.

Indeed, the highly localised nature of water-resource technology is paramount to the success, and so to the 'sustainability', of many pre-modern irrigation traditions (Sutcliffe et al., 2011), and their appeal as indigenous alternatives to modern 'big dam' developments. While several areas of influence have been considered for Sanchi's dam-based irrigation technology, we have discounted either a ready-made or a trial-and-error model. Rather, the relationship between reservoir volume, local catchments and run-off volumes suggests that they were built by a professional engineering class following a considerable period of local water balance observation (Figure 11.6). We have suggested a gradual unfolding of intensive forms of land use as monasteries became larger, and more integrated in the local economy. Evidence of a palaeotank incorporated into the design of a third-to-second-century BC reservoir at Devrajpur, for example (Shaw, 2007: 253, n. 45; Shaw et al., 2007), supports suggestions that the earliest dams were of a makeshift, temporary nature, developed only later in durable materials in response to changing socio-economic conditions (Shaw and Sutcliffe, 2003: 80–1; Agarwal and Narain, 1997: 166–75). This mirrors our suggested trajectory from adapted rock-shelter dwellings to permanent, structural monasteries, in keeping with the development from peripatetic to sedentary monasticism. As such locales became more monumentalised, so the relationship of 'entanglement' (Hodder, 2012) between monks and their socio-ecological landscape became more entrenched.

Given the lack of a recorded history of commercial rice production in the area, our suggestion of large-scale rice cultivation in central India will benefit from collaboration between, on the one hand, text-based analysis of changing religio-cultural associations of irrigated rice versus unirrigated wheat or millet (Shaw, in press) and, on the other, follow-up

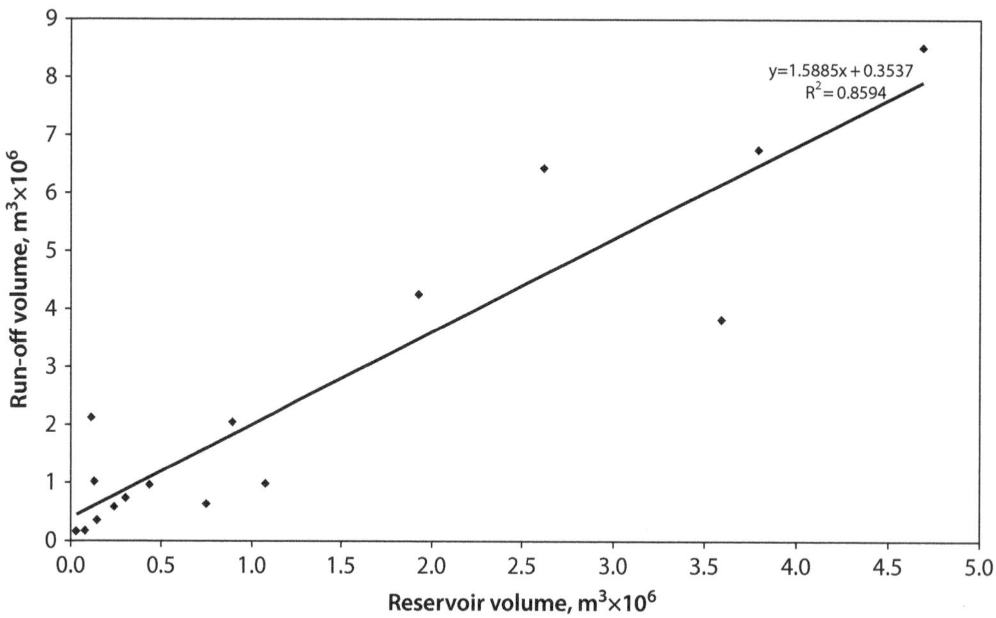

Figure 11.6 Reservoir volume and run-off. Source: author

geoarchaeological and archaeobotanical testing. Preliminary sequences from the SSP reservoir deposits revealed a predominance of spores from wet, marshland plant species, in accordance with the kind of water-logged environment expected of an upstream cropping system (Shaw et al., 2007). It is difficult, however, to identify cereal types from pollen alone, and, further, samples may not reflect the immediate environment, because of the volatility of wind-borne spores. By contrast, phytoliths, the non-organic opaline silica bodies formed within and between living plant cells, can enable the identification of individual species such as rice and wheat. The potential of this technique has been demonstrated in South Asia (Weisskopf et al., 2013; Weisskopf et al., 2015). However, identification is only reliable at the genus level, and it is often difficult to distinguish between wild and domesticated rice. Further, preliminary phytolith analysis in the SSP area (Shaw et al., 2007) shows that reservoir deposits are not generally useful in this kind of analysis because of the poor preservation of phytoliths in ancient water bodies as compared with crop post-processing sites and ceramic tempers. Sampling within habitational contexts will thus be a focus of follow-up research in the area.

The archaeobotany of Indian rice has hitherto focused on the Neolithic origins of domestication in the Gangetic valley (Fuller 2005; Kingwell-Banham et al., 2015), meaning that for later periods we are dependent on textual references to paddy fields and rice cultivation, either as backdrops to Buddhist narratives or as metaphors for Buddhist discipline (Benavides, 2005: 80; for Hindu and medical parallels see Malinar, 2016; Shaw, 2016a; Zimmermann, 2004: 377–9), which suggest that the landscape in which the 'second urbanisation' and the first Buddhist communities grew up was an overwhelmingly rice-growing one. Although archaeological samples from central India are lacking, evidence from the Deccan and South India suggests that the introduction of rice to these areas accompanied the spread of urban-based polities during the late centuries BC (Fuller, 2005; Shaw 2013a: 99). The superior yields and nutritional value of rice in relation to native millets made it an effective solution to growing urban and monastic populations, its centrality to theories linking population stress and 'agricultural involution' being well known from South East Asia (Geertz, 1963), but its introduction to central India, first as a traded ingredient and later as a cultivated crop, may also be viewed as a natural outcome of the spreading populations and new religio-cultural traditions rooted in the rice-growing Gangetic valley. Whether the eventual disuse of the Sanchi dams was also related to the long-term unsustainability of intensive rice agriculture, as suggested by the Chinese evidence (Zhuang et al., 2016), or to the changing patronage and land-tenure frameworks discussed above, warrants further investigation (Shaw, 2016a; also Gilliland et al., 2013; Lucero et al., 2015).

Morrison (2016) suggests that in medieval South India, irrigated rice, as the preferred grain of the urban elite, formed part of a two-tiered food hierarchy, with unirrigated millet remaining the staple of the masses. Morrison touches upon the possible cultural, religious and health variables in this binary set-up, particularly the high ritual status of rice in Hindu contexts. However, there were evidently very divergent attitudes, within and between religious communities and castes, towards different grains and their physical and 'energetic' impact on the body: precise classifications and taxonomies that regulated the production and consumption of different foodstuffs varied according to their different religious contexts (for example ascetic versus devotional). One must distinguish, for example, between minor ritual use of rice, as in Hindu temple worship or donation to Buddhist monasteries, and its broader religio-cultural associations, which varied according to caste, sect and locality. There are additional disparities between the perceived ritual and health properties of rice: despite its much heralded 'auspiciousness' in Hindu temple ritual (Morrison, 2016), rice, and cultivated cereals in general, have more negative associations within ascetic and medical contexts because of perceived links with violence (*hiṃsā*) and the rise of new 'urban' illnesses arguably linked to the birth of the Ayurvedic medical system (Zimmermann, 2004: 374). Such contradictions demonstrate that individual health needs, and the basic quest for survival (Ayurveda meaning literally 'the science of longevity'), can take priority over ritual dispensations (Wujastyk, 2004: 836–7), emphasising the importance of keeping human-centred preoccupations at the forefront of our understanding of Indic attitudes towards ecology. This is especially important given the emerging 'environmental paradigm' within contemporary allopathic medicine, which, in contrast to germ-based models of illness, stresses the importance of epigenetics, toxicology and nutrition as key aetiological factors in the emergence of disease (Genuis, 2012).

Conclusion

Given the growing interest in the long-term health outcomes of major socio-ecological and dietary shifts in antiquity, particularly those involving cereals (Mummert et al., 2011; Schnorr et al., 2016; Wells et al., 2016), the question of whether the supposed polarisation between rice and other cereal staples suggested in the aforementioned archaeological assessments of food change in pre-modern India (Kingwell-Banham et al., 2015; Morrison, 2016; Murphy and Fuller, 2016) was influenced by Buddhist economic or religio-philosophical dispensations obviously requires further study. Answers have hitherto been hampered by scholarship that takes literally canonical prescriptions that monks must eat whatever they receive on their begging rounds and thus have little say in dietary matters, that sedentary monasticism represents a deterioration of the ideal 'passive' model of peripatetic mendicancy geared towards individual enlightenment, and that monks must not engage directly in the ownership or management of agricultural land (*Aṅguttara Nikāya*, 2012: V, 17). At the same time, however, scholarship on ritual gift-giving demonstrates that the success of monastic–lay exchange networks depended on the *saṅgha*'s ability to disguise its direct reciprocity, with lay managerial staff dealing with transactions and activities formally prohibited to monks. Here we may recall Schmithausen's (1997) aforementioned designation of '*passive* environmental activism', not in the sense of the *saṅgha*'s 'transcendence' of nature, as suggested by Schmithausen, but in the sense of the *saṅgha*'s having to disguise its direct engagement with, and transformation of, natural resources in order to generate the patronage necessary for sustaining a non-producing monastic population, and to tackle human suffering in wider society. Thus, while the SSP data supports an 'active' model of religious change (Shaw, 2013a), the *saṅgha*'s involvement with land use only works because of its perceived passivity, with monks *seeming* to be beyond 'nature' and society. As demonstrated by historical examples of non-violent resistance as a form of civil disobedience, outward passivity matters less than the achievement of the intended aim.

The SSP data accords with 'religion as technique' models (Peel, 1968), according to which monks moved into new areas with a set of motives for local communities to extend their economic support to the monastery. Medical analogies are also relevant, the Four Noble Truths and the Eightfold Path aimed at the alleviation of suffering having close parallels with medical epistemologies (Schlieter, 2014), but also because just as no single drug can treat all illnesses, the *saṅgha* responded to individual social problems through the provision of certain practical services such as water, medicine (Zysk, 1998) and banking facilities (Schopen, 1994), without demanding the religious conversion of local populations (Shaw, 2013a). Offering as it does a human-focused model of well-being and suffering, it helps to diffuse the polarised rhetoric of the debate between the 'eco-apologists' and their critics, with its narrow focus on the question of whether or not Buddhists were particularly concerned with the suffering of animals or with the 'beauty' of 'nature', a motif which also figures independently in the 'monasteries-as-gardens' debate. And by extension, far from negating the eco-*dharma* model (Elverskog, 2014), the history and chronology of monastic landlordism and the archaeological evidence of the gradual monumentalisation of Buddhist locales in the landscape support the idea of entanglement (Hodder, 2012: 67) between monks and their physical and built environment through 'long-term relationships of material investment, care and maintenance' (Hodder, 2012: 67), the absence of which leads to decay and disrepair. While the question of whether Buddhism supported attitudes towards nature and its resources that can be described as 'eco-friendly' requires further interdisciplinary research, archaeological correlates for a socially and environmentally engaged 'Buddhist economics' (Harvey, 2000) may be instructive for the modern ecological movement in offering 'non-violent' examples of collective, ideology-based models of land ownership and management in which 'states within states' act as alternative agents of socio-environmental change, in contrast to the monetary outlook of modern development-based governmental agendas (see Shaw, 2016a for Hindu parallels). This approach sits well with modern calls for communities to respond to the social and health challenges associated with our current environmental and climate-change crisis (Deprez and Thomas, 2016).

Acknowledgements

Fieldwork was funded by the British Academy, the Society for South Asian Studies and Merton College, Oxford, under the sanction of the Archaeological Survey of India and the Department of Archaeology, Museums and Archives, Madhya Pradesh. More recent research has been funded by a British Academy Mid-Career Fellowship (2014–15) and a British Academy/Leverhulme Small Research Grant (2016–17). Grateful thanks to Yijie Zhuang, Mark Altaweel and participants at the UCL Comparative Water Technologies and Management conference (2016) for stimulating discussion.

References

Agarwal, A. and S. Narain. 1997. *Dying Wisdom: Rise, Fall, and Potential of India's Traditional Water Harvesting Systems*. New Delhi: Centre for Science and Environment.

Agrawal, A., C. Catalini and A. Goldfarb. 2014. 'Some simple economics of crowdfunding.' *National Bureau of Economic Research Working Paper* No. 19133. doi: 10.3386/w19133.

Ali, D. 2003. 'Gardens in Early Indian court life.' *Studies in History* 19 (2), 221–52.

Aṅguttara Nikāya. 2012. *The Numerical Discourses of the Buddha: A Translation of the Aṅguttara Nikāya*. Translated by Bhikkhu Bodhi. Somerville, MA: Wisdom Publications.

Bailey, G. and I. Mabbett. 2003. *The Sociology of Early Buddhism*. Cambridge: Cambridge University Press.

Benavides, G. 2005. 'Economy.' In D. S. Lopez, Jr (ed.), *Critical Terms for the Study of Buddhism*, 77–102. Chicago/London: University of Chicago Press.

Boivin, N. L., M. A. Zeder, D. Q. Fuller, A. Crowther, G. Larson, J. M. Erlandson, T. Denham and M. D. Petraglia. 2016. 'Ecological consequences of human niche construction: Examining long-term anthropogenic shaping of global species distributions.' *Proceedings of the National Academy of Societies* 113, 6388–96.

Braje, T. J. 2016. 'Evaluating the Anthropocene: Is there something useful about a geological epoch of humans?' *Antiquity* 90 (350), 504–12.

Brown, R. L. 2009. 'Nature as utopian space on the early stūpas of India.' In J. Hawkes and A. Shimada (eds), *Buddhist Stūpas in South Asia: Recent Archaeological, Art-Historical, and Historical Perspectives*, 63–80. New Delhi: Oxford University Press.

Carson, R. 1962. *Silent Spring*. Boston: Houghton Mifflin.

Chatterjee, I. 2015. 'Monastic "governmentality": Revisiting "community" and "communalism" in South Asia.' *History Compass* 13 (10), 497–511.

Chester, D. K. 2005. 'Theology and disaster studies: The need for dialogue.' *Journal of Volcanology and Geothermal Research* 146 (4), 319–28.

Clement C. R., W. M. Denevan, M. J. Heckenberger, A. B. Junqueira, E. G. Neves, W. G. Teixeira and W. I. Woods. 2015. 'The domestication of Amazonia before European conquest.' *Proceedings of the Royal Society B: Biological Sciences* 282 (1812), 20150813.

Cohen, R. S. 1998. 'Naga, Yaksini, Buddha: Local deities and local Buddhism at Ajanta.' *History of Religions* 37 (4), 360–400.

Coningham, R. A. E. 1995. 'Monks, caves and kings: A reassessment of the nature of early Buddhism in Sri Lanka.' *World Archaeology* 27 (2), 222–42.

Coningham, R. A. E. and P. Gunawardhana. 2013. *Anuradhapura. Volume 3: The Hinterland*. BAR International Series. Oxford: Archaeopress.

Coningham, R. A. E., P. Gunawardhana, M. Manuel, G. Adikari, M. Katugampola, R. Young, A. Schmidt, K. Krishnan, I. Simpson, G. McDonnell and C. Batt. 2007. 'The state of theocracy: Defining an early medieval hinterland in Sri Lanka.' *Antiquity* 81, 699–719.

Crutzen, P. and E. Stoermer. 2000. 'The "Anthropocene".' *Global Change Newsletter* 41:17–18.

Dalby, S. 2016. 'Re-evaluating the Anthropocene.' *Antiquity* 90 (350), 514–15.

Dalglish, C. 2012. 'Archaeology and landscape ethics.' *World Archaeology* 44 (3), 327–41.

Dalton, J. 2004. 'The early development of the Padmasambhava legend in Tibet: A study of IOL Tib J 644 and Pelliot tibétain 307.' *Journal of the American Oriental Society* 124 (4), 759–72.

Davison-Jenkins, D. J. 1997. *The Irrigation and Water Supply Systems of Vijayanagara*. Vijayanagara Research Project Monograph Series, 5. New Delhi: Manohar.

Deb, D. 2017. 'Folk rice varieties, traditional knowledge and nutritional security in South Asia.' In G. Poyyamoli (ed.), *Agroecology, Ecosystems and Sustainability in the Tropics*. New Delhi: Studera Press, pp. 117–34.

DellaValle, C. 2016. 'The Pollution in People: Cancer-causing Chemicals in Americans' Bodies.' Environmental Working Group Original Research. http://www.ewg.org/cancer/the-pollution-in-people.php.

Deprez, R. and R. Thomas. 2016. 'Population health improvement: It's up to the community – not the healthcare system.' *Maine Policy Review* 25 (2), 44–52.

Dixit, Y., D. A. Hodell and C. A. Petrie. 2014. 'Abrupt weakening of the summer monsoon in northwest India ~4100 yr ago.' *Geology* 42 (4), 339–42.

Dorje, O. T. (H.H. 17th Gyalwang Karmapa). 2006. 'Walking the path of environmental Buddhism through compassion and emptiness.' *Conservation Biology* 25 (6), 1094–7.

Douglas, M. 1966. *Purity and Danger: An Analysis of Concepts of Pollution and Taboo*. London: Routledge & Kegan Paul.

Dupré, J. 2016. 'A postgenomic perspective on sex and gender.' In D. L. Smith (ed.), *How Biology Shapes Philosophy: New Foundations for Naturalism*, 227–46. Cambridge: Cambridge University Press.

Edgeworth, M. 2014. 'Archaeology of the Anthropocene [forum discussion]: Introduction.' *Journal of Contemporary Archaeology* 1 (1), 73–7.

Ellis, E., M. Maslin, N. L. Boivin and A. Bauer. 2016. 'Involve social scientists in defining the Anthropocene.' *Nature* 540, 192–3.

Elverskog, J. 2014. 'Asian studies + Anthropocene[4].' *Journal of Asian Studies* 73 (4), 963–74.

Evans, D. 2016. 'Airborne laser scanning as a method for exploring long-term socio-ecological dynamics in Cambodia.' *Journal of Archaeological Science* 74, 164–75.

Fogelin, L. 2006. *Archaeology of Early Buddhism*. Lanham, MD: AltaMira Press.

Fuller, D. Q. 2005. 'Ceramics, seeds and culinary change in prehistoric India.' *Antiquity* 79 (306), 761–77.

Fuller, D. Q. 2007. 'Contrasting patterns in crop domestication and domestication rates: Recent archaeobotanical insights from the Old World.' *Annals of Botany* 100 (5), 903–24.

Geertz, C. 1963. *Agricultural Involution: The Process of Ecological Change in Indonesia*. Berkeley: Published for the Association of Asian Studies by University of California Press.

Genuis, S. J. 2012. 'What's out there making us sick?' *Journal of Environmental and Public Health*, art. no. 605137, 10 pp. doi:10.1155/2012/605137.

Gilliland, K., I. A. Simpson, W. P. Adderley, C. I. Burbidge, A. J. Cresswell, D. C. W. Sanderson, R. A. E. Coningham, M. J. Manuel, K. M. Strickland, P. Gunawardhana and G. Akikari. 2013. 'The dry tank: Development and disuse of water management infrastructure in the Anuradhapura hinterland, Sri Lanka.' *Journal of Archaeological Science* 40 (2), 1012–28.

Gombrich, R. F. 1988. *Theravāda Buddhism: A Social History from Ancient Benares to Modern Colombo*. London: Routledge & Kegan Paul.

Green, R. M. 1992. 'Buddhist economic ethics: A theoretical approach.' In R. F. Sizemore and D. K. Swearer (eds), *Ethics, Wealth, and Salvation: A Study in Buddhist Social Ethics*, 215–34. Columbia: University of South Carolina Press.

Gunawardana, R. A. L. H. 1971. 'Irrigation and hydraulic society in early medieval Ceylon.' *Past and Present* 53, 3–27.

Gunnell, Y., K. Anupama and B. Sultan. 2007. 'Response of the South Indian runoff-harvesting civilization to northeast monsoon rainfall variability during the last 2000 years: Instrumental records and indirect evidence.' *The Holocene* 17 (2), 207–15.

Harris, Ian. 1997. 'Buddhism and the discourse of environmental concern: Some methodological problems considered.' In M. E. Tucker and D. R. Williams (eds), *Buddhism and Ecology: The Interconnection of Dharma and Deeds*, 377–402. Cambridge, MA: Harvard University Center for the Study of World Religions: Distributed by Harvard University Press.

Harvey, P. 2000. *An Introduction to Buddhist Ethics: Foundations, Values, and Issues*. Cambridge: Cambridge University Press.

Harvey, P. 2007. 'Avoiding unintended harm to the environment and the Buddhist ethic of intention.' *Journal of Buddhist Ethics* 14, 1–34.

Hawkes, J. 2014. 'One size does not fit all: Landscapes of religious change in Vindhya Pradesh.' *South Asian Studies* 30 (1), 1–15.

Hodder, I. 2012. *Entangled: An Archaeology of the Relationships between Humans and Things*. Oxford: Wiley-Blackwell.

Holm, Poul, Joni Adamson, Hsinya Huang, Lars Kirdan, Sally Kitch, Iain McCalman et al. 2015. 'Humanities for the environment – a manifesto for research and action.' *Humanities* 4 (4), 977–92.

Hulme, M. 2010. 'The idea of climate change: Exploring complexity, plurality and opportunity'. *Gaia* 19, 171–4.

Imp. Gaz. 1908. *The Imperial Gazetteer of India. Volume 9: Bomjur to Central India* (new edn). Oxford: Clarendon Press.

Jenkins, W., M. E. Tucker and J. Grim (eds). 2016. *Routledge Handbook of Religion and Ecology*. Abingdon/New York: Routledge.

Jung, S. J. A., G. R. Davies, G. M. Ganssen and D. Kroon. 2004. 'Synchronous Holocene sea surface temperature and rainfall variations in the Asian monsoon system.' *Quaternary Science Reviews* 23, 2207–18.

Kingwell-Banham, E., C. Petrie and D. Fuller. 2015. 'Early agriculture in South Asia.' In G. Barker and C. Goucher (eds), *The Cambridge World History*, 261–8. Cambridge: Cambridge University Press.

Lane, P. J. 2015. 'Archaeology in the age of the Anthropocene: A critical assessment of its scope and societal contributions.' *Journal of Field Archaeology* 40 (5), 485–98.

Latour, B. 2013. 'Facing Gaia: A new enquiry into natural religion.' The Gifford Lectures, Edinburgh. https://www.ed.ac.uk/arts-humanities-soc-sci/news-events/lectures/gifford-lectures/archive/series-2012–2013/bruno-latour (accessed 21 June 2018).

Lee, L. M. 2017. 'A bridge back to the future: Public health ethics, bioethics, and environmental ethics.' *American Journal of Bioethics* 17 (9), 5–12.

Lucero, L. J., R. Fletcher and R. Coningham, 2015. 'From "collapse" to urban diaspora: The transformation of low-density, dispersed agrarian urbanism.' *Antiquity* 89 (347), 1139–54.

Macer, D. R. J. 2017. 'We can and must rebuild the bridges of interdisciplinary bioethics.' *American Journal of Bioethics* 17 (9), 1–4.

Malinar, A. 2016. 'Notions of "productive matter" and the classification of beings in Indian philosophy.' Paper presented at the conference 'Human : Non-Human – Bodies, Things, and Matter across Asia and Europe', URPP Asia and Europe, University of Zurich, 6–8 October.

Mcdermott, J. P. 1989. 'Animals and humans in early Buddhism.' *Indo-Iranian Journal* 32 (4), 269–80.

Morrison, K. D. 1995. 'Trade, urbanism, and agricultural expansion: Buddhist monastic institutions and the state in early historic Western Deccan.' *World Archaeology* 27 (2), 203–21.

Morrison, K. D. 2010. 'Dharmic projects, imperial reservoirs, and new temples of India: An historical perspective on dams in India.' *Conservation and Society* 8 (3), 182–95.

Morrison, K. D. 2012. 'Great transformations: On the archaeology of cooking.' In S. R. Graff and E. Rodríguez-Alegría (eds), *The Menial Art of Cooking: Archaeological Studies of Cooking and Food Preparation*, 231–44. Denver: University Press of Colorado.

Morrison, K. D. 2015. 'Archaeologies of flow: water and the landscapes of Southern India past, present, and future.' *Journal of Field Archaeology* 40 (5), 560–80.

Morrison, K. D. 2016. 'From millets to rice (and back again?): Cuisine, cultivation, and health in southern India.' In G. R. Schug and S. R. Walimbe (eds), *A Companion to South Asia in the Past*, 358–73. Chichester: Wiley Blackwell.

Morrison, K. D. and M. T. Lycett. 2014. 'Constructing nature: Socio-natural histories of an Indian forest.' In S. B. Hecht, K. D. Morrison and C. Padoch (eds), *The Social Lives of Forests: Past, Present, and Future of Woodland Resurgence*, 148–60. Chicago: University of Chicago Press.

Mostafalou, S. and M. Abdollahi. 2013. 'Pesticides and human chronic diseases: Evidences, mechanisms, and perspectives.' *Toxicology and Applied Pharmacology* 268, 157–77.

Mummert, A., E. Esche, J. Robinson and G. J. Armelagos. 2011. 'Stature and robusticity during the agricultural transition: Evidence from the bioarchaeological record.' *Economics and Human Biology* 9 (3), 284–301.

Murphy, C. and D. Q. Fuller. 2016. 'The transition to agricultural production in India: South Asian entanglements of domestication.' In G. Schug and S. Walimbe (eds), *A Companion to South Asia in the Past*, 344–57. Hoboken, NJ: John Wiley & Sons.

Murphy, C. and D. Q. Fuller. 2017. 'The future is long-term: Past and current directions in environmental archaeology.' *General Anthropology* 24 (1), 1, 7–10.

Naess, A. 2003. 'The deep ecology movement: Some philosophical aspects.' In A. Light and H. Rolston III (eds), *Environmental Ethics: An Anthology*, 262–74. Malden, MA: Blackwell Publishing.

Ñāṇamoli, B. and B. Bodhi, 1995. *The Middle Length Discourses of the Buddha: A Translation of the Majjhima Nikāya*. Somerville, MA: Wisdom Publications.

Neumayer, E. 2011. *Prehistoric Rock Art of India*. New Delhi: Oxford University Press.

Olivelle, P. 2006. 'The beast and the ascetic: The wild in the Indian religious imagination.' In P. Olivelle (ed.), *Ascetics and Brahmins: Studies in Ideologies and Institutions*, 91–100. Florence: Firenze University Press.

Oliver, P. 2004. 'Buddhism'. In S. Krech III, J. R. McNeill and C. Merchant (eds), *Encyclopedia of World Environmental History*, 173–6. New York: Routledge.

Olivieri, Luca M., Massimo Vidale, Abdul Nasir Khan, Tahir Saeed, Luca Colliva, Riccardo Garbini et al. 2006. 'Archaeology and settlement history in a test area of the Swat Valley: Preliminary report on the AMSV Project (1st phase).' *East and West* 56 (1/3), 73–150.

Parry, S. and J. Dupré. 2010. *Nature after the Genome*. Oxford: Wiley-Blackwell.

Peel, J. D. Y. 1968. 'Syncretism and religious change.' *Comparative Studies in Society and History* 10 (2), 121–41.

Pryor, F. L. 1990. 'A Buddhist economic system – in principle.' *American Journal of Economics and Sociology* 49 (3), 339–50.

Rajani, M. B. 2016. 'The expanse of archaeological remains at Nalanda: A study using remote sensing and GIS.' *Archives of Asian Art* 66 (1), 1–23.

Randall, A. 2016. 'Time, agency and the Anthropocene.' *Antiquity* 90 (350), 516–17.

Ray, R. 1994. *Buddhist Saints in India: A Study in Buddhist Values and Orientations*. New York: Oxford University Press.

Riede, F., P. Andersen and N. Price. 2016. 'Does environmental archaeology need an ethical promise?' *World Archaeology* 48 (4), 466–81.

Riede, F. and A. Klevnäs. 2016. 'Anthropology, weather and climate change.' *European Archaeologist* 49, 24–7.

Ruegg, D. S. 1980. 'Ahimsa and vegetarianism in the history of Buddhism.' In S. Balasooriya, A. Bareau, R. Gombrich, S. Gunasingha, U. Mallawarachchi and E. Perry (eds), *Buddhist Studies in Honour of Walpola Rahula*, 234–41. London: Gordon Fraser.

Schlieter, J. 2014. 'Endure, adapt, or overcome? The concept of suffering in Buddhist bioethics.' In R. M. Green and N. J. Palpant (eds), *Suffering and Bioethics*, 309–36. New York: Oxford University Press.

Schmithausen, L. 1997. 'The early Buddhist tradition and ecological ethics.' *Journal of Buddhist Ethics* 4, 1–74.

Schnorr, S. L., K. Sankaranarayanan, C. M. Lewis and C. Warinner. 2016. 'Insights into human evolution from ancient and contemporary microbiome studies.' *Current Opinion in Genetics and Development* 41, 14–26.

Schopen, G. 1994. 'Doing business for the lord: Lending on interest and written loan contracts in the Mūlasarvāstivāda-Vinaya.' *Journal of the American Oriental Society* 114, 527–54.

Schopen, G. 2006. 'The Buddhist "monastery" and the Indian garden: Aesthetics, assimilations, and the siting of monastic establishments.' *Journal of the American Oriental Society* 126 (4), 487–505.

Schumacher, E. F. 1973. *Small is Beautiful: Economics as if People Mattered*. London: Blond and Briggs.

Sen, S., A. K. M. Syfur and S. Ahsan. 2014. 'Crossing the boundaries of the archaeology of Somapura Mahavihara: Alternative approaches and propositions.' *Pratnatattva* 20: 49–79.

Shaw, J. 2004. 'Nāga Sculptures in Sanchi's archaeological landscape: Buddhism, Vaiṣṇavism and local agricultural cults in central India, first century BCE to fifth century CE.' *Artibus Asiae* 64 (1), 5–59.

Shaw, J. 2007. *Buddhist Landscapes in Central India: Sanchi Hill and Archaeologies of Religious and Social Change, c. Third Century BC to Fifth Century AD*. London: British Association for South Asian Studies, British Academy/Leftcoast Press.

Shaw, J. 2011. 'Monasteries, monasticism, and patronage in ancient India: Mawasa, a recently documented hilltop Buddhist complex in the Sanchi area of Madhya Pradesh.' *South Asian Studies* 27 (2), 111–30.

Shaw, J. 2013a. 'Archaeologies of Buddhist propagation in ancient India: "Ritual" and "practical" models of religious change.' *World Archaeology* 45 (1), 83–108.

Shaw, J. 2013b. 'Archaeology of religious change: Introduction.' *World Archaeology* 45 (1), 1–11.

Shaw, J. 2013c. 'Sanchi as an archaeological area.' In D. K. Chakrabarti and M. Lal (eds), *History of Ancient India. Volume 4: Political History and Administration (c. 200 BC–AD 750)*, 388–427. New Delhi: Vivekananda International Foundation and Aryan Books.

Shaw, J. 2015. 'Buddhist and non-Buddhist mortuary traditions in ancient India: Stūpas, relics and the archaeological landscape.' In C. Renfrew, M. J. Boyd and I. Morley (eds), *Death Rituals, Social Order and the Archaeology of Immortality in the Ancient World: Death Shall Have No Dominion*, 382–403. Cambridge: Cambridge University Press.

Shaw, J. 2016a. 'Religion, "nature" and environmental ethics in ancient India: Archaeologies of human:non-human suffering and well-being in early Buddhist and Hindu contexts.' *World Archaeology* 48 (4), 517–43.

Shaw, J. 2016b. 'Archaeology, climate change and environmental ethics: Diachronic perspectives on human:non-human:environment worldviews, activism and care.' *World Archaeology* 48 (4), 449–65.

Shaw, J. (ed.). 2016c. *Archaeology and Environmental Ethics*. Special issue of *World Archaeology* 48 (4). http://www.tandfonline.com/toc/rwar20/48/4?nav=tocList. Accessed 22 June 2018.

Shaw, J. (in press). 'Buddhist engagement with the "natural" environment.' In K. Trainor and P. Arai (eds), *Oxford Handbook of Buddhist Practice*. Oxford Handbooks. Oxford: Oxford University Press.

Shaw, J. and J. V. Sutcliffe. 2001. 'Ancient irrigation works in the Sanchi area: An archaeological and hydrological investigation.' *South Asian Studies* 17 (1), 55–75.

Shaw, J. and J. V. Sutcliffe. 2003. 'Water management, patronage networks and religious change: New evidence from the Sanchi dam complex and counterparts in Gujarat and Sri Lanka.' *South Asian Studies* 19, 73–104.

Shaw, J. and J. V. Sutcliffe. 2005. 'Ancient dams and Buddhist landscapes in the Sanchi area: New evidence on irrigation, land use and monasticism in central India.' *South Asian Studies* 21, 1–24.

Shaw, J., J. V. Sutcliffe, L. Lloyd-Smith, J.-L. Schwenninger and M. S. Chauhan, with contributions by O. P. Misra and E. Harvey. 2007. 'Ancient irrigation and Buddhist history in central India: Optically stimulated luminescence and pollen sequences from the Sanchi dams.' *Asian Perspectives* 46 (1), 166–201.

Shimada, A. 2012. 'The use of garden imagery in early Indian Buddhism.' In D. Ali and E. Flatt (eds), *Garden and Landscape Practices in Pre-colonial India: Histories from the Deccan*, 18–38. London: Routledge.

Skilling, P. 2014. '"Every rise has its fall": Thoughts on the history of Buddhism in Central India (Part I).' In J. Soni, M. Pahlke and C. Cüppers (eds), *Buddhist and Jaina Studies: Proceedings of the Conference in Lumbini, February 2013*, 77–122. Lumbini: Lumbini International Research Institute.

Smith, B. K. 1990. 'Eaters, food, and social hierarchy in ancient India: A dietary guide to a revolution of values.' *Journal of the American Academy of Religion* 58 (2), 177–205.

Sontheimer, G.-D. 1964. 'Religious endowments in India: The juristic personality of Hindu deities.' *Zeitschrift für vergleichende Rechtswissenschaft* 67, 45–100.

Stewart, J. 2014. 'Violence and nonviolence in Buddhist animal ethics.' *Journal of Buddhist Ethics* 21, 623–55.

Strain, C. 2016. 'Engaged Buddhist practice and ecological ethics: Challenges and reformulations.' *Worldviews* 20 (2), 189–210.

Sutcliffe, J., J. Shaw and E. Brown. 2011. 'Historical water resources in South Asia: The hydrological background.' *Hydrological Sciences Journal* 56 (5), 775–88.

Swearer, D. 2006. 'An assessment of Buddhist eco-philosophy.' *Harvard Theological Review* 99 (2), 123–37.

Tewari, R. 2003. 'The origins of iron-working in India: New evidence from the Central Ganga Plain and the Eastern Vindhyas.' *Antiquity* 77 (297), 536–44.

United Nations. 2015. Transforming Our World – The 2030 Agenda for Sustainable Development: Resolution Adopted by the General Assembly on 25 September 2015. Seventieth session, Agenda items 15 and 116. http://www.un.org/ga/search/view_doc.asp?symbol=A/RES/70/1&Lang=E.

Venkayya, V. 1906. 'Irrigation in southern India in ancient times.' *Archaeological Survey of India: Annual Report 1903–4*, 202–11.

Wasson, R. G. and W. D. O'Flaherty. 1982. 'The last meal of the Buddha.' *Journal of the American Oriental Society* 102 (4), 591–603.

Watt, G. 1889–93. *A Dictionary of the Economic Products of India*. 6 vols. London: W. H. Allen and Co./Calcutta: Office of the Superintendent of Government Printing India.

Watts, N., N. W. Adger, S. Ayeb-Karlsson, Y. Bai, P. Byass, D. Campbell-Lendrum, T. Colbourn et al. 2017. 'The Lancet Countdown: Tracking progress on health and climate change.' *The Lancet* 389 (10074), 1151–64.

Weisskopf, A., E. Harvey, E. Kingwell-Banham, M. Kajale, R. Mohanty and D. Q. Fuller. 2013. 'Archaeobotanical implications of phytolith assemblages from cultivated rice systems, wild rice strands and macro-regional patterns.' *Journal of Archaeological Science* 51, 43–53.

Weisskopf, A., L. Qin, J. Ding, P. Ding, G. Sun and D. Q. Fuller, 2015. 'Phytoliths and rice: From wet to dry and back again in the Neolithic Lower Yangtze.' *Antiquity* 89 (347), 1051–63.

Wells, J. C. K., E. Pomeroy, S. R. Walimbe, B. M. Popkin and C. S. Yajnik. 2016. 'The elevated susceptibility to diabetes in India: An evolutionary perspective.' *Frontiers in Public Health* 4 (145). doi:10.3389/fpubh.2016.00145.

Whitmee, S., A. Haines, C. Beyrer, F. Boltz, A. G. Capon, B. F. de Souza Dias, A. Ezeh et al. 2015. 'Safeguarding human health in the Anthropocene epoch: Report of the Rockefeller Foundation–*Lancet* Commission on planetary health.' *The Lancet* 386: 1973–2028.

Willis, M. D. 2009. *The Archaeology of Hindu Ritual: Temples and the Establishment of the Gods*. New York: Cambridge University Press.

Willis, M. D. 2013. 'Avalokiteśvara of the six syllables: Locating the practice of the "Great vehicle" in the landscape of central India.' In C. A. Bromberg, T. J. Lenz and J. Neelis (eds), *Evo Ṣuyadi: Essays in Honor of Richard Salomon's 65th Birthday. Bulletin of the Asia Institute* 23, 221–29.

Wujastyk, D. 2004. 'Medicine and Dharma.' *Journal of Indian Philosophy* 32 (5/6), 831–42.

Zhuang, Y., H. Lee and T. R. Kidder. 2016. 'The cradle of heaven human induction idealism: Agricultural intensification, environmental consequences and social responses in Han China and Three-Kingdoms Korea.' *World Archaeology* 48 (4), 563–85.

Zimmermann, F. 2004. '"May godly clouds rain for you!" Metaphors of well-being in Sanskrit.' In E. Ciurtin (ed.), *Du corps humain, au carrefour de plusieurs savoirs en Inde: Mélanges offerts à Arion Roşu par ses collègues et ses amis à l'occasion de son 80ème anniversaire*, 371–84. Studia Asiatica, 4–5. Bucharest: Centre d'Histoire des Religions de l'Université de Bucarest/Paris: De Boccard.

Zysk, K. G. 1986. 'The evolution of anatomical knowledge in ancient India, with special reference to cross-cultural influences.' *Journal of the American Oriental Society* 106 (4), 687–705.

Zysk, K. G. 1998. *Asceticism and Healing in Ancient India: Medicine in the Buddhist Monastery*. New Delhi: Motilal Banarsidass.

12
Water for the state or water for the people? Wittfogel in South and South East Asia in the first millennium

Janice Stargardt

Abstract

It is a striking fact that in Sri Lanka, South India and mainland South East Asia in the first millennium AD, many of the earliest polities were located in dry zones despite the availability in every case of well-watered lands near the coasts. Consequently, early irrigation and water-storage works seem to have been associated with the development of such complex societies. This chapter presents a number of case studies which illustrate the technical variety exhibited by ancient hydraulic works in Sri Lanka, South India, Cambodia and Thailand. It considers whether they were managed in a top-down or a bottom-up manner, and their implications as a critique of the famous Wittfogel hypotheses about societies with major hydraulic works as exemplars of 'total power'.

Introduction

I begin with a summary of Wittfogel's hypotheses (Wittfogel, 1957). He termed societies whose agriculture depended on large-scale works of irrigation and flood control 'hydraulic civilisations', presenting them as different in kind from Western societies. He postulated the existence of absolutist managerial states, whose bureaucrats monopolised political (and by extension economic) power through the centralised control of irrigation works. He argued that in hydraulic civilisations labour was massed and forced and its products confiscated by the state. He identified ancient Egypt, Mesopotamia, India, China and pre-Columbian Mexico as 'pre-eminent hydraulic civilisations'.

I now examine the reality of actual case studies of ancient hydraulic works in South and South East Asia, to summarise their technology and recover information about how they functioned.

Case study one: Anuradhapura and its surroundings in north-west Sri Lanka

The annual average rainfall of the area around Anuradhapura is around 1,250–1,850 mm. While modern, short-season, rain-fed hybrid rices can be cultivated under this regime, being sown towards the end of the wet season and coming to harvest less than 12 weeks later, this rainfall was marginal or inadequate to the needs of the long-season, photosensitive rices preferred in antiquity. Thus irrigation and water-storage works of great antiquity dot the area around the ancient city, notably the Tissa wewa and the Nuwara wewa, both of which have short bunds in relation to the volume of water stored behind them (Figure 12.1). This implies expert knowledge of the local terrain. Ancient, small-scale

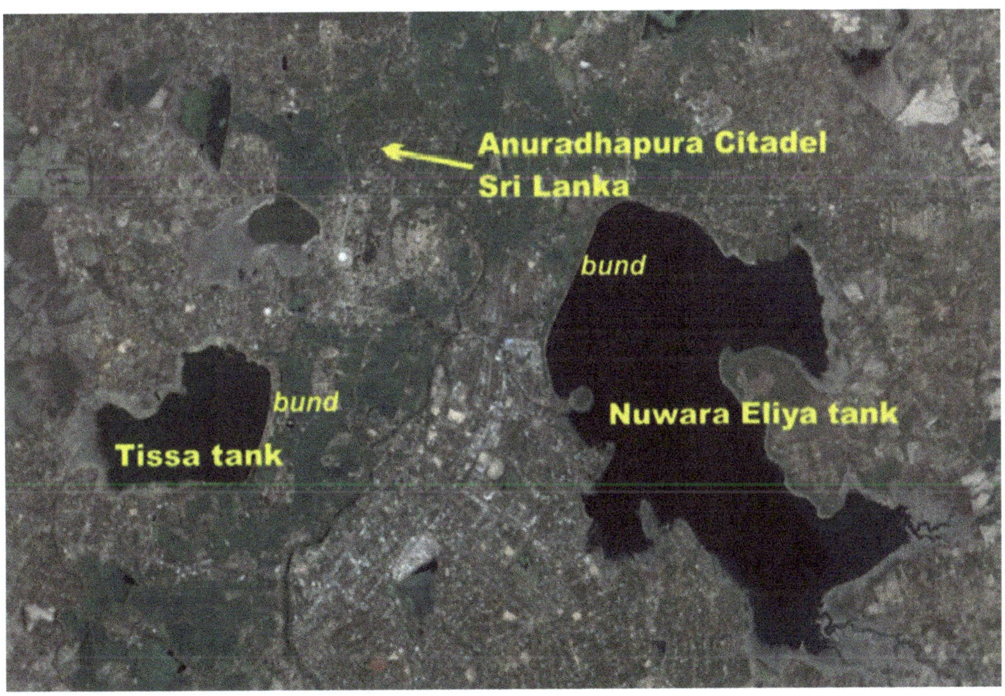

Figure 12.1 Satellite image of Anuradhapura and surrounding irrigation tanks (Google Earth, accessed April 2016, annotated by Janice Stargardt)

village tanks have rarely been studied, but 3,340 were recorded in 1984 in Anuradhapura province (Takaya and Jayawardena, 1984). The first historically recorded medium-to-large-sized tank, located a short distance away from Anuradhapura, is the Minneriya tank, said in the fifth century *Mahavamsa* chronicle (Mahānāma, 1912) to have been built by king Mahasena in the third century AD, along with 15 other tanks and two major canals. The key features of the Minneriya tank are these: its backwater covered 1,890 ha but its dam wall was only just over 2 km long and 13 m high. Whereas Wittfogel focused on the magnitude of the water stored, and constructed from that his hypotheses concerning the labour involved and the conditions under which the work took place, I wish to emphasise something he overlooked: that an *inverse ratio* existed between the scale of this dam wall and the volume of water it captured. Moreover, that inverse ratio characterises other traditional tanks of the dry zone of Sri Lanka. To achieve it, the ancient experts skilfully selected places where two natural ridges could be joined by a short dam wall, while the ridges themselves functioned as natural retaining walls for the water stored behind the dam.

Although it was apparently created on royal initiative, it is clear that a deep knowledge of the local landscape and local water forces lay behind the successful construction and operation of the Minneriya tank. It and other tanks in the same area were provided with networks of feeder and distributary canals, sluices and, sometimes, above- or underground channels linking tanks, so that they could be operated in concert. It is likely that small-scale rural tanks and canals were the places where such expert technical and environmental knowledge was accumulated, tested and transmitted over centuries, and that tanks and canals on many scales co-existed at the same time. While royal munificence might indeed determine the scale of a water project, it was essential to involve local expertise and local economic interest for its successful realisation and maintenance.

Case study two: Tanks and history in South India

The creation of the Mahendratataka (Lake Mahendra) in today's northern Tamil Nadu, in the early seventh century AD, is attributed to the Pallava king Mahendra. The tank occupied a natural depression, closed by a short dam linking two hills in the manner just described. When fully operational, the Mahendratataka stored enough water to irrigate the surrounding lands up to a distance of 13 km. The royal patron recorded his work in an inscription found in 1897 inside a cave temple (the Mahendravadi) dedicated to Vishnu, located beside the tank. The royal capital, Mahendrapura (Mahendra city), also named after him, lay less than 5 km to the south-east of the dam wall.

Here we have an apparent embodiment of the kind of top-down model Wittfogel envisaged: king, tank, temple and city all linked by the same royal name, prestige and claims to agency. Yet the tank still functioned in 1903, some 1,300 years later, despite the fact that king, court and royal city had quit the stage of history many centuries before. The census of 1891 recorded the presence of a village of 827 people where the royal city had been. What conclusions can be drawn from these facts? The necessary attributes for the maintenance and orderly use of the tank – expert technical knowledge, effective management and maintenance rules, the ability to assemble and employ the essential funds and resources – were all to be found in the village, passed down through largely oral traditions over that long period of time. In other words, for many centuries, and perhaps from the beginning of the tank's history, agency resided in the village and its institutions in ways that were to prove more important and much more enduring than those of the royal benefactor.

Case study three: Inscriptions about village irrigation committees in India and their powers

Written evidence of how these committees worked is rare, but not entirely absent. For instance, at Uttaramallur in 1898 eight inscriptions from the beginning of the ninth century were found that concerned the great tank Vayiramegatataka in Tamil Nadu (*Annual Report, Government Epigrapher,* 1898–9; Venkayya, 1906: 206; Stargardt, 1983: 187–90). Inscriptions 42, 65, 84 and 90 provided details of gifts, large and small, from many ranks of society – court and village – for the maintenance of this tank (mostly through silt removal and associated costs such as boat hire, wages and food). The inscriptions show that local tank committees were elected to be responsible for investing the gifts of gold and silver, and for the wise use of the interest accruing (at a rate of about 15 per cent according to my calculations). The tank committees collected and paid the water taxes, and imposed fines for non-payment. The inscriptions record that village assemblies appointed a committee to supervise the tanks, consisting of six members holding office for 360 days. One inscription records that the village tank committee paid the water taxes owing to the king by certain defaulting farmers, took proceeds from their land for three years and spent them on the upkeep of the tank. If the debts were cleared, the lands were to be returned to the owners, but if not, the lands were to be sold (Venkayya, 1906: 204). These were real powers, real responsibilities, and must have had a real impact on the local economy, not only directly through the wages paid to those engaged in silt removal (the norm being 140 baskets per day), and to the carpenters and blacksmiths who maintained the boat, but more significantly through the fact that the gifts were invested in the local economy and only the interest was used, thus ensuring their long-term impact. Finally, effective irrigation, the aim of all these activities, brought enhanced and more secure food production to the local population and larger taxes to the ruler, thus engaging the interest of both elite and village levels of society.

It emerges from this epigraphical evidence that the scale of the gifts to a tank determined the scale of work that could be done on it, but that gifts were made both by people of the highest rank and by village families. The powers to invest those gifts, to have access to the interest and to use it for the management of all the labour and resources required to create and maintain the tank, together with the indispensable hydraulic expertise, were all local. Details of this kind enable us to understand how it was possible for the Mahendratataka to function for over 1,300 years.

Case study four: The barays of Angkor and other Cambodian tanks of the tenth to thirteenth centuries AD

The monsoon rains in Cambodia around Angkor arrive in two phases on the south-west winds: in July–August and then in September–October (Delvert, 1961). The rest of the year is dry. This may explain the scale of the barays of Cambodia, the largest of which is the West Baray at Angkor, which measures about 8 × 2 km. It covers a regular east–west rectangular space immediately to the west of the walled and moated urban area of Angkor Thom (Figure 12.2). Work on this tank is said to have begun in the reign of Suryavarman I in the eleventh century, and to have reached completion under his successor Udayadityavarman. It appears to epitomise the kind of huge royal enterprise envisaged by Wittfogel (1957) that could only be accomplished by a massed labour force working under compulsion. The West Baray was understood by many scholars in the twentieth century as a massive royal project to irrigate downstream

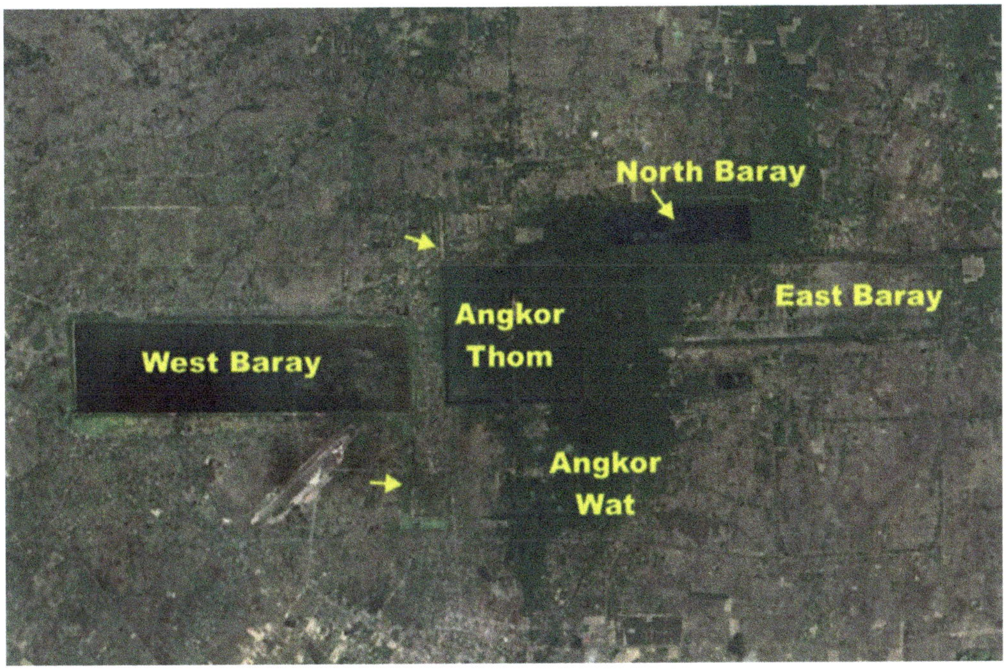

Figure 12.2 Satellite image of the West Baray, with inlet and outlet points (Google Earth, accessed April 2016, annotated by Janice Stargardt)

lands between it and the great lake, Tonle Sap (Goloubev, 1936; Groslier, 1979). Later, the practical function of the baray for irrigation was challenged in favour of a symbolic, ritual role, supported by the presence of the West Mebon temple in the centre of the baray (Van Liere, 1980). In both cases, it was considered the embodiment of 'oriental despotism', representing the work of large numbers of forced labour, commensurate with the *c.* 80 million m^3 of water the tank could potentially store (Groslier, 1979: 189–220).

I wish to emphasise aspects of this vast enterprise that contrast with that impression. Firstly, a baray was not an excavated tank, but was created *above ground* by experts who possessed an exact knowledge of the local slopes and of the seasonal water forces that descended from the Kulen Mountains to impact on the River Siem Reap, and in turn on the Angkorian plain. This basic fact immediately transforms our assumptions about the amount of labour involved. Using the prevailing north-east–south-west slope of the Angkorian plain, a ditch was dug to define the extent of the future baray. The earth dug up was used to create an earthen bund around its future area. The bund was kept low on the northern side, but increased in height down the slopes and was highest along the southern side. In Figure 12.2, a water intake point can be seen at the upper north-east corner, and an outlet point in the south-east corner which is capable of releasing excess water directly into the rectangular outer moat of the Angkor Thom urban complex. Thus with a very limited amount of labour and material but great local knowledge, it was possible to capture and store a huge body of water.

Though technically very different from the dams of Sri Lanka and South India, the Angkor barays exhibited a similar inverse ratio between the volume of water gained and the amount of labour and material involved. The baray technology did not consistently succeed in providing enduring bodies of water at Angkor: the present water retention in the West Baray is due to restoration work funded by the Australian government. A careful scrutiny of Figure 12.2 reveals the dry traces of other, smaller barays to the north and east of Angkor Thom, which no longer hold any traces of moisture, for example those surrounding the Neak Preah and the East Mebon temples. There are indications of other attempts that failed in antiquity (conversation with B.-P. Groslier, April 1977).

A second type of tank technology existed in both urban and rural areas in the Angkor period which, until recently, has received relatively little attention, but should be mentioned here. These were relatively small tanks, about 150 × 150 m, excavated into the alluvium and maintained by and for household use. As they were constantly repaired and renewed, they are difficult to date, except through the occasional ancient artefact found in their sedimentary deposits. Using the observation of the Chinese Ambassador to Angkor, Zhou Daguan, in the late thirteenth century, that such tanks served the needs of about three households, Hanus and Evans have been accumulating LiDAR-based data on these tanks in order to estimate the size of Angkor's ancient population (Hanus and Evans, 2016; Bishop et al., 2003; Stargardt, 1983: figure 52).

Case study five: The field/tank/canal system of South Thailand

Whereas most of South Thailand receives two abundant wet seasons per annum, the Satingpra peninsula – the land between the great lakes and the Gulf of Thailand – lies in a rain shadow caused by the mountain range of the isthmus. It nonetheless witnessed a thriving civilisation from about the sixth to the fourteenth centuries AD which was involved in both long-distance maritime trade and irrigated agriculture (Figure 12.3; Stargardt, 1983, 2001, 2014). The focus here is on the characteristics of

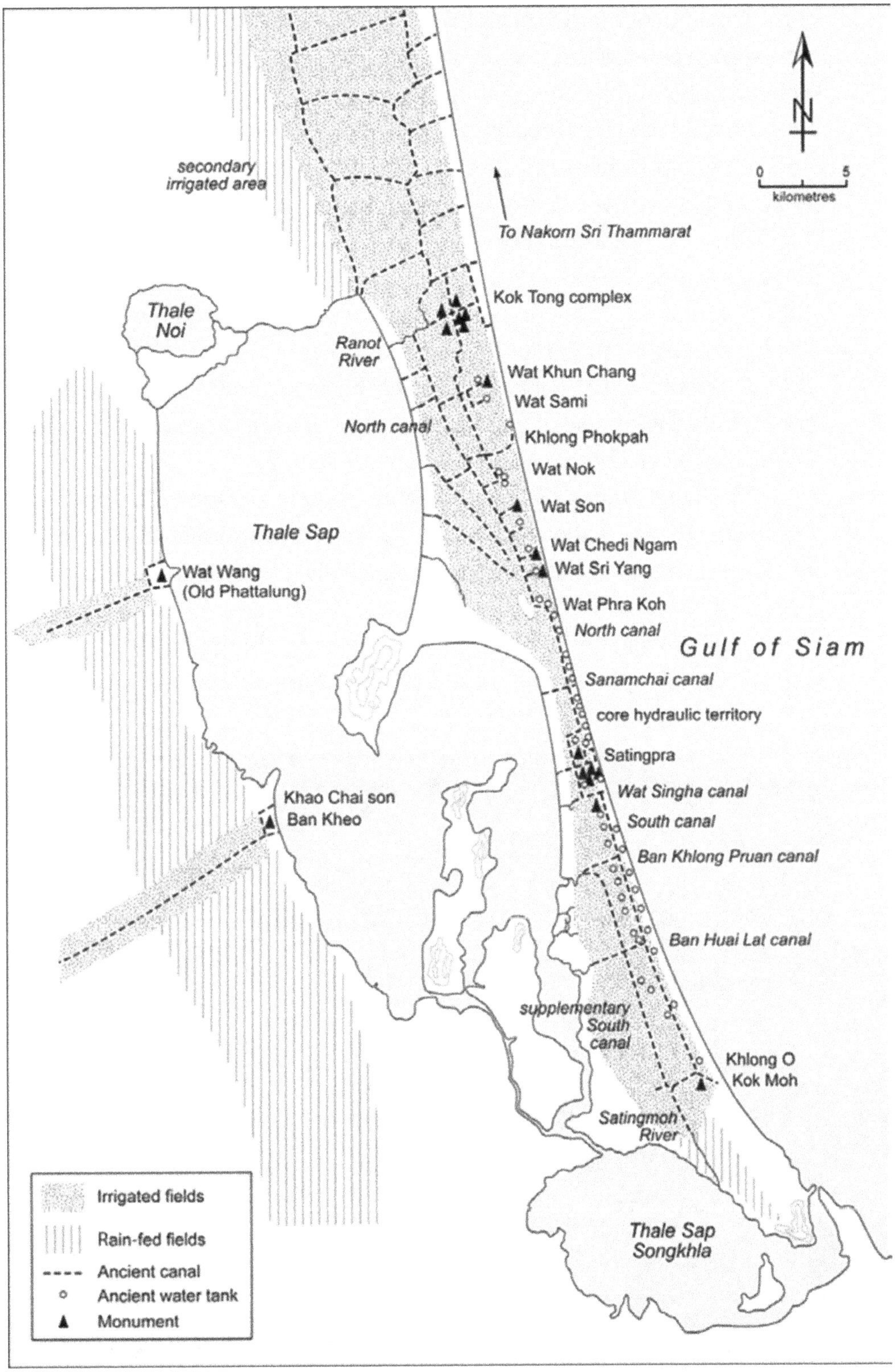

Figure 12.3 Map of the Satingpra peninsula, Isthmus of Kra, South Thailand, with ancient irrigation works (© Janice Stargardt)

the environment on the southern part of the Kra Isthmus and the irrigation system created there, which comprised tanks working together with ponded fields and a network of larger canals and smaller field channels.

Ponded fields are relatively small, aligned along even minimal slopes and separated from each other by massive bunds, usually formed so as to be well above the maximum level of water in the fields (Figure 12.4). The bunds were, and still often are, planted with secondary crops such as the sugar palm (*Borassus flabellifer* L.), whose roots strengthen the bunds (Stargardt, 2014: 117–21). Fields of this type are still found elsewhere in Thailand and also in Cambodia, in areas where the rainfall is marginal – or insufficient – to the needs of wet rice, because they lend themselves to the highly effective use of limited quantities of water, transferred frugally from field to field by seasonally piercing the bunds.

The water tanks of the Satingpra complex, South Thailand, varied in size between about 100 m × 100 m (medium) and 200 m × 200 m (large). They were studied and selectively cored by the Cambridge–Prince of Songkla research team in the 1970s and 1980s (Stargardt, 1983: 133–84 and figures 16–22). In appearance both ponded fields and tanks resembled the small tanks of ancient Cambodia just discussed, but

Figure 12.4 Photograph of surviving remnant of the ponded fields of the Satingpra peninsula, surrounded by massive bunds with sugar palms growing on them (© Janice Stargardt)

there are some important differences. The field tanks of the Satingpra peninsula were only about 3 metres above sea and lake level, whereas the plains around Angkor were about 16–18 m above mean sea level (amsl) and located above the maximum level of Tonle Sap in flood. It is noteworthy that the Satingpra tanks exposed only a small surface area to evaporation – a significant source of water loss for the great barays of Angkor.

All the water tanks tested on the Satingpra peninsula were initially dug to depths of about 3 m, well below the local water table, which was fresh. The lowest level to which the water table sank, for a short time, in the dry season was Datum minus 1.50 m. Our coring programme showed that tanks were maintained to a depth of about 3 m until the second half of the thirteenth century AD. As shown below, the water table was above 1.5 m depth for most of the year, and was the main source of water in the tanks. Water drawn from the tanks was replaced from the water table within 24 hours, this being the time it took for water to percolate through the blocky clay of marine origin into which most of the tanks and canals were dug. Tanks and canals were unlined, as lining would have inhibited the replenishing movement of the groundwater.

The environmental factor governing the level of the water table on the Satingpra peninsula, and thus the freshwater content of the tanks and canals, was the volume of water in the chain of great lakes – Thale Sap Songkhla, Thale Sap and Thale Noi – which separated the Satingpra peninsula from the rest of the Isthmus of Kra. Although the Satingpra peninsula itself received only a single season of, relatively low, rainfall from the north-east monsoon, this was not true of the surrounding areas of the isthmus and the lakes. The lakes received some of the 2,000–2,500 mm of rain per annum falling in two seasons around Phattalung, and also some of the rain from the south-west monsoon that fell on the isthmian ranges and drained down their eastern slopes into the lakes.

Although the barays and canals of Angkor were constructed on a much larger scale than those of the Satingpra system, they may have been more vulnerable than the latter to any weakening of the summer monsoon, because they depended on a single season of rainfall in the catchment area, from the Kulen Mountains down to Angkor. In spite of the small size of the individual Satingpra tanks, collectively they may have been a more significant source of irrigation water than at first appears. Moreover, as noted above, the limited quantities of water they held were a resource that renewed itself from the water table within the space of 24 hours. Thus, if managed within these environmental constraints, they supported repeated applications (Stargardt, 2014: 119–20, and table 4.1).

To sum up, the central section of the Satingpra peninsula, between the Sanamchai Canal and the Wat Singha Canal, was the most intensely irrigated area, covering 31.96 km^2, or 3,196 ha (Figure 12.4). Into this area a minimum volume of irrigation water of 1,402,200 m^3 had been introduced by means of tanks and large canals (the volume is based on the modest assumption of an average depth of 3 m, whereas most large canals were in fact somewhat deeper, see below). This can be expressed as *c.* 439 m^3 of water per ha or an average of 1.4 m^3 per m^2 of land in any 24 hours (Stargardt, 1983: 144–8, table 25).

The canals of the Satingpra peninsula were of more than one type and responded to complex environmental conditions: the larger canals averaged several standard widths and depths; the central canal (which was also the south moat of the Satingpra citadel) was 15 m wide and 4.4 m deep; the other three moats of the citadel were 15 m wide and 3.9–2.9 m deep; the lateral canals were 10 m wide and 3.4 m deep; branch canals were 6 m wide and 3.4 m deep; longitudinal canals were 10 m wide and 3.6 m deep (Stargardt, 1983: 148–59, and tables 26–29). In this system of linked lateral and longitudinal canals, the lateral canals were of crucial importance. Through their openings into the lakes they provided channels along which the rising lake waters could flow into the flat western lands of the Satingpra peninsula, assisted not only by the increasing volume of water brought by the south-west monsoon, at a time of aridity on the Satingpra peninsula, but also assisted by the wind force driving the waters into every opening along the lake front. Thus to a considerable extent the lateral canals seasonally recharged the content of the larger canals, both lateral and longitudinal.

Another type of canal in the Satingpra peninsula can be seen in Figure 12.4. More correctly described as channels than canals, they performed an important function by leading limited amounts of irrigation water from ponded field to field through clefts cut seasonally into the bunds and closed again seasonally. Such channels worked as distributaries together with both tanks and the larger canals. They add to the repertoire of technical diversity among the hydraulic works of South and South East Asia. Again, technically similar techniques are practised in rural Cambodia.

In this case study, the details of an apparently small-scale system have been revealed, which contrast with the scale of Angkor's barays and canals. The impact of the small-scale system, however, was greater than at first appears, owing to the capacity of the system to renew its contents, and to the skill with which the canals were integrated into the water cycle of the lakes. Thus the Satingpra hydraulic system also incorporated expert environmental knowledge which multiplied the impact of an apparently small-scale system.

Conclusion

The case studies above have presented glimpses of the impressive technical diversity displayed by a selection of the ancient irrigation works of South and South East Asia. Equally striking are the economical means by which great impacts were achieved, which brought their creation, operation and maintenance within the reach of village-level groups. While elite patronage may at times have initiated and determined the scale of such works, it should not be forgotten that gifts emanated from many ranks of society. Finally, written and ethnographic evidence converge in showing that the vital technical knowledge, the ability to assemble and employ the required materials and coordinate irrigation labour, all existed on the local, village level and were a precious resource that was transmitted orally over long periods.

While scholarly attention until now has usually been concentrated on the reciprocal relations between the state and hydraulic works in Asia, I have drawn attention here to the existence of an important degree of agency among village-level irrigation committees and societies in the creation and operation of hydraulic works over large areas and periods of time. In particular, I have probed the way in which an inverse ratio was often achieved between the scale of their impact and the economical means by which that was brought about. This feature remained constant, in spite of the great technical diversity among the case studies presented here, and implies the existence of a profound local knowledge of the environment that lay behind the successful creation and operation of each hydraulic system.

References

Annual Report Government Epigrapher 1898–9. Calcutta: Archaeological Survey of India.

Bishop, Paul, David C. W. Sanderson and Miriam T. Stark. 2003. 'OSL and radiocarbon dating of a pre-Angkorian canal in the Mekong delta, southern Cambodia.' *Journal of Archaeological Science* 31 (3), 319–36.

Delvert, Jean. 1961. *Le Paysan cambodgien*. Paris: Mouton.

Goloubev, Victor. 1936. 'Reconnaissances aériennes au Cambodge.' *Bulletin de l'École française d'Extrême-Orient (BEFEO)*, 1936 (2), 465–77.

Groslier, Bernard-Philippe. 1979. 'La cité hydraulique angkorienne: Exploitation ou surexploitation du sol.' *Bulletin de l'École française d'Extrême-Orient (BEFEO)* 66, 161–202.

Hanus, Kaspar and Damian Evans. 2016. 'Imaging the waters of Angkor: A method for semi-automated pond extraction from LiDAR data.' *Archaeological Prospection* 23 (2), 87–94.

Mahānāma. 1912. *The Mahavamsa or The Great Chronicle of Ceylon*. Trans. W. Geiger. London: Published for the Pali Text Society by Henry Frowde, Oxford University Press.

Stargardt, Janice. 1983. *Satingpra I, The Environmental and Economic Archaeology of South Thailand*. BAR International Series, 158. Oxford: BAR. (Available on www.Academia.edu Janice Stargardt.)

Stargardt, Janice. 2001. 'Behind the shadows: Archaeological data on two-way sea-trade between Quanzhou and Satingpra, South Thailand, 10th–14th century.' In Angela Schottenhammer (ed.), *The Emporium of the World: Maritime Quanzhou, 1000–1400*, 309–93. Leiden: Brill.

Stargardt, Janice. 2014. 'Irrigation in South Thailand as a coping strategy against climate change: Past and present.' In Barbara Schuler (ed.), *Environmental and Climate Change in South and Southeast Asia: How are Local Cultures Coping?*, 105–37. Leiden: Brill.

Takaya, Y. and G. Jayawardena. 1984. 'Agricultural transformation in Maningamuwa, a village in Dry Zone Sri Lanka.' In S. D. G. Jayawardena and N. Maeda (eds), *The Transformation of the Agricultural Landscape in Sri Lanka and South India*, 19–62. Kyoto: Center for Southeast Asian Studies.

Van Liere, W. J. 1980. 'Traditional water management in the lower Mekong Basin.' *World Archaeology* 11 (3), 265–80.

Venkayya, V. 1906. 'Irrigation in southern India in ancient times.' *Archaeological Survey of India, Annual Report 1903–4*, 203–11.

Wittfogel, Karl. 1957. *Oriental Despotism: A Comparative Study of Total Power*. New Haven: Yale University Press.

13

Agricultural development, irrigation management and social resilience in ancient Korea

Heejin Lee

Abstract

The Three Kingdoms–Unified Silla period (fourth to tenth centuries AD) of the Korean peninsula is a good example of the increasingly complex interactions between human society and the environment, with different sectors of the society becoming more and more closely intertwined. During this period, agricultural innovations, especially the development of large-scale irrigation, contributed greatly to an increase in agricultural productivity, and stimulated societal reforms. Archaeological and historical research into this early historical period of ancient Korea, as discussed in this chapter, has emphasised the popularisation of iron tools, the expansion of field systems, and the establishment of a large-scale irrigation system.

These new economic changes are placed in the context of the short- and long-term climatic changes of this time. Using this new evidence I illustrate the dynamic relationships between humans and their environment in ancient Korea and call for a reassessment of the long-term effect of new technology on society and how this contributed to social resilience.

Introduction

The Three Kingdoms (fourth to seventh centuries AD) and Unified Silla (seventh to tenth centuries AD) periods in the Korean peninsula (Figure 13.1) witnessed the formation of complex relationship between environmental

Figure 13.1 Time frame of the Three Kingdoms and Unified Silla periods from the archaeological perspective, and approximate location of each state and the capital of Silla and Unified Silla. Source: author

variables such as climate changes, in particular major (centennial) and minor (decadal) oscillations, and agricultural development in a highly complex society. The Three Kingdoms and the succeeding Unified Silla were kingdoms characterised by centralised administrative authorities, regulated taxation, an enlarged urban centre subject to systematic planning, the establishment of local administrative offices, and shared religious beliefs and ideologies (Barnes, 2015; Kwon, 2008). The Three Kingdoms practised advanced agriculture, with the rise of wet-rice farming through innovative irrigation schemes. Agricultural production increased significantly, which further stimulated the overall economic growth of the society. From the perspective of environmental archaeology, this changing relationship between the environment and the expanding rice-farming society formed a key foundation of the development of social resilience at a time of precarious climate conditions.

This chapter investigates the economic and social aspects in ancient Korea related to irrigated rice farming by synthesising a suite of archaeological and environmental evidence. This evidence includes discoveries of ancient irrigation systems, palaeo-environmental data and historical records indicating decadal oscillations that led to natural disasters.

The development of irrigated agriculture

Archaeological evidence has demonstrated that the Three Kingdoms states (Goguryeo, Baekje, Silla and Gaya, merged into Silla in the sixth century AD), the earliest kingdoms established on the Korean peninsula, arose from statelet-like confederacies in the Three Kingdoms period, in which the application of advanced iron technologies played a key role (D. H. Kim, 2009). The Three Kingdoms period is also known as an era of agricultural reforms, including the development of large-scale, state-controlled irrigation schemes (Kwon, 2008). This is a great development advance on the Bronze Age, when the general agricultural system in the Korean peninsula was a mixed cultivation of rice, wheat, barley, millets and legumes, rather than sole reliance on rice or any other crop, according to some recent archaeobotanical evidence (Ahn, 2010). The construction of irrigation infrastructures, which produced a significant improvement to the water supply, was beneficial to farmers in general, and rice farmers in particular, as wet-rice farming needs more water.

Technological advancements during the Three Kingdoms period not only greatly promoted agriculture in general, but also revolutionised rice-farming practices such as weeding, planting, ploughing methods and irrigation regimes. Such technological developments include the construction of reservoirs and levees (as we know mainly from historical accounts, including *Samguk Saki* (*Sagi*) (B. Kim, 2012) and *Samguk Yusa* (Ilyeon, 2012), compiled in the twelfth and thirteenth centuries AD), as well as the use of wooden irrigation pipes, the spread of iron farming tools, and the increased prevalence of animal-drawn ploughing. The most significant consequence of these changes was the establishment of a crop rotation system, on either a seasonal or a perennial basis. This is evidenced by the discovery of the dry-field ridges and furrows superimposed on paddy field plots at many sites, such as the Pyunggeo site (H. P. Yoon, 2013). Even before the Three Kingdoms period, rice was becoming more than a staple crop, and its special social roles were recognised in the production and consumption processes. H.-K. Lee (2010), for example, pointed out the higher proportion of rice in archaeobotanical crop assemblages discovered in the capital fortress sites, which were the centres of regional networks and therefore indicated high social rank, such as the earthen fortification of Pungnap (in particular the third-century

AD phase) and the Taemokri site. This suggests a more important role for rice at such places than in rural settlements during the proto-Three Kingdoms period.

Following the unification wars, two-thirds of the Korean peninsula was under the control of a single dynasty known as the Unified Silla (AD 668–935). This dynasty inherited the existing agricultural system from the previous political regimes and further developed it across the Korean peninsula. While there was a pronounced reduction in external military threats, the Unified Silla, especially during the eighth and ninth centuries AD, suffered from an increased frequency of natural disasters (famine, drought, insect plague, frost and floods). This high frequency of disasters is thought to have contributed to the division of the Unified Silla into the short-lived post-Three Kingdoms in the tenth century AD. The following section synthesises archaeological discoveries about agricultural systems, related historical documents and paleo-environmental data on climate changes to illustrate further the processes just described during the Three Kingdoms and Unified Silla periods.

Archaeological evidence

Farming tools

During the proto-Three Kingdoms and Three Kingdoms periods, the use of iron farming tools became popular (Table 13.1). Research on iron tools buried in graves confirms that there was a clear diversification of tool types, which are suited to certain functions such as tillage, irrigation, weeding and harvesting, based on their morphologies. One important

Table 13.1 Types of iron tools in the Three Kingdoms period (after Kim, 2009)

Function	Type of tool
Tillage	iron axe
	ard (*ddabi*)
	U-shaped shovel
	spade blade
	rake/scraper
Irrigation	*salpo*
Tillage and weeding	hoe
Harvesting	sickle

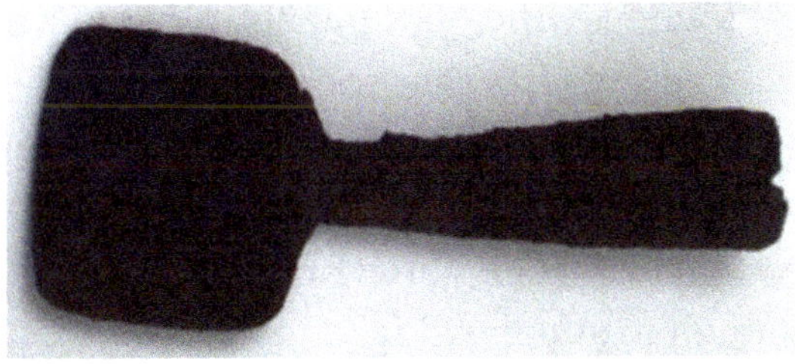

Figure 13.2 *Salpo* from Uiseong, Gyeongsangbuk province (National Museum of Korea: http://www.emuseum.go.kr/main)

type of tool was the *salpo* (Figure 13.2), which looks like a poker with a small square plate on the lower end (D. H. Kim, 2009). Its main use was to open and close the watergate between the wet-field and irrigation canals (D. H. Kim, 2009) by breaking one corner of the bund surrounding the field plot.

D. H. Kim (2009) argues that the sudden appearance of a higher frequency of *salpo* (3 out of a total of 23 tools, or 13 per cent) at the archaeological sites in the southern region (Busan and Gimhae) during the fourth to seventh centuries AD might be indicative of the establishment of wet-rice farming throughout the Korean peninsula. Despite their small quantity compared with other types of iron tools (7 out of 384 in total, *c.* 0.02 per cent), their ubiquity, together with a general increase in iron tools from the fifth century AD onwards, clearly points to a change of main farming tool assemblage (D. H. Kim, 2009). Associated with such changes was the new agricultural scheme focused more on paddy rice cultivation (D.H. Kim, 2009). Wooden farming tools excavated from wetland sites such as the Shinchangdong site are rarer. They include shovels, and ards with a square plate suitable for digging wet surfaces; such tools indicate an increased need to clear and dig the flooded land surface (D. H. Kim, 2016). Artefacts discovered in waterlogged contexts also point to the fact that irrigated rice farming was the predominant agricultural activity from the Early Iron Age (300–100 BC) onwards (D. H. Kim, 2016).

The fact that burial of digging tools is limited to tombs belonging to people of higher social rank may show that the possession of such iron tools was at the time still quite limited. They were either communally held or controlled by the elites of the settlements in the south-eastern Korean peninsula. In the western region a *salpo* appeared in the iron tool burial assemblage during the mid-fourth to early fifth century AD. More common types of iron tool such as hoes and harvesting cutter-blades, together with rarer iron tools such as spade and ploughshares, are found in the capital region of the Three Kingdoms, while spades and ploughshares are found in the settlements regarded as local centres (H. K. Lee, 2018). Lee suggests that the central government of Baekje, one of the Three Kingdoms located in the south-western region, controlled the production of iron tools and strategically distributed them to the local centres. Even in the late Three Kingdoms period, iron tools were not numerous enough to supply all farming households because of this central control. Many iron farming tools from the late Three Kingdoms period have been found that originated from state-controlled military posts such as the Achasan site (Choi, 2013), indicating long-lasting centralised management and control of the production and distribution of iron farming tools (J. H. Kim, 2005).

Dam construction

The transition to large-scale agricultural systems might be inferred from the scale of regional irrigation facilities. Although historical documents such as the *Samguk Saki* often mention the king's decrees concerning the construction of levees and water-management infrastructures (Table 13.2), little material evidence of these activities survives. One example is a large reservoir, Chejongje, which has an inscribed stele that mentions the dates of its construction and repair (Kang, 2006). Other important evidence comes from newly discovered ancient dam structures (SCPI, 2015). Judging from remains excavated at reservoirs, and from computational simulations and written records, the labour investment in the construction and management must have been enormous.

Beokgolje is a Three Kingdoms-period reservoir with related written inscriptions that mention 7,000 men deployed for the initial construction and 14,140 for later repair work (Kang, 2006). Computational simulation and examination of historical documents raised the possibility that it was capable of irrigating a large area (45/46–51 ha) (Jang et al., 2015). The historical and stele records are complemented by archaeological evidence. In particular, archaeological excavations have shed light on the locally operated irrigation systems that were used in concert with dams. Examples of such remains have been found at both the Gayari and Yaksadong reservoirs; two excavated dams show similar patterns, which comprise built levees that block the mouth of a natural valley. In these cases, water is thought to have been retained inside and released when needed (J. Kim and Son, 2015). The scale and shape of these local dams can be inferred from an example at Yaksadong, which is estimated to have been up to 12 m in height and to have had a trapezoidal levee (J. Kim and Son, 2015). Notably, these structures confirm the application

Table 13.2 Articles from *Samguk Saki* and stele inscriptions mentioning irrigation facilities and irrigation-related activities (note: the modern calculation of the ancient measurement unit Bo is under debate)

State (date – AD)	Content	Region and river	Size	Estimated labour input	Estimated irrigated area*	Source
Baekje (222)	'ordered to repair levee'	Unknown		Unknown	Unknown	*Samguk Saki* *The Book of Baekjebonki* King *Goosoo* 9
Baekje (242)	'constructed fortress and levee'	Seoul (?)		Unknown	Unknown	*Samguk Saki* *The Book of Baekjebonki* King *Goi* 9
Silla (330)	'constructed the *Beokgolje* reservoir'	Gimje	circumference 1,800 *Bo* (3.42 km?)	Unknown	124,250 tons of water; 9,000–10,000 ha of land	*Samguk Saki* *The Book of Sillabongi* King *Heulhaeisaguem (Biru)* 21
Silla (429)	'constructed the Sije reservoir'	Unknown	circumference 2,170 *Bo* (3.96 km?)	Unknown	Unknown	*Samguk Saki* *The Book of Sillabongi* King *Nulgimaripgan* 13
Baekje (475)		Seoul?	levee along the Han river	Unknown	Unknown	*Samguk Saki* *The Book of Baekjebonki* King *Gaero* 21

(Continued)

Table 13.2 (Continued)

State (date – AD)	**Content**	**Region and river**	**Size**	**Estimated labour input**	**Estimated irrigated area***	**Source**
Baekje (510)	'repair the nationwide irrigated reservoirs'			Unknown	Unknown	*Samguk Saki* *The Book of Baekjebonki* King *Muryeong 10*
Silla (531)	'issued a decree to the administrative branch in charge of repairing levees'			Unknown	Unknown	*Samguk Saki* *The Book of Sillabongi* King *Beobheung* 18
Silla Ojak stele (578)		Daegu Reservoir, location unknown	width 20 *Bo* (36 m); height 5 *Bo*, 4 *Cheok* (10 m); length 50 *Bo* (90 m)	312 men for 13 days	Unknown	Ojak stele
Silla Chengje stele (416, 476, 536, 596 and 798)		Yongcheon	circumference 5 km	7,000 workers for the construction; 136 axe-men and 14,140 soldiers for 60 days for repairs	*c.*590,000 tons of water (estimate based on hydrological analysis)	Chengje stele

Note: * based on Jang et al., 2015

of the so-called 'leaf paving technique' in the construction, a thin layer consisting of branches and leaves laid in the middle of the earthen wall structure. This organic layer might have been used as drainage, as it enables moisture inside the structure to flow out, as well as to reduce the overall weight. It may also enhance the durability of structures. The use of this technique can be seen in these two reservoirs as well as in other earthen wall structures (including the fortress) from the late Three Kingdoms period (Shin, 2014).

Although no canals for diverting water into fields have been excavated from this period to date, the effect of large-scale irrigation can be postulated from other archaeological remains, including the expanded paddy fields. Such improved irrigation conditions in the Three Kingdoms period are indirectly manifested by the altered paddy soils. At the microscopic level, internal change and some anthropogenic pedofeatures appear to be due to intensive and multiple irrigation schemes, in contrast with the Bronze Age paddy soils, which benefit from a high groundwater table that enables them to retain moisture for a long time (H. Lee et al., 2014).

Expanded paddy fields

By 2013, more than 30 paddy fields sites belonging to the Three Kingdoms–Unified Silla period across South Korea had been excavated (H. P. Yoon, 2013). In general, these ancient fields are located on the floors of valleys and in floodplain areas along major rivers and their large tributaries. A facility common to paddy fields from this period is a kind of wooden gate to store or control water flow. Water was diverted from irrigation canals and flowed to each field plot from these gates; it was often sourced from small streams nearby or from moisture collected from the valley bottom (Jeon, 2010; Kwak, 2010). Yoon (2013) has suggested that the extension of field systems commonly observed in the Three Kingdoms and Unified Silla period is probably due to an increased water supply. It appears to be more advanced than the Bronze Age paddy fields, where water from small tributaries was used and rain-fed cultivation was normal, although the structure of the wooden watergates looks similar. These reservoirs are thought to have had the capacity to distribute water widely across the transformed agrarian landscape, including to sites far from natural streams that had previously been infertile or non-arable (Jeon, 2006). However, it should be noted that the latter scenario is largely based on information from historical documents. The archaeological evidence, such as canals connecting large reservoirs to large field systems, remains to be discovered.

Excavation at the Pyunggeo (Pyeonggeo) site located on the floodplain of the River Namgang (Figure 13.3) provides clear evidence of the application of advanced agricultural practices. Floodplains along this river began to be reclaimed for arable fields from the Bronze Age onwards, when densely concentrated paddy and dry fields were built (HCRC, 2010). Ancient land use patterns involving sophisticated management of micro-environmental conditions were preserved at this site. Paddy fields from the Three Kingdoms and Unified Silla periods are typically distinguished from one another by the fact that the soil texture exhibits hydrological soil degradation, which is due to a practice of intensive irrigation, with multiple flooding sessions (H. Lee et al., 2014). They comprise expanded field plots characterised by enlarged field units. These large fields exhibit a high level of landscape modification. Settlements were usually built at the highest point of old natural levees, mainly to protect them from flooding, while dry fields were typically located within the sandy deposit area. Paddy fields of limited size were constrained to lands adjacent to swamps and on the side of the levee closer to river channels, where water collected from the slope and the soil remained waterlogged for long periods. Notably, paddy fields were much more extensively distributed at this time than during the Bronze Age, and the surfaces of the fields often show foot and hoof marks. A part of the paddy fields with a clayey sandy loam was later converted into a dry field, which indicates the use of a crop rotation system.

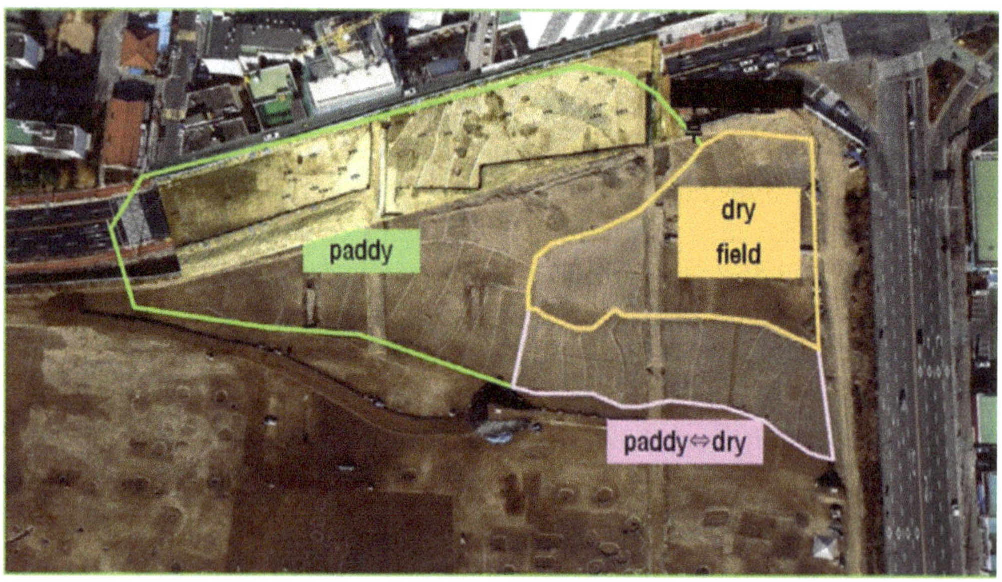

Figure 13.3 Three Kingdoms period fields from the Pyunggeo site (after HCRC, 2010)

The physical churning up of the ridges and their subsequent superimposition on older structures appear to be a strategy for replenishing the soil nutrients necessary for plant growth (H. P. Yoon, 2013).

Irrigation in historical documents

Historical documents such as the *Samguk Saki* and the *Samguk Yusa* from the Three Kingdoms and Unified Silla periods provide more detailed information about ancient agricultural systems and their maintenance (Table 13.2). In particular, these records offer valuable insights into methods of water management, and how such practices were associated with changing socio-economic backgrounds. Much historical research on the economic system that functioned during the Three Kingdoms and Unified Silla periods indicates that the construction of an irrigation system at the regional level was closely related to the centralisation of political power. The operation of this reformed agricultural economy under a reformed political organisation depended on land ownership and the right to collect taxes. The latter oscillated between local elites, who later developed into aristocrats, and the centralised government. Jeon (2006) has argued that local elites controlled the land and that a kingship emerged from the statelets of the proto-Three Kingdoms period. Thus, some of these landowners subsequently became aristocrats who were endowed with a seigneurial right to collect taxes. It is believed that the tax system for commoners, *Nokeup*, was similar to a poll tax; estates generated substantial incomes for local elites and aristocrats as a result of the arable fields which they owned and inherited. During the later phases of the Three Kingdoms period, however, each state tried to develop its own centralised government, which led to the construction of many more large irrigation systems within each state and to warfare between the different regions. The close attachment to central government was developed through the adoption of the heaven–human induction idealism (Zhuang et al., 2017) for dealing with affairs of governance, which supplanted long-practised shamanistic rituals for mitigating natural disasters (Jeon, 2013b).

Following unification by Silla, officials were paid by the central government, which also removed or reduced the right of aristocrats to collect taxes from their manors. This bureaucratic system is called *Sejo* (direct salary payment for civic officials) and *Munmukwanryojeon* (conferring barren and abandoned land instead of paying salary according to the status of governmental post). Thus taxes from commoners, including farmers, were paid to the government, and both land and salary were given to officials in return for their services rather than to aristocrats; in AD 757 these measures were reversed and the earlier *Nokeup* system was revived. The latter was symbolic of central government losing its control over governance measures. It can be surmised that the commoners' reliance on central government for their water supply had greatly increased during the early Unified Silla period. In contrast, centralised construction and management of irrigation facilities in some regions (Kang, 2006) had become inextricably engaged with society, forming a complex internal network of entanglements with the economic system. Further evidence on water-control measures can be inferred from articles in the *Samguk Saki* and the *Samguk Yusa* which show that many temples were constructed at this time along the stream that flowed through the capital, Gyeongju, in particular during the seventh to ninth centuries AD (Kang, 2009; G. J. Yi, 2010).

Documentary evidence from this period notes that floods were rare within the capital because of increased embankment area and improved flood prevention (Jeon, 2009). At the same time, the construction and management of large reservoirs, such as Cheongje, was the responsibility of the government (Kang, 2006). The background to the reversion to *Nokeup*, which appears to be a sign of social decline, has long been debated. The assumption previously widely accepted is that it resulted from political conflicts over the throne among the royal family and *Jingol* (the second-highest level of inherited social status) aristocrats. More recent studies, however, have indicated that this period of fractional strife within the ruling elites was relatively short and that the succession to the throne throughout the rest of the Unified Silla period was stable (Kwon, 2014).

An alternative view emphasises that the political turmoil was the consequence of severe natural disasters that are known to have occurred, at an unusually high frequency, during this period (Jeon, 2013a). Reduced tax revenues due to a succession of serious famines had imposed a burden on the central government, which paid salaries to officials. What may have been worse is that aristocrats had exploited those farmers even in the time of low production. This exploitation is regarded as having incited riots against the government, and the consequent political chaos led to the collapse of the Unified Silla through dynastic change. The relationship between socio-political changes and rice production indicates that the development of wet-rice farming is closely intertwined with the irrigation methods controlled by central government and with the tax system in Korea (Jeon, 2006). The influence of central government over common farmers was weakened around this time, even though the ideological expectation that the king should be held responsible for climatic anomalies remained strong throughout the early Unified Silla period. The repeated failure of agricultural production, combined with overexploitation of the population, led to an increase in the number of thieves; groups of them went on plundering raids, began to foment subversion, and mutinied against the government (Jeon, 2006, 2013a).

Pollen records suggest that there was a cooler period from AD 200 to 800 and a warmer period between AD 800 and 1300, parallel with the changes observed in north-east Asia (Park, 2013). Ironically, more natural disasters were recorded in historical documents during the warmer period, which raises the issue of the reliability of early historical documents. In contrast with the high reliability of modern meteorological records, records of the frequency of disaster occurrence in historical texts are often unreliable. It is nevertheless assumed that these documents do reflect the trends in real climatic events and their social consequences, to some extent. Detailed examination of the *Samguk Saki* suggests a possibility of decadal-scale (20–30 years' duration) climatic oscillations that resulted in a higher frequency of severe natural disasters in the eighth century AD (H. Lee, 2017).

Records show that, during the eighth century AD, social riots resulting from the reduction of agricultural yields because of natural disasters (in particular drought, famine and flood, early frost) were concentrated in a short period, while during the ninth century their occurrence patterns were more dispersed. Because of the discrepancy between the eighth- and ninth-century records, it remains an open question whether or not frequent outbreaks of famine and other natural disasters were due to climatic changes. If they were, why were the patterns of natural disaster occurrence in the eighth and ninth centuries different? If not, were they symptoms of failures in agricultural governance, or causes of political instability?

Jeon (2013b) stated that a pronounced aridification tendency could be seen during the second half of the eighth century in Japan; it caused natural disasters such as famine and subsequent social riots. Jeon suggested that a similar and contemporary climatic trend may have developed in ancient Korea. We can infer that the slight increase of temperature demonstrated by the pollen data (Park, 2011, 2013) may point to climatic oscillations and a trend to a warmer climate after the seventh to eighth centuries. Climatic oscillations during the ninth century are hard to ascertain from the current historical evidence. I would like to conjecture that the clearer decadal-scale of climatic oscillations in the eighth century is associated more closely with climatic vagaries than are ninth-century events. On the other hand, a slight temperature rise does not necessarily have adverse effects. It would favour the growth of rice, especially in the Korean peninsula, which is much cooler than other rice production regions. In addition, the agricultural system was well equipped with various irrigation facilities that mitigated the shortage of water in that period. Considering these advantages, it is all the more peculiar that the society and its agricultural system appear to have been so vulnerable to the climatic events of the Unified Silla period. Was resilience not developed along with the larger and more complex economic system?

Human-induced environmental degradation

In this section, I further synthesise social and environmental evidence of the Three Kingdoms period and explore how the society shaped and was shaped by environmental changes in a wider context. This period is marked by advanced agricultural technologies, especially the application of intensive modes of cultivation (M. Kim et al., 2013; H. Yi, 2012). In other words, the society invests more infrastructure and cultivation management in order to significantly retain agricultural productivity. An estimated population increase between the fifth and seventh centuries AD exacerbated the problem of land shortage and rendered the need to increase productivity even more acute. At the same time, methods of mitigating and relieving disasters were developed, so that it was easier to cope with crises. On the other hand we also could suspect the disadvantage of an extended agricultural system. In cases where climatic oscillations were manifested in droughts and anomalous precipitations, their effects were probably also exacerbated by the 'newly extended agricultural systems' established during the Three Kingdoms period. This is just a conjecture at present, but there is a hint of supporting contextual evidence in Gyeongju.

Gyeongju was one of the capitals of the Three Kingdoms period, and also became the capital in the Unified Silla, with a large population. To sustain such a large population, large swathes of land would have been cleared for farming and construction, and trees felled for fuel. This large-scale landscape modification would have caused severe disturbance to local ecosystems. Archaeological surveys have shown that the residential towns and roads constructed during the late Unified Silla period did not conform to the usual urban pattern characterised by square block districts (*bangli*) (I. H. Hwang, 2015). The more random development leads us to assume that poor refugees from the countryside were moving to Gyeongju and the expansion of the city had moved out of government control. It is also very likely that Gyeongju suffered at the time from the overexploitation of natural resources. A key outcome was the rapid siltation of streams that resulted from increased sedimentation, caused by extensive deforestation, which elevated the river bed (S. Hwang and Yoon, 2013). The Buckcheon stream flowed through the capital and fed a number of large temples and other buildings (Kang, 2009). But its lower part was raised as a result of the sedimentation of fluvial deposits in the sixth to seventh centuries AD (S. Hwang and Yoon, 2013; S. Yoon and Hwang, 2004). Investigations in the area along the Buckcheon stream, have found that there were Unified Silla buildings on the lower river terrace, destroyed by fast-running water there. On the higher river terrace, cultural deposits from the period lasting from the Bronze Age to the Three Kingdoms period are well preserved, without signs of severe floods. This contrast may indicate an expansion of settlement territory to the lower terrace, and a high groundwater level during the Unified Silla period (S. Hwang, 2007). It can therefore be hypothesised that extensive agricultural systems also led to environmental damage and to rigidity in social resilience, and to what appears to have been a case of overly interconnected subsystems of environment, socio-politics and economy, which may have made the entire society vulnerable.

Did more advanced agriculture lead to greater social resilience? The role of natural disasters in the fall of the Unified Silla Dynasty

The Unified Silla inherited agricultural systems from the Three Kingdoms period, and these were further developed into large-scale extensive irrigation systems. The additional impacts of these new systems have seldom been discussed. However, this large-scale agriculture-oriented society appears to have faced a number of new challenges, including natural disasters in the eighth century AD. The impacts of natural disasters and related famines were enormous, and may have upset the newly established socio-economic systems. The failure of agricultural production during the Unified Silla period had often been linked to rebel uprisings in rural areas. It has been noted above, for example, that one critical administrative decree was the revival of *Nokeup*, which granted aristocrats the right to collect tax from locals on their privately owned land instead of being paid a salary by central government; a number of scholars previously argued that this was the moment that tipped the Unified Silla to its decline. This critical moment of the enforced revival of an old political system can be interpreted as demonstrating that the Unified Silla society had reached the threshold at which the social resilience to maintain the status quo was lost (H. Lee, 2017). Indeed, the government lost the control over local affairs, in particular taxation, that was crucial in difficult economic situations, while the expectations of the commoners concerning the central government's role in famine relief remained unchanged. Even worse, as mentioned above, aristocrats exploited locals even more harshly at the same time as environmental oscillations were wreaking havoc (Jeon, 2006). In addition, the large population of this area was infected by various epidemic diseases around this time; disease outbreaks are thought to be attributable to the sudden mixture of ethnically and culturally heterogeneous groups that arose from the presence of foreign troops, and to impoverished living conditions during the unification wars (H. S. Lee, 2003).

All the above factors combined made the society very fragile. Natural disasters were a key trigger inciting the rebels against the government during the late ninth century. Although it is questionable whether the natural disasters recorded were simply the consequences of climatic events or economic failures, the increased social instability and the power shift to aristocrats had profoundly affected the centralised irrigation management, along with the tax system. Eventually the authority became less respected, which might also explain why irrigation facilities were so useless at preventing famines: not much investment was made in maintaining the basic functioning of these systems.

Correlations between advanced irrigation and social vulnerability, and resilience

A large-scale irrigation scheme at this time seems to have increased the total amount of water supplied to arable land for rice farming. Previously neglected barren land and floodplain were reverted to arable fields. There is no doubt that these synergistic effects contributed to increased agricultural production. However, the increased capacity of this newly established system to maintain a stable water supply appears to have had some disadvantages. Extended fields would require more water to sustain. Problems can arise during extreme droughts and prolonged periods of bad weather with low precipitation. When the increased demand for water was not met, the lack of water to irrigate extensive fields would have been a serious hazard, leading to a reduction in agricultural produce. Societal disruptions triggered by climatic events may have induced environmental degradation and amplified the negative consequences of those events. Expanded irrigation schemes may have experienced similar negative effects from climatic vagaries, and may have contributed further to increasing social conflicts in the complex social-economic network.

Conclusions

This chapter integrates current archaeological discoveries and historical research concerning irrigation systems in the ancient kingdoms of South Korea. The archaeological surveys have generated new evidence of local irrigation facilities that were not recorded in historical texts. This evidence contributes fresh insights into the role irrigation systems play in agricultural production and the beneficial effects they have on it. Large-scale irrigation facilities enabled the flexible control of water alongside multiple drainage schemes and led to improvements in the productivity of rice farming. Such intensified agricultural practices stimulated a series of fundamental socio-economic transformations over more than a thousand years and formed the foundation of the Korean national economy (centred on rice farming) up to the nineteenth century. The more complex consequences of this transformation, driven by innovative irrigation, related agricultural schemes and significant landscape modification, were long overlooked.

I illustrate the possibility that the excessive interconnection of the ecological system with the socio-economic system through agriculture at this time may have drawn the society into a new kind of vulnerability and threatened the long-term resilience of the ancient dynasties. But, as we have seen, only dynastic changes took place, not the collapse of the entire agricultural system, as rice farming continued to develop during the succeeding dynasties. Crop failures exacerbated social conflicts and were linked to serious rebellions against the central government at this time. A key contributing factor, as well as a potential driving force, was probably oscillations in climate, while another was the fact that the agricultural system was deeply embedded within social networks. As a result, the expanded agricultural system, boosted by irrigation renovations and fostered by social development, may have caused adverse effects in times of drought. In the later dynasties (the Goryeo and Joseon kingdoms), from the eleventh to the nineteenth centuries, the similarly large-scale agricultural systems were sustained through improvements in socio-economic conditions, such as reforms of the tax system and technological advancement. But this does not mean that they were exempt from the impact of environmental changes, as we see the continuous occurrence of natural disasters derived from decadal oscillations. Further comparison between the Three Kingdoms and Unified Silla, and the Goryeo and Joseon, periods may reveal how irrigation systems were developed side by side with changing socio-economic and socio-political situations and climatic vagaries.

Another possible avenue for future research is to expand our synthesis by incorporating additional historical accounts with archaeological evidence at the advent of modern agriculture. The case studies presented in this chapter call for more attention to be paid to decadal climatic oscillations and for a nuanced understanding of the complex social-environmental network of agricultural production in ancient Korea.

References

Ahn, S.-M. 2010. 'The emergence of rice agriculture in Korea: Archaeobotanical perspectives.' *Archaeological and Anthropological Sciences* 2, 89–98.

Barnes, G. L. 2015. *Archaeology of East Asia: The Rise of Civilization in China, Korea and Japan*. Oxford: Oxbow Books.

Choi, J. 2013. *Achasanhwa Koguryouinambangjeongchaek* (Acha mountain fortresses and the southward expansion of Koguryo). Seoul: Sekyeongsa Press.

HCRC (History and Culture Research Center of Gyeongnam Development Institute). 2010. *Jinjupyeongeodong 3–1 jigu* (Excavation report of Pyeonggeodong, 3–1 District, Jinju). Jinju: Gyeongnam Development Institute.

Hwang, I. H. 2015. 'Sillawanggyeong jungsimbuui dosihwagwajeong mlt bangli gujo gochal' (A study on the urbanisation process and Bangri structure of the centre of Silla capital). *Journal of Ancient Korean History* 90, 65–91.

Hwang, S. 2007. 'Godae gyeongju jiyeokui hongsu ganungseongwa inganhwaldong' (The possibility of flooding and human activities of Gyeongju Area in ancient times). *Journal of the Korean Geographical Society* 42, 879–97.

Hwang, S. and S. Yoon. 2013. 'Jayeonjaehaewa inwuijeok hwangyeongbyeonhwaga toilsilla bunguie michi yeonghyang' (Influences of changes in natural environments by natural hazards and human activities in ancient times in Korea on collapse of the Unified Silla Dynasty). *Jayeonjirihakhoiji* (*Journal of the Korean Association of Regional Geographers*) 19 (4), 580–99.

Ilyeon. 2012. *Samkuk Yusa* (Memorabilia of Three Kingdoms). Seoul: Institute of Korean Classical Humanities.

Jang, C. H., H. J. Kim and J. Y. Sung. 2015. 'Jihyeonghwa sumunhakjeokbunseokul tonghae bon godaesurisiseolui nonggyeongsangsanryeok yeongu' (Study of agricultural productivity for ancient irrigation facilities using topographical and hydrological analysis). *Journal of Korean Ancient History* 89, 71–86.

Jeon, D. 2006. *Hangukgodaesahoigyeongjesa* (The economic history of ancient Korea). Paju: Taehaksa.

Jeon, D. 2009. *Silla wangjeonghui yeoka* (The history of Silla Capital). Seoul: Saemoonsa Press.

Jeon, D. 2010. 'Samguk mit tongilsillaui surisiseol' (Irrigation facilities of Three Kingdoms and Unified Silla). In J. Y. Sung (ed.), *Paddy Rice Farming and Facilities of Ancient Korea*, 316–39. Seoul: Seokyeong.

Jeon, D. 2013a. 'Samguksidae' (Drought occurrences and government precautions during the Three Kingdoms Period and the Unified Silla Period). *Journal of Korean Historical Studies* 160, 1–46.

Jeon, D. 2013b. 'Three Kingdoms period'. In Board of Seoul History Publication (ed.), *Seoul Jaehae-sa* (The history of the natural disaster of Seoul). Seoul: Seoul City Press.

Kang, B. W. 2006. 'Large-scale reservoir construction and political centralization: A case study from ancient Korea.' *Journal of Anthropological Research* 62 (2), 193–216.

Kang, B. W. 2009. 'Gyeongju bukcheonui suriyeo gwanhan yeoksa mit gogohakjeok gochal' (A study of success and failure in water management of the Buk Chun in Kyongju: Historical and archaeological approaches). *Sillamunhwa (Journal of the Center of Research for Silla Culture)* 25, 337–60.

Kim, B. 2012. *Samguk Saki* (History of the Three Kingdoms). Seoul: Institute of Korean Classical Humanities.

Kim, D. H. 2009. 'Seonsa·Godae nongguui soyuhyeongtae geomto: Yeongnamjiyeokuljungsimuiro' (Review on possession type of prehistory and ancient farming tools: Focusing on Yeongnam Region). *Hanguksanggosahakbo (Journal of Korean Ancient History)* 64, 25–60.

Kim, D. H. 2016. 'Godaeui mokje gigyeongu yeongu' (A study on wooden ploughing tools of ancient times). *Junganggogo (Studies of Joongang Archaeology)* 21, 81–121.

Kim, J. H. 2005. 'Goguryeoui cheolje nonggiguwa nonggyeonggisului baljeon' (Iron-based agricultural utensils and agricultural techniques in Goguryeo). *Dongbukayeoksanonchong (Journal of East Asian History)* 2005, 57–96.

Kim, J. and S. W. Son. 2015. 'Engineering approach to original shape restoration of ancient water reservoir embankment' (in Korean). *Journal of Korean Ancient History* 89, 53–69.

Kim, M., S.-M. Ahn and Y. Jeong. 2013. 'Rice (*Oryza sativa* L.): Seed-size comparison and cultivation in ancient Korea.' *Economic Botany* 67 (4), 378–86.

Kwak, J. 2010. 'Cheongdonggisidae-chogicheolgisidaeui surisiseol' (Irrigation facilities from Bronze Age to Iron Age). In J. Y. Sung (ed.), *Hangukgodaeui sujeonnongeipgwa surisiseol* (Paddy rice farming and facilities of ancient Korea), 231–309. Seoul: Seokyeong.

Kwon, O. 2008. 'Samgguksidae gaegwan' (Overview of the Three Kingdoms period). In Korean Archaeological Association (ed.), *Hangukgogohakgangui* (Lecture on Korean archaeology), 229–32. Seoul: Sahoipyeongron.

Kwon, Y. O. 2014. 'Silla hadewa sillamal' (The Hade period and the closing years of Silla). *Yeoksawa gyeonggyeo (History and Boundaries)* 90, 39–79.

Lee, H. 2017. 'Samguk-tongilsillasidae hwangyeonggwa inganui sanghojakyongui yihae' (Understanding human–environment interaction during the Three Kingdoms–Unified Silla period: Mechanism of sustainability of a society against climatic oscillations). *Hangukyeongu (Journal of Korean Studies)* 60, 235–76.

Lee, H., C. French and R. I. Macphail. 2014. 'Microscopic examination of ancient and modern irrigated paddy soils in South Korea, with special reference to the formation of silty clay concentration features.' *Geoarchaeology* 29 (4), 326–48.

Lee, H.-K. 2010. 'Wonsanguksidae jungbujibangui jakmuljoseong yeongu' (A study on the establishment of the crop assemblage of the proto-Three Kingdoms period in central Korea). *Hangukgogohakbo (Journal of the Korean Archaeological Society)* 75, 98–125.

Lee, H. K. 2018. 'Baekjegukgahyeongseonggiui cheolje nonggigu culto yangsang yeongu' (Baekje state formation and iron agricultural implements). *Hanguksangosahakbo (Journal of Ancient Korean History)* 99, 102–36.

Lee, H. S. 2003. 'Silla tongilgi jeonyeombyengui yuhanggwa daeungchak' (Epidemics and government policies for their control in Unified Silla Korea (668–936 AD)). *Hanguksangosahakbo (Journal of Ancient Korean History)* 31, 209–56.

Park, J. 2011. 'A modern pollen–temperature calibration data set from Korea and quantitative temperature reconstructions for the Holocene.' *The Holocene* 21 (7), 1125–35.

Park, J. 2013. 'Namhan jiyeokui holocenejunghugi gihubyeonhwa' (Mid- and late-Holocene climate change in South Korea). *Gihuyeongu (Journal of Climate Research)* 8, 127–42.

SCPI (Seongrim Cultural Property Institute). 2015. *Ulsan yaksadongujeok* (The ancient site at Yacksa-dong, Ulsan). Gyeongju: Gyeongsangbukdo Province.

Shin, H. 2014. 'Samguksidae tochuk gujomului buyeobbeob yeongu' (A study on the 'Leaves paving technique' of earthen structures in the Three Kingdoms Period). *Baeksanhakbo* 98, 357–89.

Yi, G. J. 2010. 'Sillawanggyeongui Hyeongseonggwa sawon' (The process of the Silla's capital city construction and Buddhist temples). *Dongakmlsulsahak* (*Dongak Art History*) 11, 105–23.

Yi, H. 2012. 'The formation and development of Samhan.' In M. E. Byington (ed.), *Early Korea. Volume 2: The Samhan Period in Korean History*, 17–60. Cambridge, MA: Korea Institute, Harvard University; distributed by University of Hawai'i Press.

Yoon, H. P. 2013. 'Gyeongjakyuguruel tonghae bon gyeongiyiyongbangsikui byeoncheon yeongu' (Changes in land use schemes through cultivation features). In S.-M. Ahn (ed.), *Nongeuipui gogohak* (Archaeology of agriculture), 153–83. Seoul: Sahyoepyeongron.

Yoon, S. and S. Hwang. 2004. 'Gyeongju mit cheonbuk jiyeokui seongsangji jihyeongbaldal' (The geomorphic development of alluvial fans in the Gyeongju city and Cheonbuk area, south-eastern Korea). *Daehanjirihakhoiji* (*Journal of the Korean Geographical Society*) 39, 56–69.

Zhuang, Y., H. Lee and T. R. Kidder. 2017. 'The cradle of heaven–human induction idealism: Agricultural intensification, environmental consequences and social responses in Han China and Three-Kingdoms Korea.' *World Archaeology* 48 (4), 563–85.

14

Quoting Gandhi, or how to study ancient irrigation when the future depended on what one did today

Maurits W. Ertsen

Abstract

Irrigated agriculture and societal complexity remain prominent concepts in archaeology. Recently, simulation techniques have allowed new ways of studying the relation between these two. Based on the argument that simulated actions by model agents need to be consistent with temporal and spatial aspects of these actions, this chapter explores issues of short-term, small-scale interactions between agents in human–nature interactions. The short term is the relevant scale for studying irrigation systems in order to build an understanding of an irrigation system's creation, the users' understanding, and the predictability of emerging realisations within irrigation. Despite uncertainties about how to represent physics and behaviour in simulations, all actions that are looked at happened in specific places at specific times. The way material limits are produced is a result of human and non-human agencies interacting in local settings that can be defined in terms of time and space.

Introduction

In 1989, one of the so-called 'big three' authors in the Netherlands, novelist and writer Willem Frederik Hermans, published *Au Pair*, his last important novel before he passed away in 1995 (Hermans, 1989). In the book, Hermans presented the story of a Dutch girl who becomes an au pair – how could it be otherwise, given the title? – in Paris, the city where the author lived at the time. The book received rather mixed reviews, to say the least. Some reviewers considered it to be a great piece of storytelling and even to offer theoretical insights into human societies and relations, others considered it rather vague. At least one reviewer complained about the author's lack of focus, using an example of immediate relevance to the point I would like to make (Van den Bergh, 1990).

On pages 274 to 276 – at least in the first Dutch edition – the author sets the main character Paulina and her friend on a walk to their car. A reconstruction of their thoughts and conversation suggests that the walk from the entrance of their apartment building to the actual car should have taken at least 10 minutes. The exact time frame of the walk is not explicitly mentioned in the book, so this 10-minute period is a reconstruction based on the evidence provided in the text of the novel. What is explicitly mentioned in the book text is that the car Paulina and her friend were walking towards was parked right in front of the building entrance – or rather exit. Obviously, one can discuss what 'right in front' actually entails, but this reviewer considered that a 10-minute walk in Paris from a door to a car could not be anything close to 'right in front'. The reviewer complained that the textual account of the actions 'talking on the streets of Paris' by the novel characters was not consistent with the (expected) coverage in time and space of those same actions of the two persons 'walking on the streets of Paris'.

I argue that the field of archaeology has a similar issue to solve as the problem that the reviewer suggested Hermans's novel had. In the archaeological narrative – the 'novel' – the role of irrigation features prominently. Many early civilisations were based on the additional food production that irrigation could bring. This food would allow the emergence in society of non-food producing groups, with associated changes in hierarchies and power relations. How irrigated agriculture and societal complexity link has always been a prominent question in archaeology. In recent years, the use of simulation techniques has become an important way of studying that same relation. The rich data records of irrigation-based societies – both ancient and current – are used to construct more or less elaborate models. As much as the actions of Paulina in *Au Pair* need to fit with the time–space projections of these actions, simulation models need to ensure that the actions of model agents are consistent with temporal and spatial aspects of these actions. One cannot simply ascribe the task 'maintain the canal' to a model agent without considering the spatial coverage of such a task in a given time period.

In line with such reasoning, I have argued elsewhere that simulation models should be based on short-term agencies because that would be how social relations and emerging changes need to be understood (Ertsen, 2016; Ertsen et al., 2014). Clustering the short term into lumped decisions or ignoring the short term by applying longer-term time steps in a model world – things most models nowadays do for reasons of simplicity and computational time – leads to circular reasoning. Clustering decisions that human agents make every day suggests that those decisions would not change anyway. Even decisions that might result in one action per year – like the next holiday destination for example – might have to be conceptualised as a series of discussions and decisions: at least, that would be representative of my own situation. Clustering individual agents into organisations or societies and granting these clustered entities decision power is problematic as well. The articulated entities themselves are the result of individual agencies, which are lost in clustering. Furthermore, clustering usually results in the association of certain predefined features to the larger-scale entities according to the patterns that the (archaeological) record suggests. This creates a major problem of circularity, as the same features that are observed are used as input in the model that should study how those features emerge.

Building on these earlier publications, I aim to explore issues of short-term, small-scale interactions between agents of different kinds as I study human–nature interactions. I discuss why and how the short term – in my case days merging into years – is the relevant scale at which to study irrigation systems – in my case the irrigation of the Hohokam in what we now know as Arizona – in order to review how we can understand irrigation systems' creation, users' understanding of these systems, and issues of predictability when those systems are used. My discussions are theoretically supported by actor–network approaches, with specific emphasis on work from Bruno Latour, the French sociologist and philosopher, who argues that human decision making and development of societal institutions is always a local activity. I will elaborate on these notions for the Hohokam irrigation systems, but first I will dive into the issues of scale(s) and perception(s) a little more. Let us consider an alien coming from outer space to destroy the Earth.

Alien scales?

A perfect illustration that the relevant scale is not a given, but depends rather heavily on the observer, the observed and all kinds of interactions between the two, is provided by the children's book by Tony Ross *I'm Coming to Get You!* (Ross, 1984). The story unfolds as, after wreaking havoc on several planets of its own galaxy by eating them or blowing them up, a still-hungry monster directs its spaceship towards a pretty blue planet. On this planet, called Earth, the monster plans to attack the little boy Tommy Brown. Tommy is actually pretty scared of monsters. The alien monster arrives during the night and hides behind a rock with a huge beam it has found. In the morning, the monster jumps ahead to attack Tommy, only to find out that Tommy is quite a bit bigger than the monster. Tommy does not really seem to notice the monster at all, as the monster does not even reach the height of Tommy's shoes. The reader discovers that the rock was nothing more than a pebble and the beam was just an old match. The take-home message from this short story is that it is tempting to select a scale after one knows the outcome. In terms of (spatial) scales, for almost the whole story the relevant scale was the alien. However, on the last pages the relevant scale seems to turn to Tommy, the giant compared to the alien. Does this mean that the story needs to be retold, with the alien being tiny all the time, as its effect on humans and their planet was so small?

Telling Tony Ross's story that way may ignore the other aliens that visited the same planet Earth in 1991, some years after the small alien's trip was told to us; the drawing published in the series 'Ayer y Hoy' (Yesterday and today) by Matt in the newspaper *La Nación* is available as evidence for the 1991 close encounter (Figure 14.1). We see one alien who has landed on the planet reporting back to two aliens still in the spaceship. The message is clear: 'No civilización. Nada. Lo único es que está lleno de hormigas.' ('No civilisation. Nothing. It is just full of ants.') In this account, the question turns the other way: should the story be told from the perspective of the (huge) alien, which is actually bigger than the planet itself? In that story the role of humans might be less important than humans would like, which would be a reason not to select the alien scale in the first place.

For those who think that reasoning on scales with aliens is a little far-fetched, consider the micro-scales within our own bodies: the average adult human gut is home to approximately 1 kg of bacteria, about the weight of a human brain (Dinan et al., 2015). Apparently, these microbiota have influenced brain development in mammals and are partially regulating behaviour, such as social interaction and stress management. Perhaps an outside observer would conclude that human beings are nothing but containers for bacteria and other microbiota. Would an observer without a mindset predefined to define the spatial scale of humans – and the associated skill set – select humans as the main focus of study when so many other species are around?

Figure 14.1 Aliens visiting Earth, from *La Nación*, 1991

One account we have of such an observer from outside is the friend of Arthur Dent, one of the main characters in *The Hitchhiker's Guide to the Galaxy* (Adams, 1979). At the start of the story, the whole planet Earth is about to be destroyed by a Vogon constructor fleet to create a hyperspace bypass. In an attempt to respond to cries and protests (in the movie version) the head of the Vogon crew explains that humans have nothing to complain about. The plans for the bypass had been available for quite some time at the local planning office on Alpha Centauri, a mere four light years away. The humans just had not cared to consult them, another fine example illustrating that selecting appropriate temporal and spatial scales is rather influential for the outcome of stories.

We meet the two main characters of the stories when (human) contractors want to demolish Arthur Dent's house to make way for a (human) bypass on Earth and his friend tries to warn him about the Vogons about to demolish the Earth. Arthur's friend calls himself Ford Prefect, after what (some) humans know as a British car manufactured from 1938 to 1961. When arriving from his home planet somewhere in the vicinity of

Betelgeuse, Ford had simply tried to ensure he would blend in. Therefore, he decided to adopt one of the most common names he could find on planet Earth. In the 2005 movie version, Ford is even depicted as almost being run over by his name-giver while greeting an actual Ford Prefect. Especially the movie version stresses the point that an outsider would not necessarily distinguish between humans and non-humans in terms of species or acting agents.

What may have appeared as a slightly confusing and crazy detour involving aliens, bacteria and hitchhikers actually brings me to an important issue. I would argue that because we can imagine that outside observers must have a hard time distinguishing between humans and non-humans, we are forced to rethink that same distinction which is typically so easily made by human observers. I imagine outsiders seeing me struggling with my computer and keyboard to convey my message. I wear glasses, a friend of mine has a new hip joint, many people use walking sticks, even external skeletons become available. What is a mobile phone other than an extended communication-related sensory agent, as closely linked to the human body as shoes or skin – or microbiota in our bodies? Why distinguish a priori between entities that think they are special because they think that they can think and those entities that are thought of as not being capable of thinking, or reading or reasoning? Recent studies on animals suggest that there may be more to the language capacities of animals than we have thought for a long time. Many of these new insights may have been defined because the researchers explicitly selected research protocols that allowed those non-human agents to engage with the research on a scale that was more relevant to them, including by allowing elephants to look into large mirrors when studying whether they could recognise themselves in a mirror (De Waal, 2017).

In line with such approaches and findings, my answer to the general question of how to deal with the predefining that we are vulnerable to is that we as human scholars studying humans should be extremely careful when making distinctions between humans and non-humans in studying what we want to study. That is not an argument in favour of some kind of new objectivity we have to follow. Scholars make choices according to certain interests. Actually, I would suggest that we embrace the principle that there is no escape from our own intentionalities when we do mobilise our historical and archaeological records to answer those questions we want to have answers to. Acknowledging such a position should entail that we are only more explicit about our selections of time and space in our studies. When archaeologists – and scholars associating themselves closely with archaeology like me – study the landscapes emerging from archaeological data, the research should clarify how that landscape is emerging from a double process of '"thinging" entities together' (Latour, 2007: 140). In a first process, the historical agents created a landscape. In a second process, the landscape is created by archaeologists – using the material remains in the data records and the perceptions of the archaeologist(s).

Time, space and relations

To clarify this double process of creating reality and its close relation to issues of time and space, and to show how important it could be to have clarity on the dimensions that one uses, let us assume an irrigation canal. For simplicity we consider only two users along that canal. Let our starting point be that the downstream user is allowed to start irrigating at a certain moment. The water that is to be used downstream needs to pass the upstream user first. Although the upstream user may not have the right to irrigate at that moment, she may not be able to resist the temptation to take some of it. Perhaps this upstream user wants to open her gate in order to guide a certain amount of water to her field with crops, which are in need of some water. Let us assume she does so for one hour, after which she decides to close her gate, as the crops have had enough water. The original opening of the gate will have caused a change in the amount of water flowing to our second user downstream. The closing of the gate again changes the flow that moves downstream. Now, we can have different options, depending on the properties of the gates and the canal, and depending on the willingness or ability of the downstream user to act. These are just a few of those options:

1. The downstream user sees the upstream act herself, which must mean that the two users are fairly close to each other. The spatial setting of the interaction is small. This does not necessarily mean that the downstream user will act upon the observation. The two users may not be within sufficiently easy reach to communicate, the downstream user may not feel the need to act, or may be less inclined because of other reasons: she may like the neighbour, or fear her.
2. The downstream user cannot observe the acts of the other user, but notices a change in canal flow fairly quickly, for example after 20 minutes. The observation may allow the downstream user to act – say, to move upstream to investigate – but whether

this act results in the problem being solved depends on the time needed to investigate, which is dependent at least on issues like the time needed to walk the distance, the type of terrain to be crossed, and the fitness of the downstream actor.

3. It may be the case that the downstream actor is able to observe any acts of the upstream user only after that upstream user has closed her gate, after the one hour we assumed. Thus, the downstream user has not been able to observe the acts of the other user directly. She will have fewer (or at least different) possible ways of relating the changes in water flow to the acts of others than in a situation of direct observation. The upstream field may be wet, but proving that someone deliberately took water out of turn is not that easy, as I can confirm, because this third case is exactly what happened during my traineeship in an Argentinian irrigation system.
4. Perhaps the downstream user notices the change in flow and responds directly by changing her own gate. This in turn will change the flow going downstream. Hydraulically speaking, however, there will also be effects upstream. Without going into technical details, this means that the action of the upstream user in opening her gate may cause a secondary change in the water level at her own gate through the action of the downstream user. Perhaps the upstream user wants to change her own gate again, setting in motion another series of changes.

Within an archaeological setting, the material remains that are available to us would not typically allow us to study these relations between users, flows and canals. The temporal resolution is simply not dense enough. Therefore, any analysis of these relations would typically involve mental or model-based reconstructions. A specific reason why a mathematical model would be required is that the flow patterns that may be created by the actions of users cannot be computed by hand, and even analytical solutions may not be available. The changes in flows, water levels and energy gradients are to be computed by a numerical model – that is, only when we assume we are interested in these details. I argue that we should be interested in them, as, in a relatively short amount of time – in our theoretical case possibly hours – the different water levels that our users experience may be related to the actions of one (or both) of these users. If one selected a modelling time step of 12 hours, which would still be rather short compared with archaeological timescales of hundreds of years, this chain of action–reaction would be missed. However, the same chain may have ruined the social relation between the users, with potential consequences for other issues of cooperation or decision making … Let us move to the Hohokam and their landscapes for an illustration.

The Hohokam

The name 'Hohokam' refers to a culture known through archaeology that occupied areas along the middle Gila and Lower Salt rivers in the Sonoran Desert roughly between AD 1 and the middle of the fifteenth century (Figure 14.2). Research is conducted on settlements and irrigation on both rivers, but the excellent data sets from the middle Gila area have proved to be extremely valuable for reconstructing possible Hohokam realities, including settlements, climatic conditions, food production and community organisation (see Woodson, 2016; Ertsen et al., 2014; Zhu

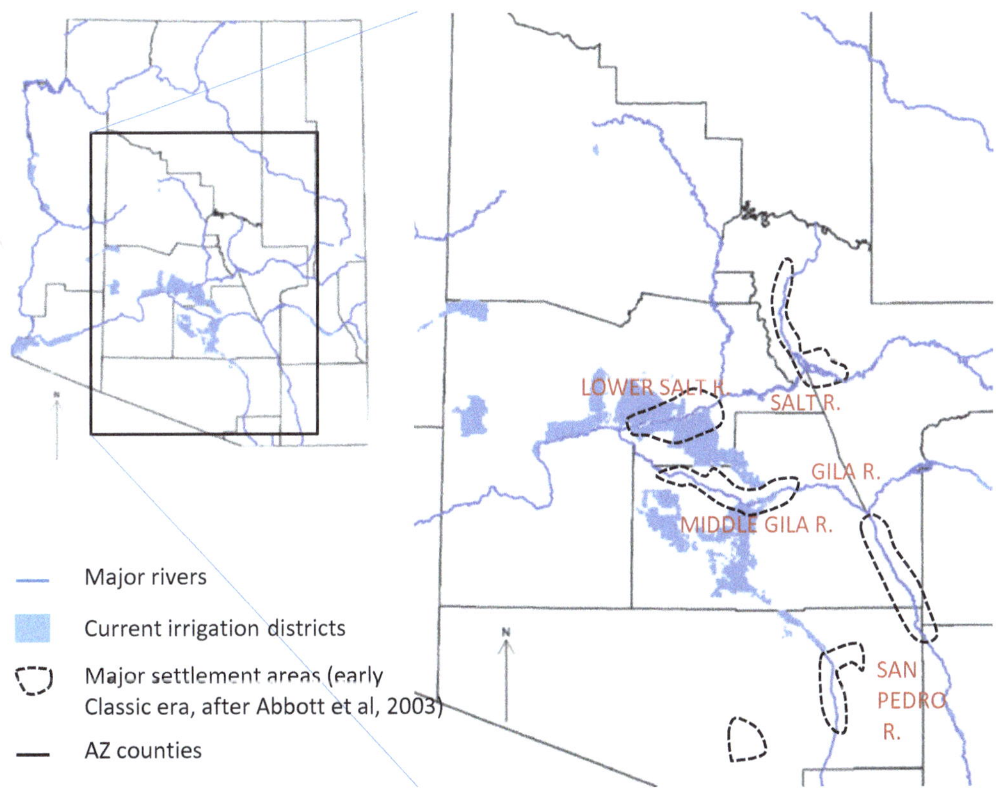

Figure 14.2 Overview of Hohokam settlements in the Salt and Gila rivers areas (early Classic area). Data source: Arizona State Land Department, Arizona Land Resources Information System, Major Rivers & County; Arizona Department of Water Resources, Irrigation districts; 11 July 2013 in Arizona Geospatial Data and Maps. Image source: author

et al., 2015). Hohokam culture was rediscovered by the European settlers who came to the region, when they encountered extensive irrigation canals that had not been in use for some time, but were still clearly visible. As the people using these canal systems were not found any more, the name Hohokam ('they that disappeared') was given to the group of people that must have constructed the water features. Given the impressive canal sizes, irrigation must have been important when 'they' were still around, but still, it is highly likely that Hohokam irrigation was of a supplemental nature. In contrast to using irrigation to meet full crop demands, the River Gila discharge and its flowing conditions may not have allowed for the tight water control that would have ensured exact crop water delivery. It is far more reasonable to assume that the Hohokam used their canal systems to store moisture in the soil of their fields a few times in a growing season.

In addition, the archaeological record suggests that exploiting desert resources yielded as much as 50 per cent of Hohokam needs (Hunt and Ingram, 2007). Irrigated agriculture, with its maize, beans, squash and cotton, must have been important for the Hohokam, but they harvested wild plants and hunted animals too. Agave was important for both fibre and food and would have (been) grown along irrigation canal banks, basically wiping out the borders between irrigation, nature and agriculture even more. In maintaining their resource use, the Hohokam had to balance irrigation, which brought with it tasks like water use and control, canal maintenance and crop cultivation, with many other tasks related to resource collection and production, like hunting, walking and storing. Recent research suggests that in Hohokam irrigated agriculture, a large bottleneck would have occurred between late June and early August, as this period saw activities like crop harvesting and transplanting in irrigation overlapping with non-irrigated-crop-related tasks like canal cleaning and collecting resources from other sources (Zoric, 2015).

Another potential major bottleneck in terms of labour mobilisation would have been encountered at the start of a new agricultural cycle in (Arizona) spring. This period would have been in March–April, when the River Gila discharges, associated water levels have become sufficiently high to bring water to the fields, and winter frost is no longer a risk to the young plants. However, that same River Gila could still bring flood peaks during this period, which could damage (and potentially even destroy) canals. Depending on the magnitude of the flood and the water levels, some floods may even have been able to enter (lower-lying) fields and destroy the crops (Zoric, 2015). To complicate matters of labour division and success in terms of food availability, flood events that might destroy irrigated fields would have been beneficial for resource collection from the desert later in the year, as a flood event can be associated with higher moisture availability in the larger environment.

Research suggests that the frequency of these higher discharges in spring may have been greater than previously assumed, with floods possibly occurring as often as every five to seven years (Zoric, 2015). This new insight is important: it may change our perspective on Hohokam society, because it changes our ideas on extremes and vulnerability in that same society. When extreme events occur more often, it is not unreasonable to assume that those events will no longer be perceived as extreme. After all, something is 'extreme' when it is 'not normal'. Our current reconstructions of past climates have an immediate impact on how we have to think about the environmental perceptions of contemporary (Hohokam) agents.

Thinking about what counts as extreme in the meaning of 'not normal' has a longer-term dimension as well. Societal response to periods of drought is particularly discussed – for the Hohokam, but also for example for the Maya (Scarborough, 2003) – but slow processes like increasing salinity in Mesopotamian irrigated agriculture also serve as examples of what have been called increasingly extreme conditions leading to collapse (see Wilkinson et al., 2013). The decline of the Hohokam as a recognisable social group from AD 1150 onwards, leading to its ultimate disappearance after about 1450, has been often explained by increasing aridity in the region. Without suggesting that drought was not an issue, we can make two important observations. First, the archaeological record suggests that the Hohokam have gone through earlier periods of (even more extreme) drought without disappearing, but not without change, in settlements for example (Woodson, 2016; Pande and Ertsen, 2014). Second, such a long-term approach does not allow us to construct a narrative of change. With slowly but steadily decreasing water availability, when does one decide to change or leave? Why would one decide something is extreme, when the change is so gradual that one doesn't notice it?

New ideas about Hohokam environmental conditions allow us to develop new narratives about Hohokam 'collapse' as well. Rainfall reconstructions for the period between 1000 and 1500 (Figure 14.3) suggest that the last 300 years of known (more accurately 'claimed') Hohokam presence saw a high incidence of both droughts and floods, even in periods that on average would be classified as dry or wet. These new reconstructions of Hohokam reality would allow a narrative of change that could be tested in short-term models, and analysis along short-term lines, that are

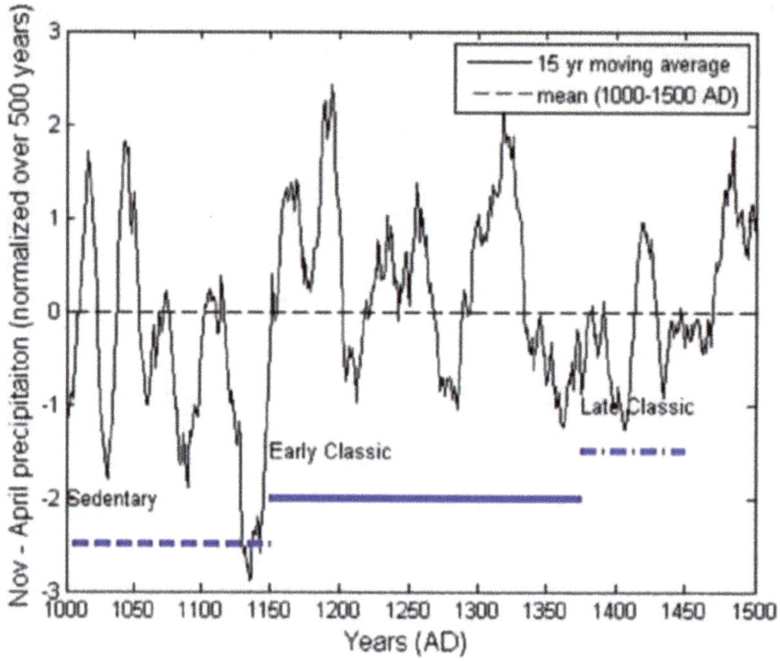

Figure 14.3 Reconstructed and standardised (i.e. subtract the mean and divide by the standard deviation of the time series) winter precipitation (south-western Arizona, US) for the period AD 1000–1500. First presented in Pande and Ertsen, 2014. Data from http://www.ncdc.noaa.gov/paleo/pubs/ni2002/az6.html; see also Ni et al., 2002

crucial to understanding why change and continuity happened. Relatively quick successions of droughts and floods – low and high seasonal river discharges alternating in periods of years – would have built up stress within a community. The Hohokam seem to have witnessed a dispersal of the larger cooperative networks in their last centuries of existence in the archaeological record. It may not be possible to generalise for the whole Hohokam area, but there is evidence that in wet periods people moved away from settlements, which makes sense, as the surrounding landscape would have provided ample resources. In dry periods, many people would have moved back to the areas closer to the river.

Now, let us think this through a little more. If such population movements away from and towards the river area took place in periods with overarching droughts, but with the occasional wetter period allowing desert resource use and destroying the canal system in spring as well, what would happen? A community dependent on irrigation would need to repair the canal systems after a flood, for food production in a year in which the desert would also have yielded resources. In the years to follow, however, the repaired irrigation system may have been less able to secure irrigated crops, or at least not able to irrigate all fields, because of lower river discharges. The floods might have demanded more energy to keep the systems working than the gains from cooperative irrigation could sustain in the drier years to follow. This would have put stress on cooperative efforts: why invest time in repairing a system after a flood when it did not pay off the last time because of drought?

The Hohokam may have disappeared from the archaeological record because it turned out to be very difficult for that group of people to deal with risky environments. As I suggested, the meaning of 'risk' may have changed as much as the environment itself. Does one's judgement of extreme events change once those events are seen as occurring more often or as part of a pattern? Does that gradual change escape notice because it is not a series of strong events? When does an event cease to be an event and become a slow process of change? How are the effects of events distributed over people and spaces? We are only at the beginning of answering such questions for the Hohokam, but to answer them we need to consider the close relations between time and space as perceived by us and the Hohokam.

Predefined landscapes

What could we take from this? My Hohokam story shows human agents trying to create and promote 'correct behaviour' in non-human actors – rivers, fields and crops – but we encounter non-human agents refusing to be caught exactly the way that human agents want. Matter is not necessarily obeying human orders – water may decide not to flow, as it did many times for the Hohokam. In both versions of the reconstruction process I referred to in the section on alien scales, things (matter, non-human agents) matter: the shape of a river, the slope of the land, the grain size of the soil; but how could we include such properties in our analysis, retain their influence, allow meaningful conclusions, and not fall into the trap of defining a new orthodoxy, this time one of matter dominating the discourse or the like?

We find several responses to the question of how much agency we should assign to matter. In Hodder's recent work, the concept 'entanglement' emphasises the 'limited unfixed nature of things in themselves and their relationships with each other' (Hodder, 2014: 24). Entanglement suggests that because things exist on their own, with their 'material possibilities', not everything is still possible in terms of the realisations of realities involving matter and humans. Things create 'potentials and constraints' (Hodder, 2014: 25). In line with such reasoning, Strang argues that 'material things and their agentive effects' constitute 'a form of potentiality' that is to be 'harnessed' (Strang, 2014: 142) by human agents. Although Hodder and Strang clearly argue that relations between matter and human are key, the way they frame the position of matter, with its apparent present potentiality to be freed by humans in those relations, comes pretty close to positions such as those defended by a scholar like Elder-Vass (2008, 2015), who assumes that the mechanisms revealed by science are actual mechanisms existing somewhere out there, presumably just waiting to be discovered.

In his ideas of New Realism, the German philosopher Gabriel (2013) argues that although we can know things and facts, those things and facts cannot exist on their own. Meaning is mobilised within several different perspectives ('*Sinnfelder*'). Therefore, 'the world' does not exist, as nothing exists on its own. Although Gabriel and Latour would not necessarily agree with each other's exact phrasing – Gabriel uses the forbidden word 'context' – both scholars argue that the knowledge that humans develop concerning the objects that shape their world depends on how humans connect to those objects. In shaping those connections, objects have agency too, and are multiple in the sense that other agents can connect to similar objects in different ways. Following Latour, who criticises the idea that one can extract oneself from matter by projecting the material as an external world outside of oneself, I would not defend any position that suggests that we can study material entities independently of what human agents do or think.

The claim that matter exists outside of knowing human subjects may be reasonable, but is also meaningless, as in our analysis we can deal only with the representations of any external reality developed by these subjects. Apparently, even starting from a position of entities existing and events occurring independently of what humans think or say about them, Elder-Vass considers that there could be entities that depend on human concepts or actions (Elder-Vass, 2015). This poses the question of who decides which entities and events are made up by humans and which are truly discovered (or whose potential is realised). I struggle with this issue. The assumption that there is a world (problematic as this term is) outside of knowing subjects may not be unreasonable in itself. Matter would be independent of humans. For example, the hypothesis that the Earth will continue to exist when the last human has disappeared is defendable. At the same time this proposition is nothing more than an interesting thought experiment: when humans are no longer there, there will be no human reference to the Earth any more, and the Earth will have ceased to exist as something relevant to humans.

Knowledge is not necessarily just made up and therefore totally free (or fake) for all, but that in itself is no evidence of any knowledge being a true representation of any external reality (see Latour, 1999, 2002). As soon as knowing subjects relate to earthly matter, any given external Earth ceases to exist, because the Earth is defined in human terms. That is not the same as saying that knowledge is not useful or is always irrational, but it does mean that knowledge production is always a social activity. As an answer to the problem of the double process of constructing reality, our research needs to offer options for our scholarly reconstructions that allow the same 'possibility of holding society together as a durable whole' (Latour, 1991: 103) as the original constructions did for those who were involved in the first process, in my case the Hohokam.

We should clarify how the actor–network of landscape making along the River Gila was shaped by human and non-human agents alike (Ertsen, 2016; see Latour, 1991). Biggs (2010) shows how one of those non-human agents – tidal movements in the Mekong Delta – effectively blocked many of the shipping canals with sediments very soon after these waterways had been dug – without the tide making a difference to whether those interventions were Vietnamese, French or American. Materiality matters as much as the goals and actions of human agents. Environmental agents with their agency influence the survival strategies of those (human) agents who engage with that environment and so with the agents themselves, like the bacteria in human guts (see Kendal et al., 2011; Ertsen et al., 2016).

Models

Irrigation is an expression of, and a basis for, emerging social relations. Negotiations between human and non-human agencies shape social relations, as Latour shows with his example of the hotel key, which acts upon the hotel guests (Latour, 2000). Not too long ago, when we checked in, we would receive a hotel key with a weight. As we usually receive a key card nowadays, the role of the key has slightly changed, but we still encounter a social relation that pre-dates our arrival and is expressed in the key – in material stuff. Social relations without a material component would have a hard time surviving anyway, as they need to be renegotiated all the time (Ertsen, 2016). Matter provides temporal and spatial links between agents in infrastructure, in institutions. Social relations do not float around before being manifested, they are only manifest when realised in matter, and matter includes books, statues and language.

As much as canals are produced through engagements with the material, in that same process of thinging entities together that I referred to as the double production process, such engagements only offer possibilities of keeping the (re)constructions as stable entities, both for the original agents and for the archaeological researcher. In my own work on the Hohokam, I have a fair amount of control of those short-term processes in irrigation that should be studied, but only by making several assumptions, including the idea that gravity worked similarly thousands of years ago, a suggestion one is never sure of when following Latourian logic. After all, in 1999, Latour argued that because an element is defined by its associations, a new event is created on the occasion of each of these associations (Latour, 1999). Take microbes again. Latour argues that if microbes existed before 1864, they only did so after 1864 when Pasteur defined entities as microbes. Could this simply mean that the agency of those entities we now call microbes has been around all the time, but that we only call these entities microbes since more recent times? Please consider what our successors may think of our ideas on the matter; why would our current ideas hold?

To make a very simple statement, we cannot be sure that our current ideas on reality are real. However, as soon as we claim that microbes – or environments for that matter – did not exist before they were thought of by humans, we have to ask 'What does that mean?' Should climate change scholars ignore historical climate change because the concept was unknown at the time? Intuitively this makes no sense either. Nevertheless, by defining climate change as an entity, historians change the climate in a progressive construction of reality. Climate is not something out there that is finally being discovered. Climate and its changes are translations which completely transform that which gets transported between agents, whether these agents are historical or current.

I do not think this means we can no longer study any archaeological or historical reality. I do think, however, that we need to be careful in our phrasing and framing. Water is not a form of potentiality that needs to be manipulated by human agents to come to reality. There may be much to be grasped by any human actor, but the grasping and its consequences are constructed in close negotiations with the water itself, and with its former harnessing agents which we encounter through water infrastructure. What is to be grasped or manipulated changes as soon as it is grasped or manipulated. If water only has to be harnessed, there seems to be little agency left for the resource itself. For some scholars, that is only good news, as for them non-humans cannot have agency anyway, as they would have no intentionality (see the discussion in Strang, 2014).

I do not want to suggest that intentionality as a concept is irrelevant, but if we define agency in Latourian terms – as having an effect upon other agents – intentionality is no longer a required property of an agent. From a Latourian perspective, we know we can never go further than making claims about reality, as reality itself does not exist. But still, which claims do we want to make? What is a more trusted way of representing agencies? Whatever that may mean, it is clear that our own changing understanding of environmental dynamics changes our archaeological reconstructions as well. It also changes our own options within our current society. Only a few years ago, the Dutch coastal defences became statistically 50 centimetres lower in terms of protection against floods because the method of incorporating the impact of waves was changed. New ideas about the physics of waves changed relative sea-level rise more than climatic change had done in the same period.

To conclude, I would suggest that despite such uncertainties, related to physics and behaviour, in representing them, we can assume certain flows of agencies of non-humans and humans alike in the archaeological cases that we study. Obviously, we are never sure whether our reconstructed reality comes close enough to the realities as experienced by the many agents of the time. We can be sure, however, that all actions happened at specific places at specific times and that there were spatial and temporal limits to all agencies. To be more precise, the way material limits are produced is a result of human and non-human agencies interacting in local settings that can be defined in terms of time and space. In those interactions, our landscapes are not passive or static backgrounds of human agency: they engage directly with other agents. The outcomes of these combined agencies are often idiosyncratic and unpredictable; I would argue that a careful selection of the scales we mobilise in our studies allows us to trace reality-building-in-action, both our own and that of our archaeological agents.

Acknowledgement

This research was undertaken in conjunction with the Gila River Indian Community's Cultural Resource Management Program and the Pima-Maricopa Irrigation Project under funding from the Department of the Interior, US Bureau of Reclamation, under the Tribal Self-Governance Act of 1994 (P.L. 103–413), for the design and development of a water delivery system utilising Central Arizona Project water.

References

Abbott, D. R., C. D. Breternitz and C. K. Robinson. 2003. 'Challenging conventional conceptions'. In D. R. Abbott (ed.), *Centuries of Decline during the Hohokam Classic Period at Pueblo Grande*. Tucson: University of Arizona Press.

Adams, D. 1979. *The Hitchhiker's Guide to the Galaxy*. London: Pan Books.

Biggs, D. 2010. *Quagmire: Nation-Building and Nature in the Mekong Delta*. Seattle: University of Washington Press.

De Waal, F. 2017. *Are We Smart Enough to Know How Smart Animals Are?* New York: Norton.

Dinan, T. G., R. M. Stilling, C. Stanton and J. F. Cryan. 2015. 'Collective unconscious: How gut microbes shape human behavior.' *Journal of Psychiatric Research* 63, 1–9.

Elder-Vass, D. 2008. 'Searching for realism, structure and agency in Actor Network Theory.' *British Journal of Sociology* 59 (3), 455–73.

Elder-Vass, D. 2015. 'Disassembling actor-network theory.' *Philosophy of the Social Sciences* 45 (1), 100–21.

Ertsen, M. W. 2016. '"Friendship is a slow ripening fruit": An agency perspective on water, values and infrastructure.' *World Archaeology* 48 (4), 500–16.

Ertsen, M. W., C. Mauch and E. Russell (eds). 2016. *Molding the Planet: Human Niche Construction at Work*. RCC Perspectives: Transformations in Environment and Society, 2016/5. Munich: Rachel Carson Center.

Ertsen, M. W., J. T. Murphy, L. E. Purdue and T. Zhu. 2014. 'A journey of a thousand miles begins with one small step: Human agency, hydrological processes and time in socio-hydrology.' *Hydrology and Earth Systems Sciences* 18 (4), 1369–82.

Gabriel, M. 2013. *Warum es die Welt nicht gibt*. Berlin: Ullstein.

Hermans, W. F. 1989. *Au Pair*. Amsterdam: De Bezige Bij.

Hodder, I. 2014. 'The entanglements of humans and things: A long-term view.' *New Literary History* 45, 19–36.

Hunt, R. C. and S. E. Ingram. 2007. 'Traditional maize production in the Southwest: Irrigation along the Middle Gila River.' Manuscript in the possession of the author.

Kendal, J., J. J. Tehrani and J. Odling-Smee. 2011. 'Human niche construction in interdisciplinary focus.' *Philosophical Transactions of the Royal Society B* 366 (1566), 785–92.

Latour, B. 1991. 'Technology is society made durable.' In J. Law (ed.), *A Sociology of Monsters: Essays on Power, Technology and Domination*, 103–32. London: Routledge.

Latour, B. 1999. *Pandora's Hope. Essays on the Reality of Science Studies*. Cambridge, MA: Harvard University Press.

Latour, B. 2000. 'When things strike back: A possible contribution of "science studies" to the social sciences.' *British Journal of Sociology* 51 (1), 107–23.

Latour, B. 2002. *War of the Worlds: What about Peace?* Trans. C. Bigg, ed. J. Tresch. Chicago: Prickly Paradigm Press.

Latour, B. 2007. 'Can we get our materialism back, please?' *Isis* 98, 138–42.

Ni, F., T. Cavazos, M. K. Hughes, A. C. Comrie and G. Funkhouser. 2002. 'Cool-season precipitation in the southwestern USA since AD 1000: comparison of linear and nonlinear techniques for reconstruction.' *International Journal of Climatology* 22 (13), 1645–62. https://doi.org/10.1002/joc.804.

Pande, S. and M. W. Ertsen. 2014. 'Endogenous change: On cooperation and water in ancient history.' *Hydrology and Earth Systems Sciences* 18, 1745–60.

Ross, T. 1984. *I'm Coming to Get You!* London: Andersen.

Scarborough, V. L. 2003. *The Flow of Power: Ancient Water Systems and Landscape*. Santa Fe, NM: SAR Press.

Strang, V. 2014. 'Fluid consistencies: Material and materiality in human engagements with water.' *Archaeological Dialogues* 21 (2), 133–50.

Van den Bergh, H. 1990. 'Een regelrecht misbaksel …: Is "Au Pair" van Willem Frederik Hermans een meesterwerk of een mislukking?' *Vrij Nederland*, 27 January.

Wilkinson, T. J., M. Gibson and M. Widell. 2013. *Models of Mesopotamian Landscapes: How Small-Scale Processes Contributed to the Growth of Early Civilizations*. Oxford: Archaeopress.

Woodson, M. K. 2016. *The Social Organization of Hohokam Irrigation in the Middle Gila River Valley, Arizona*. Tucson: University of Arizona Press.

Zhu, T., M. W. Ertsen and N. C. van der Giesen. 2015. 'Long term effects of climate on human adaptation in the middle Gila River Valley, Arizona, America.' *Water History* 7, 511–31.
Zoric, A. 2015. 'Advantages and limitations of small-scale physical and agent-based modelling for the prehistoric Hohokam canal systems: Rediscovering irrigation in Snaketown.' MSc dissertation, Delft University of Technology.

Index

www.ingramcontent.com/pod-product-compliance
Lightning Source LLC
LaVergne TN
LVHW082116200926
848484LV00008B/1963